EXPERIMENTS IN MOLECULAR GENETICS

Experiments in Molecular Genetics

Jeffrey H. Miller

Society of Fellows, Harvard University
and
Départment de Biologie Moléculaire
Université de Genève

Cold Spring Harbor Laboratory 1972

EXPERIMENTS IN MOLECULAR GENETICS

International Standard Book Number: 0-87969-106-9

Library of Congress Catalog Card Number: 72-78914

Printed in the United States of America

SECOND PRINTING, 1974

THIRD PRINTING, 1977

Cover photos: Mutator colonies stained for both constitutive β-galactosidase (blue) and constitutive alkaline phosphatase (yellow). For description of methods see Experiment 3. Photograph courtesy of G. Hombrecher from the laboratory of W. Vielmetter, Institut Genetik, Universität Köln.

Frontispiece: Mutator colony stained with Giemsa stain (see Experiment 3). Photograph courtesy of G. Hombrecher from the laboratory of W. Vielmetter, Institut Genetik, Universität Köln.

Foreword

One of the more obvious conclusions from the biology of the past twenty-five years is that biochemistry moves faster when it can utilize genetic methods. Likewise, much of what we now call genetics would not have been discovered had not biochemical methods been introduced. Though today many of the practitioners who merge the biochemical and genetic approaches call themselves molecular biologists, there are many other scientists who see no reason to change their names, seeing nothing dishonorable with the terms biochemist and geneticist. But no matter what we call ourselves today, almost no one argues for a purest approach which sees virtue in solving a problem solely by genetic crosses or by massive numbers of postdoctoral biochemists. Of course, there still exist many important problems where now only one of these approaches can be applied. Increasingly, however, we find we can remove this limitation by choosing a more appropriate organism to work with.

Now there is hardly any discussion as to the organism of first choice for probing a fundamental biological problem. *E. coli* stands so far in the forefront that it frequently provokes boredom on the part of the uninitiated, who may have started reading about it in high school and who in college found it hard to take a biology or biochemistry course where it did not sneak in. Thus, it is easy to say that the days of *E. coli*'s domination must soon pass, and so on to embryology and those organisms which have nuceli, mitochondria, and perhaps a number of chloroplasts.

But sighting a new frontier does not always mean that the time is ripe for everyone to open it up. Over and over the past decade has produced examples of key problems where the use of mutants was indispensible to their solution. This situation is still true today. Virtually each new issue of the *Proceedings of the National Academy of Science* contains one or more incisive articles whose conclusions are based on the use of specific *E. coli* mutants.

So knowledge of how to work with *E. coli* as a genetic system is likely to remain a key ingredient in the biology curriculum for many years to come. Teaching bacterial genetics, however, as a purely formal subject without integration into the mainstream of modern biochemistry and molecular biology would be a very dull job. It would fail to convey the excitement of current genetic research and leave the impression that it is an esoteric topic best suited for those who only live genetics. Instead we believe the *E. coli* genetics is best taught in the context of contemporary research where mutants are vital for solving fundamental biological dilemmas.

This is the way the bacterial genetics course at Cold Spring Harbor has been taught since its inception the summer of 1950. First taught by M. Demerec, E. Witkin, and V. Bryson, it has been given here each succeeding summer to ten to twenty students of highly diverse backgrounds, ranging from the pure physical scientist to the applied microbiologist. By now some 250 people have come here for this specific purpose. Many have been strongly influenced by this experience and quickly settled on subsequent careers in molecular genetics.

Now we have to face the fact, however, that the three-week interval in which our summer courses must be given is only sufficient for a small fraction of the experiments that make up bacterial genetics. Inevitably, the experiments which are chosen for a given summer must reflect the specific research interests of the instructors of that year. Furthermore, the number of people who need to be familiar with current tricks for doing bacterial genetics greatly exceeds those that can come here to take our course or go for a learning period to a lab that specializes on this topic. Yet the intelligent novice should have at his disposal a way to become familiar with all the new procedures, if not the "lore" that he might someday need.

The moment thus seems propitious to bring forth an all-inclusive manual where "everything you need to know" about bacterial genetics can be found. In getting Jeffrey Miller to do this job, we have been most fortunate. He is old enough already to be a master in this field, yet, when he started writing he was too young to know how much work is necessary to turn out a good book. The final result, I believe, is a superb job; one, I hope, that will find widescale use for many years to come.

Cold Spring Harbor
March, 1972

J. D. Watson

Preface

Why do we still study *E. coli*? One attraction of working with such cells is that they represent a simplified system. The possibility of harvesting a large number of cells in a short time, and the advantage offered by a haploid organism containing only one chromosome and which can double every 20 minutes, have prompted many investigators to use *E. coli* for genetics research. Their work has considerably increased our knowledge of this organism and has resulted in the development of numerous specialized techniques, many of which we use in this manual.

Most importantly, there remain a vast number of basic problems in the field of cell biology which are as yet poorly understood in even a relatively simple system like *E. coli*. DNA replication, recombination, and repair are examples. Much about the mechanism of transcriptional (mRNA) control is unclear, for instance, positive control. The question of how the synthesis of ribosomal and transfer RNA is controlled is still largely unanswered. Many details of protein synthesis have not yet been elucidated, and the DNA and RNA sequences coding for the initiation and termination of transcription and translation are just now being deduced. The problem of controlled degradation of mRNA and of proteins still is mostly unsolved, and the study of membranes and transport may also be dependent on systems such as *E. coli*. Finally, there remain many unresolved aspects of intermediary metabolism.

How can bacterial genetics help solve these problems? The answer to this question is the subject of this manual. The basic approach that this field offers is the isolation of mutants. The discovery of new control systems, the tailoring of enzymes by genetic manipulation, and the definition of genes involved in biochemical pathways and processes are all direct results of this approach. Therefore, much of the text describes methods for the induction, isolation, characterization, and mapping of different types of mutations.

To facilitate the teaching of experimental molecular biology, we have compiled a series of experiments which can be done on a class basis and which cover many of the areas of modern bacterial genetics. Many of these experiments have been performed by student groups, such as the summer Bacterial Genetics Courses at Cold Spring Harbor. We use the *lac* operon for illustration often, and a review text, *The Lactose Operon* (Cold Spring Harbor Laboratory, 1970) has recently been published which provides a valuable summary of work on this basic system. We were fortunate to able to draw on the excellent experimental manual by Clowes and Hayes (John Wiley and Sons, Inc., 1968) which served as a model for much of this book.

We have also tried to bring together into one volume as many recipes and methods as possible to enable investigators to use this text as a research handbook. Although the *lac* system is used for demonstrative purposes throughout part of the manual, almost all of the methods and techniques described are general. Thus, the description of indicator plates, mutagenesis, Hfr crosses, strain construction, hybridization, and enzyme assays are applicable to a wide variety of systems. Also, we hope that compiling these techniques will enable investigators not thoroughly acquainted with genetic manipulations in *E. coli* to form strategies for building strains and isolating mutants. One of the recent advances in bacterial genetics has been the development of techniques for incorporating bacterial genes into the DNA of certain phages. These specialized transducing phage are then used to provide DNA greatly enriched for the specific gene of interest. Experiments utilizing current methods for isolating these phage are presented in detail in this manual.

We have attempted to arrange these experiments in order of increasing difficulty and have tried to introduce the concepts of some of the later experiments in earlier sections. For instance, Experiment 42 (The Isolation of *trp-lac* Fusion Strains) is a series of steps, each of which has been covered previously. The object of the experiment is to isolate strains in which the *lac* genes are under the control of the *trp* operon. First, phage-resistant mutants are isolated and examined on lactose indicator plates. Recombination and complementation tests are then used to determine the end points of the deletions. Finally, Hfr crosses and F′ transfers are employed to prepare and test *trpR* derivatives of the fusion strains. (This experiment was successfully performed by different groups of 20 students at Cold Spring Harbor in 1969 and 1970.) We would like to emphasize, however, that there is no mandatory order to these experiments. We have included many more experiments in this manual than could possibly be accomplished in a one-semester course. This gives the students and instructors a large freedom of choice in selecting experiments and planning courses. For instance, Unit VI is certainly optional since this requires special equipment which may be unavailable in some laboratories.

In order to facilitate the use of this manual, we have made available strain kits containing the 79 strains described here and 5 lysates in small, 1-dram, agar-filled stab bottles. We also include in each kit a precision-made device, described in the Appendix, which is used for interrupted matings. Kits can be obtained from Cold Spring Harbor Laboratory for $100 to cover costs of handling, packaging, and mailing.

Acknowledgments

It is a pleasure to acknowledge the help of many people without whom this manual could never have been compiled. First of all I am indebted to Jim Watson, who conceived of the idea to produce this manual, who encouraged and advised me throughout the preparation of the book, and who spent many hours reading the final manuscript and the proofs.

This book is a collection of experiments written by several authors. I wrote the introductory material, Units I–V, Unit IX through the end of Experiment 59, and part of Unit VII. Terry Platt wrote Unit VIII, Appendix VI, and parts of Unit VII; Bill Haseltine contributed Experiments 44–47 in Unit VI; Jack Greenblatt prepared Experiments 60–62 in Unit IX; and Benno Müller-Hill wrote Experiment 43 in Unit VI and parts of Unit VII. Also, Larry Taylor and Brooks Low contributed Appendices IV and II, respectively, and Ernesto Bade wrote part of Experiment 47. I have reviewed and edited all of this material to put together the final manual, and I am solely responsible for any errors which are present.

Significant parts of this manuscript were read by Charles Yanofsky, Bob Weisberg, Frank Stahl, Joel Kirschbaum, Don Ganem, John Scaife, Larry Taylor, Ray Gesteland, Nancy Hopkins, Geoffrey Zubay, and Terry Platt. I am grateful for their comments. I am particularly indebted to David Botstein, Brooks Low, and Bill Reznikoff who read most of the first draft in detail and suggested many revisions, and also to David Zipser, Ahmad Bukhari, and Ernesto Bade for reading parts of the proofs.

During the initial preparation of the experimental protocols I benefited from the advice of many people. In particular I would like to thank Ed Lin, Bob Weisberg, Joel Kirschbaum, Don Ganem, Larry Taylor, Brooks Low, Max Gottesman, Jon Beckwith, Ethan Signer, Walter Gilbert, Dan Morse, Julian and Marilyn Gross, Lucien Caro, Klaus Weber, David Zipser, Ahmad Bukhari, Jürgen Schrenk, Mike Malamy, Ekke and Linde Bautz, Marc Van Montagu, and David Dressler. Several of these experiments were retested specifically for this manual by Joel Kirschbaum and Don Ganem. I am grateful for their help.

I would also like to thank Madeline Szadkowski for typing the final manuscript, Judy Gordon for doing the copy editing and proof reading, and Glen Lyle and Judy Gordon for preparing the design of the book. Hanna Neubauer and F. W. Taylor, Co., Northport, N.Y., contributed the art work.

I wrote this manual while a junior fellow of the Society of Fellows of Harvard University, and I am most grateful for their support. I also used the facilities of the Harvard Biological Laboratories (in the Watson-Gilbert group), the Genetics Institute of the University of Cologne (in the laboratory of B. Müller-Hill), and the Cold Spring Harbor Laboratories during this period, and am indebted to these institutions for their support.

Finally, I would like to thank all of the people in the Cold Spring Harbor community who made my past two summers at Cold Spring Harbor (where most of this manual was prepared) enjoyable. In particular I am indebted to Elfie, John, and all the Cairns; Liz and Jim Watson; Frauka, Henry, and the Westphals; and Marianne, Walter, and the Kellers.

Cold Spring Harbor
January, 1972

Jeffrey H. Miller

Table of Contents

Unit II Matings between Male and Female Cells

Unit III Mutagenesis and the Isolation of Mutants

Unit IV Strain Construction

Episome Transfers

Manipulation of Genetic Markers

Isolation of Auxotrophic Mutants

Construction of Hfr Strains

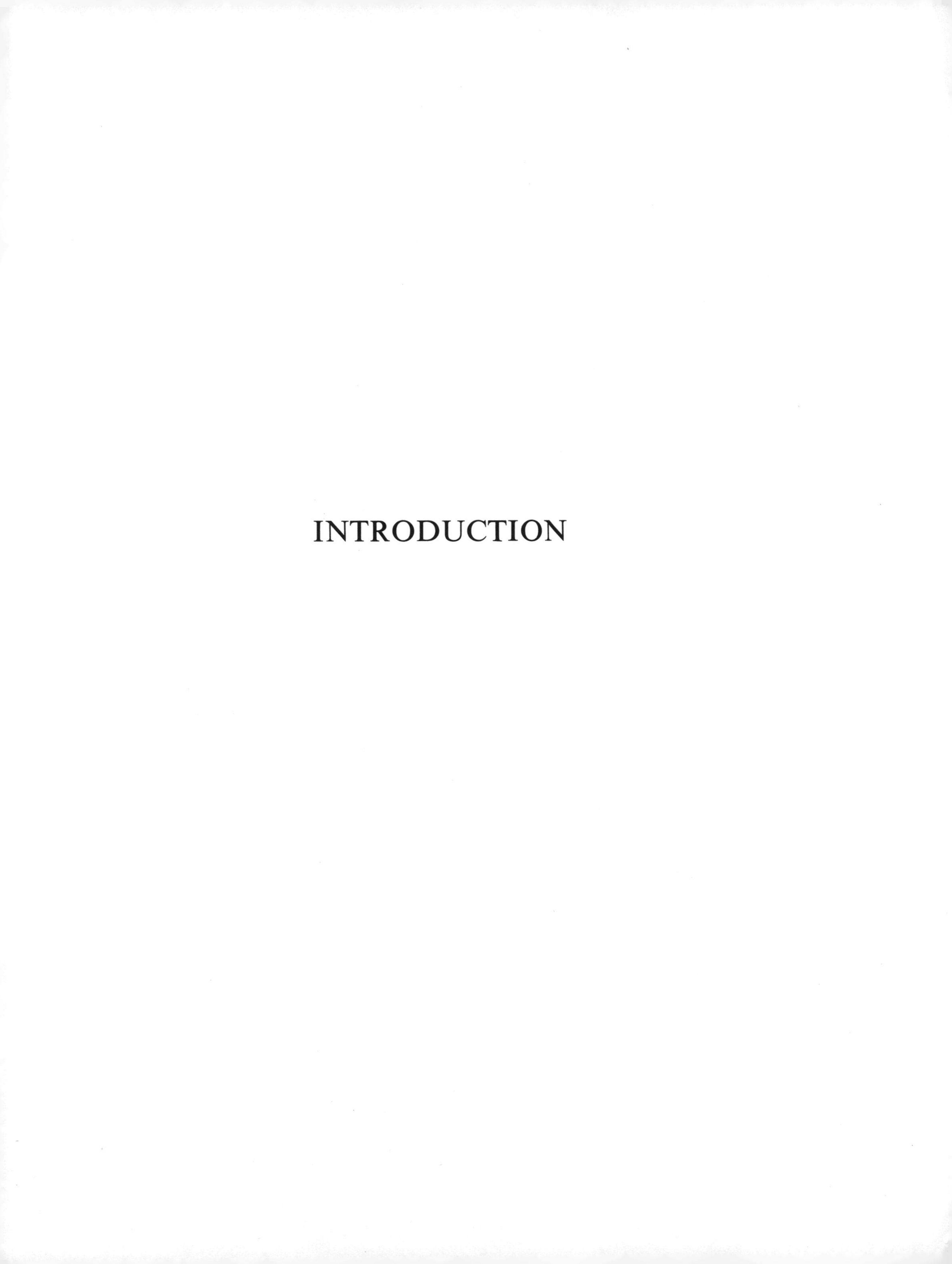

INTRODUCTION

INTRODUCTION TO THE *lac* SYSTEM

Structural Genes

In *E. coli* a cluster of three genes constitutes the lactose or *lac* operon. The first two gene products are required for the utilization of lactose as a carbon source. The *z* gene codes for the structure of the enzyme β-galactosidase (Figure 1), which catalyzes the cleavage of lactose into glucose and galactose. The *y* gene directs the synthesis of a permease required to transport lactose into the cell. A third gene, the *a* gene, codes for the structure of thiogalactoside transacetylase. Although the *in vivo* role of this enzyme is not clear, it is not essential either for cell growth

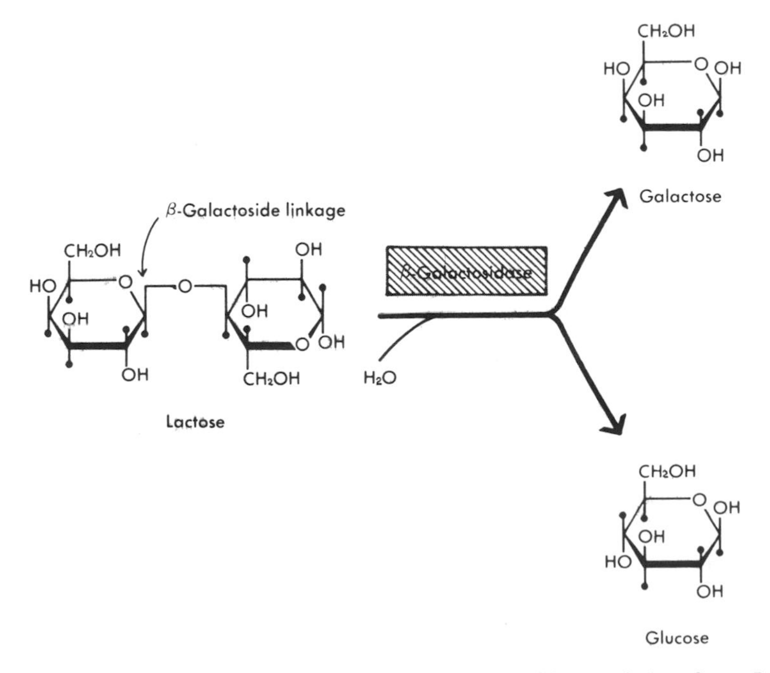

Figure 1. Cleavage of lactose by β-galactosidase. (Reprinted with permission from J. D. Watson, *Molecular Biology of the Gene*, 2nd Ed., W. A. Benjamin, Inc., New York.)

or for lactose metabolism. These three genes are transcribed into a single polycistronic messenger RNA (mRNA) molecule, and are thus said to constitute an operon. Details for the assays of these enzymes are given in Unit VII.

Repressor-Operator Control

The synthesis of *lac* mRNA is controlled by a protein molecule, termed the repressor. The structural gene for the *lac* repressor is the *i* gene. In its active form the repressor binds to the operator (*o*) region of the *lac* operon (on the DNA) and blocks the transcription of *lac* messenger RNA. In this form the *lac* operon is said to be "repressed." When the repressor is inactivated, either by mutation or binding reversibly to certain small molecules, it falls off the DNA allowing transcription of the *lac* operon. In this state the *lac* operon is "induced." Derivatives of lactose, such as the synthetic analog IPTG, which bind to the repressor and alter its DNA binding capacity, are termed "inducers." IPTG is not metabolized by the cell and is widely used as a gratuitous inducer.

Cells with mutations in the *i* gene which result in the synthesis of a defective repressor molecule have fully induced or constitutive levels of the *lac* enzymes, regardless of whether inducers are present or not. Most of these mutations are recessive. Thus, a diploid which is i^-/i^+ will have an inducible (non-constitutive) phenotype. Some *i* mutations, however, are "trans-dominant," and the corresponding i^-/i^+ diploids are partially constitutive. Another class of mutations in the *i* gene, i^s mutations, results in a Lac$^-$ phenotype. In this case, the binding site for inducers is altered and the *lac* operon can no longer be induced by lactose or IPTG. We find that i^s mutations are dominant to i^+. The operator region (*o*) can also be altered, and these lesions have been termed o^c mutations. Operator mutants are partially constitutive and only the genes on the same DNA element are affected. These mutations are therefore said to be "cis-dominant." For a more complete discussion of the diploid analysis of these mutations, see Experiment 20.

Promoter; Cyclic AMP Control

The *lac* operon is also controlled, together with several other operons, by a cyclic AMP regulated system. A protein factor (CAP) activates these operons in the presence of cyclic AMP and allows maximal levels of transcription. In the absence of this factor or cyclic AMP, the levels of the *lac* enzymes are severely reduced. Unit IX discusses this system in greater detail.

We can define genetically a region at the beginning of the operon (Figure 2) termed the promoter (*p*), in which transcription is initiated. Point mutations which map within *p* lower the maximal levels of all three of the *lac* enzymes.

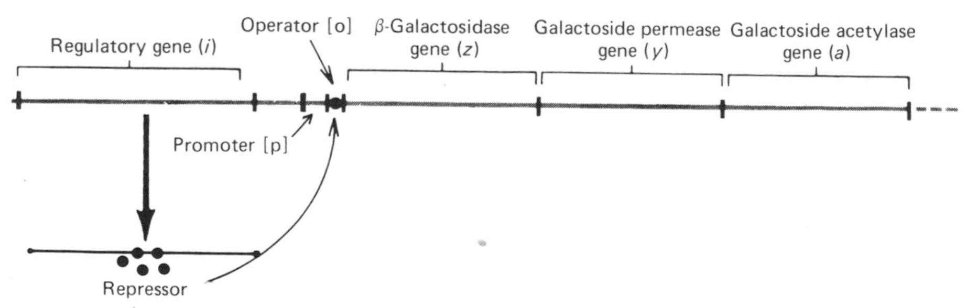

Figure 2. Genetic elements of the *lac* operon.

Reference

For detailed reading on the *lac* system, we refer the reader to

The Lactose Operon, J. R. Beckwith and D. Zipser, ed., Cold Spring Harbor Laboratory, 1970.

FORMAT; USE OF MATERIALS

Each experiment contains a theoretical introduction, followed by a list of strains used in each exercise. These are listed together with their entire genotype and also the important properties relevant to the experiment. A detailed methods section then describes each procedure step by step.

Materials are listed on a day to day basis. Unless otherwise noted, the material requirements are calculated per student or pair of students working together.

We have avoided using the term "sterile" in both the methods and materials sections. Unless otherwise stated, all materials coming in contact with bacteria (test tubes, toothpicks, velvets) are assumed to be sterile. Recipes for media and methods for preparing plates are listed in the Appendix. For rich medium, LB can be used throughout unless specifically stated otherwise. We recommend a phosphate buffer (called medium A or 1 × A) as the standard minimal medium for both liquid medium and also for agar plates (see Appendix). Usually the term "test tube" refers to a 16–18 mm × 150 mm test tube, and the term "small test tube" refers to a 13 mm × 75 mm test tube. A fresh saturated culture is often termed an "overnight culture" or simply an "overnight," and a freshly growing culture in log phase (see Experiment 1) is referred to as an exponential culture.

NOMENCLATURE

Genotype and Phenotype

It is important to distinguish between genotype and phenotype when describing strains. The genotype of a strain is a set of genetic determinants, whereas the phenotype is a set of observable properties. We follow most of the recommendations of Demerec et al. (1966) with respect to genetic nomenclature, and so use the symbols listed by Taylor (1970) to refer to genetic loci. Therefore, three-letter *italicized* symbols are used to describe genetic loci, for instance *lac* or *met*. These symbols are followed by capital letters to further subdivide and distinguish related loci, for instance *metA* and *metB*. Mutational sites are designated by numbers, thus *met A1*, *met A2*, *met A3*, etc.

The plus (+) sign is used to designate the wild-type allele of a locus. Thus, $metA^+$, or simply "+," indicates the wild-type allele of the *metA* locus. When listing the genotype of a strain, only the alleles which differ from wild type are written. Mutant alleles are normally not given a minus (−) sign. Therefore, a strain with the genotype *trp*, *his*, *metA* has mutations in each of these loci. Some authors use trp^-, his^-, and $metA^-$, respectively, to refer to this genotype. In this manual, however, we rarely use the minus sign with reference to genotypes, and then only for occasional emphasis (except for the *lac* system).

Bacterial genes carried on F′ factors or specialized transducing phage are treated in exactly the same way. When we describe F′ factors carrying particular bacterial genes, we use a plus (+) sign to emphasize that the loci on the F′ factor are in the wild-type state. Thus, $F'lac^+proA^+,B^+$ indicates an F′ factor carrying the *lac* and *proA,B* region of the bacterial chromosome; and $F'lacZ\ proA^+,B^+$ indicates an F′ factor carrying the *lac* and *proA,B* regions but also carrying a mutation in the *z* gene.

Phenotypic designations are made either with written descriptions, or by abbreviations. In the latter case, we use non-italicized symbols beginning with a capital letter. Thus, Lac^- indicates inability to grow with lactose as the sole carbon source, Met^- inability to grow in the absence of methionine, etc. The following table presents many of the genetic and phenotypic symbols used throughout this manual. For a more complete list, we refer the reader to Taylor, 1970, and to Appendix III.

Suppressors

The classification of suppressor loci is a constant source of confusion. This occurs because the wild type carries no suppressor and is genetically *sup*$^+$, where *sup* is the symbol for suppressor loci. The phenotypic symbol for the wild-type suppressor-minus state is, however, Su$^-$. Strains carrying a suppressor are phenotypically Su$^+$ but genotypically *sup*$^-$. Specific suppressors are identified by number when speaking of the phenotype, and by letter when speaking of the genotype. Thus, Su1 is indicated by *supD*. A complete list of the symbols used for each of the suppressors in *E. coli* is given in Table 22A. To avoid confusion we refrain from writing the *sup*$^+$ or *sup*$^-$ designation,, but do use *supD* or *supE* to indicate a particular suppressor locus (the Su$^+$ state) and use no symbol at all to indicate the genotype of the Su$^-$ state.

Table 1. Genetic and Phenotypic Symbols Used Throughout This Manual

Genetic locus	Phenotype symbol	Phenotypic trait affected
ampA	Pen^r, Pen^s	resistance or sensitivity to penicillin or ampicillin
ara	Ara	ability to utilize arabinose
arg	Arg	requirement for arginine
attλ		integration site for prophage λ
attϕ80		integration site for prophage ϕ80
cysB	Cys	requirement for cysteine
gal	Gal	ability to utilize galactose
his	His	requirement for histidine
ilv	Ilv	requirement for isoleucine + valine
lac	Lac	ability to utilize lactose
leu	Leu	requirement for leucine
*malA,B**	Mal (λ^r)	ability to utilize maltose (resistance to λ)
met	Met	requirement for methionine
mtl	Mtl	utilization of D-mannitol
nal	Nal^r, Nal^s	resistance or sensitivity to nalidixic acid
pro	Pro	requirement for proline
purE	Ade, Pur	requirement for adenine
pyrC,F	Ura, Pyr	requirement for uracil
recA,B,C	Rec	ultraviolet sensitivity and competence for genetic recombination

* Mutants resistant to λ have mutations either in *malA* or *malB*.

Phenotype symbols are used when referring to properties of cells or strains. The genetic symbols are applied when describing a genetic locus, a group of genes, or a particular allele.

Table 1. *(continued)*

Genetic locus	Phenotype symbol	Phenotypic trait affected
rif	Rif^r, Rif^s	RNA polymerase resistance or sensitivity to rifampicin
spc	Spc^r, Spc^s	resistance or sensitivity to spectinomycin
strA	Str^r, Str^d, Str^s	resistance, dependence, or sensitivity to streptomycin
sup	Su	suppressor of nonsense mutations—for a complete listing of all suppressors, see introduction to experiments 22–24
thi	B1	requirement for thiamine
thr	Thr	requirement for threonine
tonA	$T1,5^r$; $T1,5^s$	resistance or sensitivity to phages T1, T5, and $\phi80$
tonB	$T1^r$, $T1^s$ TonB	resistance or sensitivity to phages T1, $\phi80$, and colicins B, I, and V; active transport of Fe
trp	Trp	requirement for tryptophan
tsx	T6	resistance or sensitivity to phage T6
xyl	Xyl	ability to utilize D-xylose

F Factor

In accordance with the recommendations of Demerec et al., we use the following genetic symbols to denote the sex of the bacteria (see Unit II):

- F^- — the state of lacking the sex factor, F;
- F^+ — the state of carrying an autonomous sex factor;
- F' — the state of carrying a sex factor which carries a genetically recognizable segment of the bacterial chromosome;
- Hfr — the state of harboring a sex factor which is integrated within the chromosome.

Prophage—Drug Resistance

We designate a prophage by using parentheses. Thus, (λ) indicates "carries λ as a prophage." For phenotypic designations involving sensitivity or resistance, we use the superscripts s and r, respectively. Thus, Str^r refers to streptomycin resistance, and Str^s to streptomycin sensitivity. Unless otherwise stated, Str^r indicates resistance to 200 μg/ml streptomycin, Rif^r to 100 μg/ml rifampicin, and Nal^r to 20 μg/ml nalidixic acid.

Sample Usage

A sample description of a bacterial cross illustrates the use of these symbols in referring to phenotypes and genotypes:

Cross of an HfrH (genotype *supE*; phenotype Su^+, Str^s) with an F^- (genotype *leu lac strA*; phenotype Leu^- Lac^- Str^r)

> "An Hfr strain was crossed with an F^- recipient and Lac^+ Str^r recombinants were selected. After purification these colonies were tested for Leu^+ and it was concluded that 10% had received the *leu* region of the chromosome. Since the *leu* mutation in the F^- was an amber mutation, the Leu^+ Lac^+ Str^r recombinants were further tested to determine whether the Leu^+ phenotype was due to the presence of a nonsense suppressor, the *supE* locus. If this were the case, then a *leu supE* (or leu^- *supE*) genotype would score as a Leu^+ phenotype. To test for an Su^+ phenotype, T4 amber phage were used . . . ."

It can be seen in this example that when we refer to the ability of a strain to grow without added leucine, we are describing a phenotypic trait and use symbols such as Leu^+ or Leu^-. When we refer to the *leu* locus, or a mutation in this region, then we are describing genotype and use italicized symbols.

NOMENCLATURE FOR THE *lac* SYSTEM

Unfortunately, the field of *lac* genetics and also the text *The Lactose Operon* use a different set of symbols than that recommended by Demerec et al. and by Taylor. Instead of designating genetic loci as *lacI*, *lacO*, *lacZ*, *lacY*, and *lacA*, most of the studies in the literature use *i*, *o*, *z*, *y*, and *a*, respectively. Although it is possible to convert these symbols to the Demerec system for this manual, we find it advantageous to retain the current *lac* nomenclature and state that it is an exception to the Demerec system. There exists such a large volume of important literature which uses these symbols that it would be necessary to learn this system anyway to be able to understand these papers properly. Also, we find it simpler to use symbols such as i^s, i^{-d}, and o^c to depict genotypes than to use the Demerec system, which prescribes that alleles should give no information as to phenotype.

The diploid i^s/i^{-d} should be written as *lacI694/lacI24* with an accompanying explanation that *I694* has an $\mathrm{i^s}$ phenotype and *I24* an $\mathrm{i^{-d}}$ phenotype. This is somewhat cumbersome, and we find designations such as i^s and i^{-d}, and o^c actually clearer and more informative.

For the *lac* system only, small italicized letters are used to indicate the genetic loci *i*, *p*, *o*, *z*, *y*, and *a*. Also, minus (−) and plus (+) signs are employed to indicate mutant and wild-type alleles, respectively. The symbols *lacZ*, *lacI*, etc. are retained only in cases where the entire genotype is listed, such as the strain list at the beginning of the manual.

Since all of the *lac* enzymes are easily assayed, as is the *lac* repressor, phenotypes can be associated with each genetic locus. This would not lead to confusion if different symbols such as $\mathrm{Ind^+}$ were used to indicate "inducible," rather than the more commonly used $\mathrm{i^+}$. Unfortunately, this has not been the case, and the same italicized symbols such as i^+ and i^- are often used for both genotypes and phenotypes. For instance, statements such as "the i^-/i^+ diploid had an i^+ phenotype" appear often in the literature. We recommend using unitalicized symbols for phenotypes in the *lac* system and do so in this book. The following table lists some of the more common symbols used.

Demerec system genetic locus	*Lac* symbol (genotypic, phenotypic)	Phenotype
lacI	i^+, i^+	inducible (not constitutive) for the *lac* enzymes
lacI	i^-, i^-	constitutive (non-inducible) synthesis of the *lac* enzymes
lacI	i^s, i^s	cannot be induced by lactose or IPTG; lac^-; synthesizes very low level of the *lac* enzymes
lacI	i^{-d}, i^{-d}	dominant constitutive (i^-); partially constitutive in the presence of additional wild type *i* gene
lacO	o^c, o^c	partially constitutive synthesis of the *lac* enzymes
lacP	p^-, p^-	lowered maximal levels of the *lac* enzymes

References

DEMEREC, M., E. A. ADELBERG, A. J. CLARK and P. E. HARTMAN. 1966. A proposal for uniform nomenclature in bacterial genetics. *Genetics 54:* 61.

TAYLOR, A. L. 1970. Current linkage map of *Escherichia coli. Bacteriol. Rev. 34:* 155.

STRAIN LIST

The following list describes all of the strains used in the experiments in this manual. These are available from Cold Spring Harbor as a separate kit. The list describes both the genotype (Demerec system) and the phenotype of each strain. In cases where specific strains have been referred to in the literature, the alternative name is included. In cases where the specific cistron or allele is unknown, then no letter or number is given. Thus, *trp* indicates a mutation in either *trpA*, *trpB*, *trpC*, *trpD*, or *trpE*. The intervals referred to in strains CSH1 to CSH20 are diagrammed in Figure 21A.

Strain Number	Alternate name	Sex	Genotype / Phenotype / Comments
CSH1		F^-	**Gen:** *trp lacZ strA thi* **Phen:** $Trp^-Lac^-Str^r B1^-$ carries nonsense mutation in interval 1 of the *z* gene
CSH2		F^-	**Gen:** *trp lacZ strA thi* **Phen:** $Trp^-Lac^-Str^r B1^-$ carries nonsense mutation in interval 3 of the *z* gene
CSH3		F^-	**Gen:** *trp lacZ strA thi* **Phen:** $Trp^-Lac^-Str^r B1^-$ carries nonsense mutation in interval 4 of the *z* gene
CSH4		F^-	**Gen:** *trp lacZ strA thi* **Phen:** $Trp^-Lac^-Str^r B1^-$ carries nonsense mutation in interval 6 of the *z* gene
CSH5		F^-	**Gen:** *trp lacZ strA thi* **Phen:** $Trp^-Lac^-Str^r B1^-$ carries nonsense mutation in interval 8 of the *z* gene
CSH6		F^-	**Gen:** *trp lacZ strA thi* **Phen:** $Trp^-Lac^-Str^r B1^-$ carries nonsense mutation in interval 9 of the *z* gene
CSH7	M7047	F^-	**Gen:** *lacY strA thi* **Phen:** $Lac^-Str^r B1^-$ carries amber mutation in the *y* gene
CSH8		F^-	**Gen:** *trp lacZ strA thi* **Phen:** $Trp^-Lac^-Str^r B1^-$ carries nonsense mutation in interval 11 of the *z* gene

CSH9		F^-	**Gen:** *trp lacZ strA thi* **Phen:** $Trp^- Lac^- Str^r B1^-$ carries nonsense mutation in interval 15 of the *z* gene
CSH10		F^-	**Gen:** *trp lacZ strA thi* **Phen:** $Trp^- Lac^- Str^r B1^-$ carries nonsense mutation in interval 20 of the *z* gene
CSH11		F^-	**Gen:** *trp lacZ strA thi* **Phen:** $Trp^- Lac^- Str^r B1^-$ carries nonsense mutation in interval 26 of the *z* gene
CSH11a		F^-	**Gen:** *trp lacZ strA thi* **Phen:** $Trp^- Lac^- Str^r B1^-$ carries unmapped nonsense mutation in the *z* gene
CSH11b		F^-	**Gen:** *trp lacZ strA thi* **Phen:** $Trp^- Lac^- Str^r B1^-$ carries unmapped nonsense mutation in the *z* gene
CSH11c		F^-	**Gen:** *trp lacZ strA thi* **Phen:** $Trp^- Lac^- Str^r B1^-$ carries unmapped nonsense mutation in the *z* gene
CSH12	XA21	F^-	**Gen:** *lacZ strA thi* **Phen:** $Lac^- Str^r B1^-$ carries deletion of beginning of *z* gene; alpha acceptor
CSH13	H120	F′ *lacZ* $proA^+,B^+$	**Gen:** Δ(*lacpro*) *supE thi* **Phen:** $Lac^- B1^- Su2\ (SuII^+)$ carries deletion #H120 in the *z* gene on the F′
CSH14	H111	F′ *lacZ* $proA^+,B^+$	**Gen:** Δ(*lacpro*) *supE thi* **Phen:** $Lac^- B1^- Su2\ (SuII^+)$ carries deletion #H111 in the *z* gene on the F′

Strain List *(continued)*

Strain Number	Alternate name	Sex	Genotype Phenotype Comments
CSH15	H119	F′ *lacZ proA*$^+$,*B*$^+$	**Gen:** Δ(*lacpro*) *supE thi* **Phen:** Lac$^-$B1$^-$Su2 (SuII$^+$) carries deletion # H119 in the *z* gene on the F′
CSH16	H114	F′ *lacZ proA*$^+$,*B*$^+$	**Gen:** Δ(*lacpro*) *supE thi* **Phen:** Lac$^-$B1$^-$Su2 (SuII$^+$) carries deletion # H114 in the *z* gene on the F′
CSH17	H145	F′ *lacZ proA*$^+$,*B*$^+$	**Gen:** Δ(*lacpro*) *supE thi* **Phen:** Lac$^-$B1$^-$Su2 (SuII$^+$) carries deletion # H145 in the *z* gene on the F′
CSH18	H125	F′ *lacZ proA*$^+$,*B*$^+$	**Gen:** Δ(*lacpro*) *supE thi* **Phen:** Lac$^-$B1$^-$Su2 (SuII$^+$) carries deletion # H125 in the *z* gene on the F′
CSH19	H138	F′ *lacZ proA*$^+$,*B*$^+$	**Gen:** Δ(*lacpro*) *supE thi* **Phen:** Lac$^-$B1$^-$Su2 (SuII$^+$) carries deletion # H138 in the *z* gene on the F′
CSH20	H220	F′ *lacZ proA*$^+$,*B*$^+$	**Gen:** Δ(*lacpro*) *supE thi* **Phen:** Lac$^-$B1$^-$Su2 (SuII$^+$) carries deletion # H220 in the *z* gene on the F′
CSH20a		F′ *lacZ proA*$^+$,*B*$^+$	**Gen:** Δ(*lacpro*) *supE thi* **Phen:** Lac$^-$B1$^-$Su2 (SuII$^+$) carries unmapped deletion in the *z* gene on the F′
CSH20b		F′ *lacZ proA*$^+$,*B*$^+$	**Gen:** Δ(*lacpro*) *supE thi* **Phen:** Lac$^-$B1$^-$Su2 (SuII$^+$) carries unmapped deletion in the *z* gene on the F′

CSH21		F′ *lacZ proA*$^+$,*B*$^+$	**Gen:** Δ(*lacpro*) *supE nalA thi* **Phen:** Lac$^-$Nal^r B1$^-$Su2 (SuII$^+$) carries ochre mutation (X90) in the *z* gene
CSH22		F′ *lacZ proA*$^+$,*B*$^+$	**Gen:** *trpR* Δ(*lacpro*) *thi* **Phen:** 5MT^r Lac$^-$B1$^-$ carries deletion (M15) in the *z* gene on the F′
CSH23	E5014	F′ *lac*$^+$*proA*$^+$,*B*$^+$	**Gen:** Δ(*lacpro*) *supE spc thi* **Phen:** Spc^r B1$^-$Su2 (SuII$^+$)
CSH24	ECO	F'_{ts} *lac*$^+$	**Gen:** Δ(*lacpro*) *supE thi* **Phen:** Pro$^-$B1$^-$Su2 (SuII$^+$) carries temperature-sensitive F′ *lac*
CSH25		F$^-$	**Gen:** *supF thi* **Phen:** Su3 (SuIII$^+$) B1$^-$
CSH26		F$^-$	**Gen:** *ara* Δ(*lacpro*) *thi* **Phen:** Ara$^-$Lac$^-$Pro$^-$B1$^-$
CSH27		F$^-$	**Gen:** *trpA33 thi* **Phen:** Trp$^-$B1$^-$
CSH28		F′ *lac*$^+$*proA*$^+$,*B*$^+$	**Gen:** Δ(*lacpro*) *supF trp pyrF his strA thi* **Phen:** Su3 (SuIII$^+$) Trp$^-$Ura$^-$His$^-$Str^r B1$^-$
CSH29		F$^-$	**Gen:** *trpB thi* **Phen:** Trp$^-$B1$^-$
CSH30		F$^-$	**Gen:** *trpC thi* **Phen:** Trp$^-$B1$^-$
CSH31		F$^-$	**Gen:** *trpD thi* **Phen:** Trp$^-$B1$^-$
CSH32		F$^-$	**Gen:** *trpE thi* **Phen:** Trp$^-$B1$^-$

Strain List (*continued*)

Strain Number	Alternate name	Sex	Genotype / Phenotype / Comments
CSH33	WD5017	F′ *colV*$^+$,*B*$^+$ *trp*$^+$ *cysB*$^+$	**Gen:** *thi* **Phen:** B1$^-$
CSH34	E7089	F′ *lacZ proA*$^+$,*B*$^+$	**Gen:** Δ(*lacpro*) *supE thi* **Phen:** Lac$^-$B1$^-$Su2 (SuII$^+$) carries ochre mutation (U118) in *z* gene on F′
CSH35	E7101	F′ *lacI*^{s} *proA*$^+$,*B*$^+$	**Gen:** Δ(*lacpro*) *supE thi* **Phen:** Lac$^-$B1$^-$Su2 (SuII$^+$) carries i^s mutation on the F′
CSH36	E7074	F′ *lacI proA*$^+$,*B*$^+$	**Gen:** Δ(*lacpro*) *supE thi* **Phen:** B1$^-$Su2 (SuII$^+$) constitutive for the *lac* enzymes (i^-)
CSH37		F′ *lacO proA*$^+$,*B*$^+$	**Gen:** Δ(*lacpro*) *supE thi* **Phen:** B1$^-$Su2 (SuII$^+$) partially constitutive for *lac* enzymes (o^c)
CSH38		F′ *lacP proA*$^+$,*B*$^+$	**Gen:** Δ(*lacpro*) *supE thi* **Phen:** B1$^-$Su2 (SuII$^+$) grows on lactose but not melibiose at 42° (p^-)
CSH39		F′ *lacZ proA*$^+$,*B*$^+$	**Gen:** Δ(*lacpro*) *thi* nalA **Phen:** Lac$^-$B1$^-$Nal^r carries amber mutation (YA536) in *z* gene on F′
CSH40		F′ *lacY proA*$^+$,*B*$^+$	**Gen:** Δ(*lacpro*) *thi* **Phen:** Lac$^-$B1$^-$ carries amber mutation in *y* gene on F′

CSH41		F′ *lacI,P proA*$^+$,*B*$^+$	**Gen:** Δ(*lacpro*) *galE thi* **Phen:** Gal$^-$Lac$^-$B1$^-$ carries mutations in the *i* gene (i^-) and the *lac* promoter (p^-) on the F′
CSH42	LC173	F$^+$	**Gen:** *thr leu lac thyA mal ilv thi* (T46) **Phen:** Thr$^-$Leu$^-$Lac$^-$Thy$^-$Mal$^-$Ilv$^-$B1$^-$DNA$_{ts}$ contains mutation (T46) preventing initiation of DNA synthesis at high temperature
CSH43	BMH480	F$^-$	**Gen:** *tonA* Δ(*lac*) (λ*C*I857*St*68h80) *thi* **Phen:** $\phi 80^r$ Lac$^-$B1$^-$ carries heat-inducible prophage
CSH44	BMH479	F$^-$	**Gen:** *tonA* Δ(*lac*) (λ*C*I857*St*68h80) *thi* (λ*C*I857*St*68h80d*lac*$^+$) **Phen:** $\phi 80^r$ Lac$^+$B1$^-$ double lysogen for heat-inducible prophage
CSH45		F$^-$	**Gen:** Δ(*lac*) *thi* (λ*C*I857*S*7) *trpR* **Phen:** Lac$^-$B1$^-$5MT^r carries heat-inducible prophage
CSH46	M96	F$^-$	**Gen:** *ara* Δ(*lacpro*) *thi* (λ*C*I857*St*68h80d*lacI*,*Z*) **Phen:** Ara$^-$Lac$^-$ (z^-U118) Pro$^-$B1$^-$ contains temperature-inducible prophage which carries the *lac* operon; *lac* region contains a mutation in the *z* gene (U118) and also in the *i* promoter (SQ)
CSH47	KL25	Hfr	**Gen:** *sup* **Phen:** Su$^+$ ←——— – – – – – – – – – – – – *ilv metB leu* ... *pyrE*
CSH48	X178	F′ *colV*$^+$,*B*$^+$, *trp*$^+$	**Gen:** *his* (ϕ80) *thi* **Phen:** His$^-$B1$^-$

Strain List (*continued*)

Strain Number	Alternate name	Sex	Genotype Phenotype Comments
CSH49		F^-	**Gen:** *ara*$^-$ Δ(*lacpro*) Δ(*trpA-D*) *str A* *thi* (φ80d*lacIZYA*) *mal* (φ80*h*d*trpA*$^+$*B*$^+$*C*$^+$*D*$^+$) **Phen:** Mal$^-$Ara$^-$Pro$^-$Lac$^-$Trp$^+$B1$^-$λ$^-$ lysogenic for either φ80*h*d*trp* and φ80*h* or for φ80*hptrp*
CSH50		F^-	**Gen:** *ara* Δ(*lacpro*) *strA thi* **Phen:** Ara$^-$Lac$^-$Pro$^-$Str^r B1$^-$
CSH51	X7700	F^-	**Gen:** *ara* Δ(*lacpro*) *strA thi* (φ80d*lac*$^+$) **Phen:** Ara$^-$Pro$^-$Str^r B1$^-$
CSH52	X7700*recA*		same as CSH51 but *recA*
CSH53	X8632	F^-	**Gen:** *ara* Δ(*lacpro*) *strA thi* (φ80d*lacI*) **Phen:** Ara$^-$Pro$^-$Lac$^+$ Str^r (i$^-$) B1$^-$
CSH54	X7150	F^-	**Gen:** Δ(*lacpro*) *supF trp pyrF his strA thi* **Phen:** Lac$^-$Pro$^-$Su3 (SuIII$^+$) Trp$^-$Ura$^-$His$^-$Str^r B1$^-$
CSH55	X7026N	F^-	**Gen:** Δ(*lacpro*) *supE nalA thi* **Phen:** Lac$^-$Pro$^-$Su2 (SuII$^+$) B1$^-$
CSH56	X68c	F^-	**Gen:** *ara* Δ(*lacpro*) *supD nalA thi* **Phen:** Ara$^-$Lac$^-$Pro$^-$Su1 (SuI$^+$) Nal^r B1$^-$
CSH57		F^-	**Gen:** *ara leu lacY purE gal trp his argG malA strA xyl mtl ilv metA* or *B thi* **Phen:** Ara$^-$Leu$^-$Lac$^-$Ade$^-$Gal$^-$Trp$^-$His$^-$Arg$^-$Mal$^-$Str^r Xyl$^-$Mtl$^-$Ile$^-$Met$^-$B1$^-$

CSH58	RL250	F^-	**Gen:** *ara thr leu proA lac gal trp his nalA recA1 thyA strA xyl mtl argE thi sup thy* **Phen:** $Ara^-Thr^-Leu^-Pro^-Lac^-Gal^-Trp^-His^-Str^rXyl^-Mtl^-Arg^-B1^-Su^+Thy^-$
CSH59	X181a	F^-	**Gen:** *pyrC trp strA thi* **Phen:** $Ura^-Trp^-Str^r$ $B1^-$
CSH60	Ra-2	Hfr	**Gen:** *sup* **Phen:** Su^+ ←——— – – – – – – – – – – – – – – *metB argE leu* … *ilv*
CSH61	D7011	HfrC	**Gen:** *trpR thi* **Phen:** $5MT^r$, $B1^-$ ←——— – – – – – – – – – – – – – – *purE lac leu* … *gal*
CSH62	CA8000	HfrH	**Gen:** *thi* **Phen:** $B1^-$ ←——— – – – – – – – – – – – – – – *thr leu lac* … *malB*
CSH63		HfrH	**Gen:** *val*^{r} *thi* Δ(*lac pro*) **Phen:** Val^r (resistant to greater than 50 μg L-valine ($B1^-Lac^-Pro^-$ ←——— – – – – – – – – – – – – – – *val lac* … *malB*
CSH64	KL14	Hfr	**Gen:** *thi* **Phen:** $B1^-$ ←——— – – – – – – – – – – – – – – *argA strA* … *thyA*
CSH65	XA28N	F^-	**Gen:** *leu lac strA thi thr* **Phen:** $Leu^-Lac^+Str^r$ $B1^-$ Thr^-

Strain List (*continued*)

Strain Number	Alternate name	Sex	Genotype Phenotype Comments
CSH66	M7133	F^-	**Gen:** Δ(*lac*) *thi* (λCI857S7*plac*5 $i^- z^+ y^-$) **Phen:** $B1^- Lac^-$ carries temperature-inducible prophage carrying *lac* region ($i^- z^+ y^-$)
CSH67	R5	Hfr	**Gen:** *lac gal xyl mtl malA thi* (λ) **Phen:** $Lac^- Gal^- Xyl^- Mtl^- Mal^- B1^-$ ←———— - - - - - - - - - - - - - - *leu thr ilv* ... *proA*
CSH68	Hfr6	Hfr	**Gen:** *mtl met malB* **Phen:** $Mtl^- Met^- Mal^-$ λ^r ←———— - - - - - - - - - - - - - - *tsx purE trp* ... *lac*
CSH69	KL99	Hfr	**Gen:** *thi* **Phen:** $B1^-$ ←———— - - - - - - - - - - - - - - *pyrC trp his* ... *pyrD*

CSH70	P4X	Hfr	**Gen:** *metB argE thi* **Phen:** Met⁻Arg⁻B1⁻ ← - - - - - - - - - *proA leu* … *proB*
CSH71	RW361	HfrH	**Gen:** Δ(*gal att80λ bio uvrB*) *thi* **Phen:** Gal⁻Bio⁻B1⁻ grows poorly on tetrazolium plates at 41° but grows well at 33° on these plates
CSH72		F'_{ts} *trp*⁺ *att80*	**Gen:** *ara* Δ(*lacpro*) *tonB trp strA thi* (φ80d*lacI*) *recA* **Phen:** T1ʳ B1⁻Strʳ Ara⁻Pro⁻Rec⁻ loses F′ factor at high temperature; temperature-sensitive on minimal plates
CSH73		HfrH	**Gen:** Δ*lac* Δ(*ara-leu*) *thi* **Phen:** Lac⁻Ara⁻Leu⁻B1⁻
CSH74	K46; KL16	Hfr	**Gen:** *thi* **Phen:** B1⁻ ← - - - - - - - - - *lysA thyA* … *argG*

UNIT I

INTRODUCTORY EXPERIMENTS

INTRODUCTION TO UNIT I

Notes on Simple Techniques

Dilution Methods

For this experiment as well as in all subsequent exercises it will be necessary to dilute bacterial cultures (or phage lysates) before plating. A saturated culture contains on the order of 10^9 live cells/ml. To be able to visualize several hundred isolated colonies on a plate we would have to dilute this culture 10 million-fold. One way to do this is by making a series of dilutions of either 1:10 or 1:100. A 10^{-1} dilution is achieved by pipetting 1 ml into 9 ml of dilution buffer, and a 10^{-2} dilution is achieved by adding 0.1 ml to 9.9 ml. There is no particular need to use 10 ml, and often it is more practical to use larger volumes which require fewer manipulations. Alternatively, dilution tubes with 5 ml buffer each can be employed. In this case 0.05 ml is used for a 10^{-2} dilution and 0.55 ml for a 10^{-1} dilution. The dilution buffer is either minimal salts (1 × A) or simply saline. To achieve a final dilution of 10^{-6} we would have to use a series of three 10^{-2} dilutions:

$$10^{-2} \times 10^{-2} \times 10^{-2} = 10^{-6}.$$

A 10^{-7} dilution can be achieved with an additional 10^{-1} dilution:

$$10^{-2} \times 10^{-2} \times 10^{-2} \times 10^{-1} = 10^{-7}.$$

A 10^{-6} dilution of a culture which originally contained 2×10^9 cells/ml contains 2×10^3 cells/ml or 2000 cells/ml. If we pipette 0.1 ml of this, then we are delivering 200 cells. In this manual we always refer to the dilution **before** plating 0.1 ml. The directions "plate a 10^{-4} dilution" call for the taking of an aliquot **from** a 10^{-4} dilution of the culture. Since the cells are distributed in a specified volume at random (if they are well mixed), the Poisson distribution is applicable; thus the standard deviation is the square root of the number of colonies. For the experiments described in this manual, 200–300 colonies are sufficient.

Dilution tubes can be prepared directly before use according to exact needs. When a high degree of accuracy is required, dilution tubes should be prepared after autoclaving, since evaporation during sterilization is inevitable.

Care should always be taken to wipe off the end of the pipettes (particularly 0.1-ml pipettes) before transferring to the next dilution tube. This is done either by wiping against the side of the test tube or by wiping with a small tissue. To minimize inaccuracies due to random error, important titers can be done with multiple determinations. For a discussion of sources of error in making serial dilutions and a review of other colony count techniques, see Postgate (1967) and Kubitschek (1969).

Sterile Plating and Streaking Techniques

To obtain single isolated colonies of a bacterial strain, either a wire loop or flat wooden toothpicks are used. As demonstrated in Figure 1A, cells are first applied to a sector of a plate (A) by either a sterile toothpick (usually round) or else with a wire needle or loop. The cells are then streaked across the sector (B) and after the needle is sterilized by flaming over a bunsen burner and then cooled by dipping in the agar plate, the bacteria are further streaked in a zig-zag pattern towards the center of the plate. Two of these operations (C) (sometimes three are required) usually result in isolated colonies in each sector. As many as eight sectors per plate can be streaked with little difficulty.

Small aliquots of a cell suspension can be spread over the surface of an agar plate by using a bent glass rod (Figure 1B), a bent pasteur pipette, or even a wire loop. The rod is immersed in 90% ethanol and passed through a bunsen burner directly before use. Alternatively, cells can be applied to a plate by means of a **soft agar**

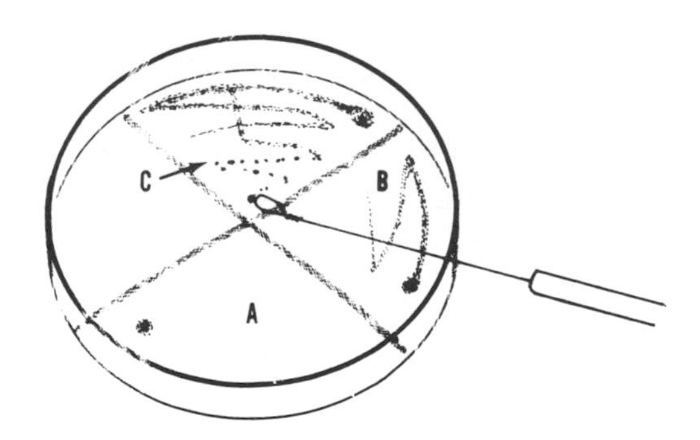

Figure 1A. Streaking for single colonies.

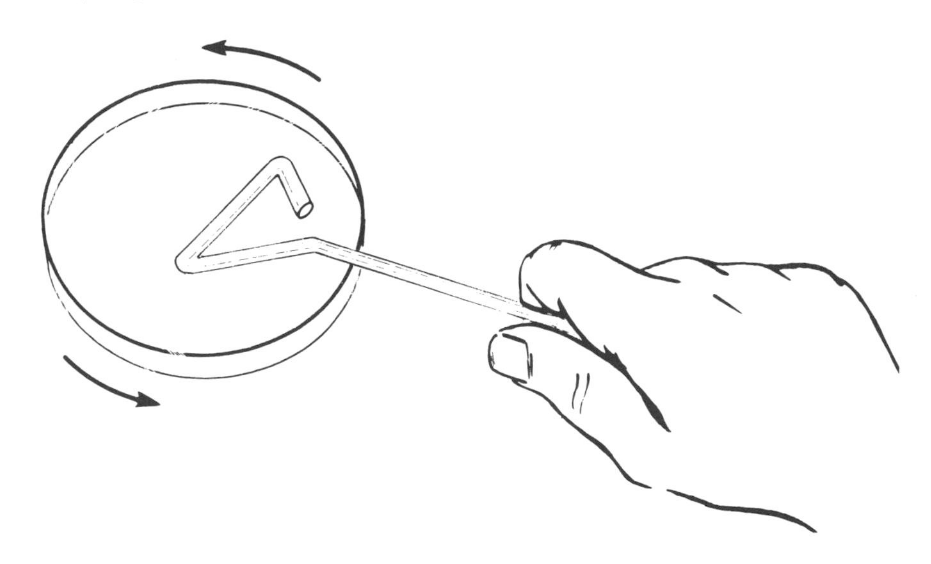

Figure 1B. Spreading liquid over the surface of an agar plate.

overlay. Usually 2–3 ml of 0.8% agar are added to 0.1 to 0.2 ml aliquots of a cell suspension and plated out directly. The soft agar is kept molten at 45°. Soft agar hardens quickly at room temperature, so manipulations should be done quickly.

For those not familiar with bacteriology, we cannot stress enough how important it is to work with pure strains. Before any strain can be used it must be free of contamination. As a general rule, a strain should go through at least two steps of single colony isolations before being stored. **Never use a strain which has been resuscitated from storage or has been received from another investigator without repurifying it and testing single colonies!**

Methods for storing strains are described in the Appendix. In general, for periods of several weeks, colonies on plates kept in a refrigerator are usable, and for several months agar slants are convenient. Stabs are recommended for longer periods, as are cultures kept in glycerol at −15°C and lyophiles. Usually cultures are prepared by inoculating several milliliters of broth with a single colony and then growing to saturation overnight. We use the term *overnight* to refer to a recently prepared saturated culture.

On the day of an experiment the overnight is subcultured by transferring several drops into a fresh broth tube. Continued subculturing should be avoided, and it is always preferable to prepare a new culture from a single colony. Otherwise, mutants and faster growing contaminants have a chance to accumulate and overgrow the original strain. Strains with unstable properties, for instance certain Hfr's, are readily lost by continued subculturing.

Aeration

For genetic experiments it is usually convenient to aerate small volumes (2–5 ml) in 16 mm or 18 mm diameter test tubes placed on a roller drum at 50–60 rpm. Aeration of larger volumes is achieved by hard shaking of an erlenmeyer flask. The volume of the bacterial culture should be less than 20% of the rated volume of the flask. Thus, 100 ml in a 500 ml flask is satisfactory. It is also possible to use a bubbler tube and bubble air through from an aeration pump (Clowes and Hayes, 1968).

References

Clowes, R. C. and W. Hayes. 1968. *Experiments in Microbial Genetics.* John Wiley and Sons, New York.

Kubitschek, H. E. 1969. Counting and sizing micro-organisms with the Coulter counter. *Methods in Microbiology*, Vol. 1, p. 593. Academic Press, New York.

Postgate, J. R. 1967. Viability measurements and the survival of microbes under minimum stress. *Adv. Microbial Physiol. 1:* 1.

EXPERIMENT 1

Determination of Viable Cell Counts: Bacterial Growth Curves

Often it is necessary to determine the number of cells in a bacterial culture at a given time. This can be done in several ways:

Cell Number (Absolute)

The number of cells can be measured with the aid of microscopes equipped with counting chambers constructed to hold standard volumes, such as a Petroff-Hausser (Mallette). Electronic counters such as the Coulter Counter (Kubitschek) have also been used for this purpose.

Viable Cells—Colony Counts

Viable cells are determined by spreading a measured aliquot of a diluted bacterial culture onto nutrient agar plates and counting the resulting colonies after a period of incubation.

Counting the cells directly measures the total number of bacteria, whereas viable cell counts measure only those cells which will form colonies. The ratio viable cell count/total cell count is often termed the plating efficiency, when colony forming ability is used to determine the number of viable cells. Usually, the plating efficiency is higher on rich medium than on minimal medium. Also, many indicator media lower the plating efficiency.

Cell Mass

The cell mass can be measured by determining the dry weight of a culture, or by recording the turbidity. In cases where most of the cells are viable, both the dry weight and the turbidity are directly related to cell number.

Bacterial cultures exhibit light scattering, which is approximately proportional to cell mass. At wavelengths where the ratio of absorbance to light scattering is low (e.g. 550–600 mμ), the optical density of a growing culture can be followed in a spectrophotometer and is proportional to the number of cells. The exact relationship of optical density at a particular wavelength to cell number varies with the particular strain used and the growth conditions, since the mass/cell number ratio varies with the medium (Maaløe and Kjeldgaard).

Logarithmic Growth

Growing bacteria, dividing by binary fission, exhibit exponential or logarithmic growth kinetics until a point of saturation of the culture is reached. During this time the increase in the number of bacteria (N) per unit time (t) is proportional to the number of bacteria present in the culture.

$$dN/dt = kN \qquad (1)$$

where k is a growth constant; on integration this yields

$$\log(N) = \log(N_o) + Kt \qquad (2)$$

where $K = k/2.3$ and N_o = the number of bacteria present when $t = 0$. If we plot log (N) or log (OD) versus the time (t), then the slope = K.

Often we are interested in the generation time of the culture, the time required for the cells to double in number. In this case $N = 2N_o$. Since from Equation 2 we now have $\log 2 = Kt$, the generation time $= t = 0.3/K$. We can thus easily determine the generation time of a culture from a plot of the logarithm of the optical density versus the time.

Growth Curves

When exponentially growing cells are transferred to fresh medium, they continue growing with logarithmic kinetics. If instead the inoculum is taken from a saturated culture consisting of half-starved, slowly-multiplying cells, there is a lag time (usually consisting of 30–60 minutes) before the cells resume rapid growth. In Figure 1C the growth curve of a strain used in this manual (CSH26) is shown. The curve was obtained by measuring the optical density at 550 mμ at various times after dilution of a full grown culture, exactly as prescribed in "Methods" for Experiment 1. Both a linear and semilog plot (Figure 1D) of the OD_{550} mμ versus time are presented. In addition, viable cell counts were taken by diluting the culture and plating on rich plates.

Time	Viable Cell Count	OD_{550} mμ
10 minutes	0.9×10^8/ml	0.420
50 minutes	2.2×10^8/ml	1.220

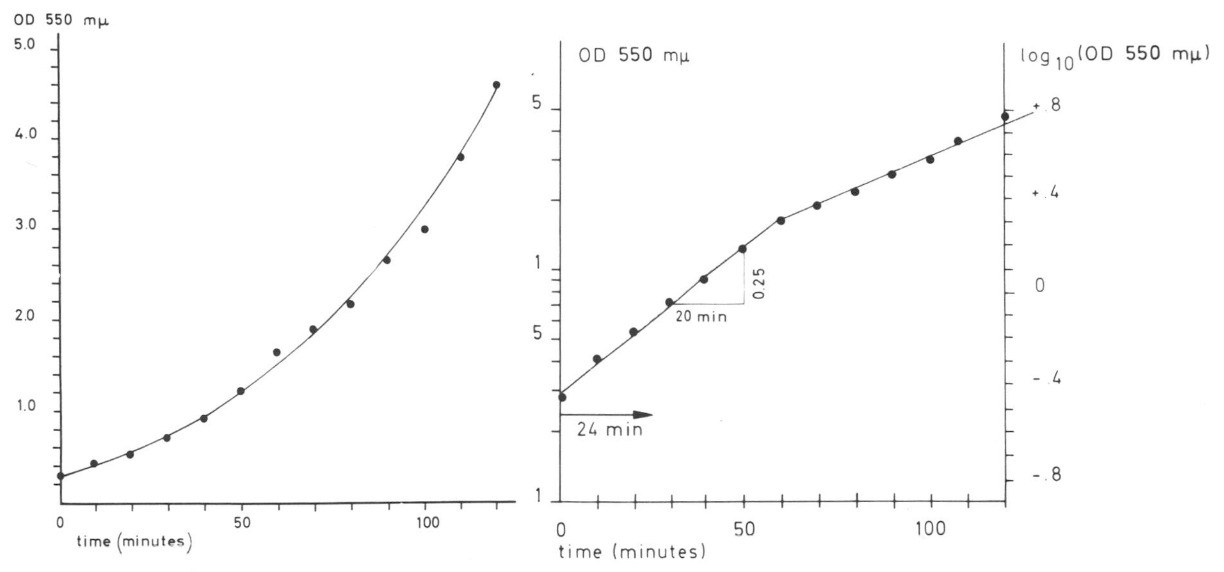

Figure 1C. Growth curve, linear plot.

Figure 1D. Growth curve, semilog plot.

It can be seen that under the conditions of the experiment and for this strain, one OD_{550} mμ unit corresponds to 1.9×10^8 viable cells/ml, or 1×10^8 cells/ml corresponds to 0.520 OD_{550} mμ units. It can also be seen that in this experiment the exponential growth rate remained unchanged until a density of $3\text{–}4 \times 10^8$ cells/ml (as determined by viable cell count) was reached. We interpret this change in slope as the end of exponential phase. Since the slope of the curve (right-hand scale) is

$$\frac{0.25}{20 \text{ min}} = \frac{0.0125}{\text{min}} = K,$$

then the generation time

$$(t) = \frac{0.3}{K} = \frac{(0.3) \text{ min}}{(0.0125)} = 24 \text{ min}.$$

In practice we usually plot the data on semilog paper (left-hand scale) rather than compute the log values. In this case we can read the generation time directly from the graph by determining the time required for the ordinate value (the OD_{550}) to double.

Effect of Nutrients—Aeration

E. coli can grow on a simple chemical medium in which glucose is the carbon and energy source, provided the medium is buffered at a pH near 7.0 and contains magnesium, phosphate, and a nitrogen source (usually ammonium chloride or ammonium sulfate). Strains of *E. coli* grow more rapidly in rich broth than in this minimal medium because the broth supplies many of the compounds which the cell would otherwise have to synthesize. Addition of several percent agar to the medium still allows growth, only now this is confined to the surface of a solid medium, convenient for use in petri dishes.

E. coli is capable of growing either in the presence of air (aerobically) or in its absence (anaerobically). However, the growth rate in the absence of aeration is significantly poorer than that achieved with good aeration. After a density of 10^7 cells is reached, air must be supplied either by shaking the liquid or by bubbling air through the cultures to allow rapid growth (Stent, 1963).

Additional Methods

Both staining and the measurement of dye uptake have been used as methods of determining cell number (Postgate, 1967), as have slide culture techniques (Postgate, 1967, 1969). Here the ratio of viable to non-viable cells is accurately determined by examining a microculture on an agar surface sealed on a slide. After several hours incubation the number of microcolonies versus single cells is viewed under a microscope.

Design of Experiment

In the following experiment the growth rate of two different *E. coli* strains will be determined. Both optical density readings and viable cell counts will be determined at different times from growing cultures. In addition to demonstrating the relationship between turbidity and number of viable cells, this experiment will demonstrate the techniques of dilution and plating.

References

CLOWES, R. C. and W. HAYES. 1968. *Experiments in Microbial Genetics.* John Wiley and Sons, New York.—A complete description of methods of microculture.

KUBITSCHEK, H. E. 1958. Electronic counting and sizing of bacteria. *Nature 182:* 234.

KUBITSCHEK, H. E. 1969. Counting and sizing micro-organisms with the Coulter counter. *Methods in Microbiology*, Vol. 1, p. 593. Academic Press, New York.—This and the previous reference review the use of electronic counters.

MAALØE, O. and N. O. KJELDGAARD. 1966. *Control of Macromolecular Synthesis.* W. A. Benjamin, Inc. New York.—Contains a complete discussion of microbial growth.

MALLETTE, M. F. 1969. Evaluation of growth by physical and chemical means. *Methods in Microbiology*, Vol. 1, p. 521. Academic Press, New York.—Reviews current physical and chemical methods used to measure growth of bacteria and includes a good discussion of light-scattering by bacteria and turbidimetry, and also of the use of chambers for microscopic counting.

POSTGATE, J. R. 1967. Viability measurements and the survival of microbes under minimum stress. *Adv. Microbial Physiol. 1:* 1.

POSTGATE, J. R. 1969. Viable counts and viability. *Methods in Microbiology*, Vol. 1, p. 611. Academic Press, New York.

QUESNEL, L. B. 1969. Methods of microculture. *Methods in Microbiology*, Vol. 1, p. 365. Academic Press, New York.—A complete description of methods of microculture.

STANIER, R. Y., M. DOUDOROFF and E. A. ADELBERG. 1970. *General Microbiology*, 3rd Ed., p. 298. Macmillan, New York.—A complete description of microbial growth.

STENT, G. S. 1963. *Molecular Biology of Bacterial Viruses.* W. H. Freeman and Co., San Francisco.

Method

Day 1

Subculture an overnight of each of the two strains to be used (CSH51 and CSH61). (Half the class could use one of these strains and half the other.) Dilute each fresh overnight 1:50 into prewarmed rich broth. We recommend pipetting 0.4 ml into 20 ml broth in a 250-ml erlenmeyer flask (or 0.3 ml into 15 ml in a 100- or 125-ml flask) which is then shaken vigorously in a 37° room or in a shaking waterbath.

Points should be taken by withdrawing 0.5 ml with a 1-ml pipette and immediately diluting 1:1 by delivering the sample into 0.5 ml chilled broth in a test tube. This should be done with as little disruption of the shaking of the cells as possible. The flask should be maintained at 37° at all times throughout these manipulations. The 1-ml sample should be read at 550 mμ in a spectrophotometer as soon as possible. If samples are to be further diluted for viable cell counts, 0.1 ml of this should be withdrawn directly before reading the optical density.

The chilled broth must be sterile as must all pipettes used in this experiment. For viable cell counts, dilutions and plating should be done promptly. Both a 10^{-5} and a 10^{-6} dilution should be plated for points after the first hour, but a 10^{-5} dilution is sufficient for points during the first hour.

Beginning 30 minutes after the subculture is made, take points for optical density readings every 10 minutes and time points for viable cell counts every 20 minutes (thus every other optical density sample will also be diluted for plating). Use as a blank an aliquot of chilled rich broth which is identical to the broth used for growth and for dilution. When the optical density reading reaches 0.6, use 1:5 dilutions for the next samples, since readings too much above this are less reliable due to the small amount of light transmitted. Continue to take points for 3 hours, or until the optical density has leveled off.

Construct a plot of optical density *vs.* time, and also the log of the cell density *vs.* time (on semi-log paper). Compute the generation time of each strain under these conditions. Incubate the plates at 37°.

Schedule of Time Points

Time	−30′ (subculture)	0	10	20	30	40	50	60	70	80	90	100	110	120	130	140	150	160	180
OD		+	+	+	+	+	+	+	+	+	+	+	+	+	+	+	+	+	+
viable cell count		+		+		+		+		+		+		+		+		+	+

Day 2 Count the colonies on each of the titer plates, and construct curves using these cell numbers. Compare the curves, and also the relationships between viable cell counts and optical densities at a given wavelength.

This experiment can be varied in many ways. The effect of medium (minimal versus rich), temperature, and aeration can be measured. For a class consisting of different groups of students, it is suggested that each group change one of these variables to determine its effect on the growth curve. The plating efficiency can also be measured by comparing the number of live cells (as judged by the viable cell count) with the actual number of cells seen in a microscope counting chamber (Petroff-Hausser). The relative plating efficiency on different types of plates (LB versus minimal, etc.) can also be included as an exercise. Finally, the growth rates of different strains can be compared, for instance CSH51 with its *recA*$^-$ derivative (strain CSH52).

Materials

Day 1

36 dilution tubes
16 LB plates
42 0.1-ml and 10 1-ml pipettes for titering
18 1-ml pipettes for withdrawing samples
125-ml or 250-ml erlenmeyer flask
shaking waterbath at 37°
overnight of CSH51 and CSH61
18 test tubes with either 0.5 ml or 0.8 ml broth kept in ice bucket
2 test tubes with 10 ml broth each (for blank); all broth tubes should contain the identical broth as the growth medium
18 pasteur pipettes
spectrophotometer set at 550 mμ

Flasks containing side arms which conveniently fit into Klett reading devices are available from Metalloglass.

EXPERIMENT 2

Preparation and Plaque Assay of a Phage Stock

Lytic and Lysogenic Cycles

A variety of different viruses or **bacteriophage** are capable of infecting and lysing *E. coli*. These phage, consisting of a nucleic acid molecule enclosed in a protein sheath or "head" (see Figure 2A), adsorb to receptor sites on the bacterial cell wall. Infection begins when the phage DNA is injected into the cell. Viral genes are then expressed in an ordered temporal sequence. Phage enzymes direct the synthesis of new phage DNA, the assembly of new virus particles, and the disruption of the cell wall (lysis), allowing release of several hundred new phage. The process beginning with a single phage genome and ending with lysis and the release of new phage progeny is often termed the **lytic cycle**.

Certain bacteriophage can undergo, in addition to the lytic response, a second process referred to as the **lysogenic cycle** (Figure 2B). This occurs when the injected DNA is integrated into the bacterial chromosome and specific phage proteins, termed **repressors**, act to prevent the autonomous replication of the phage DNA and the expression of phage functions which are required for the lytic cycle. Normally, the insertion of phage DNA is mediated by phage-encoded enzymes. These direct the integration of the phage genome at specific sites on the bacterial

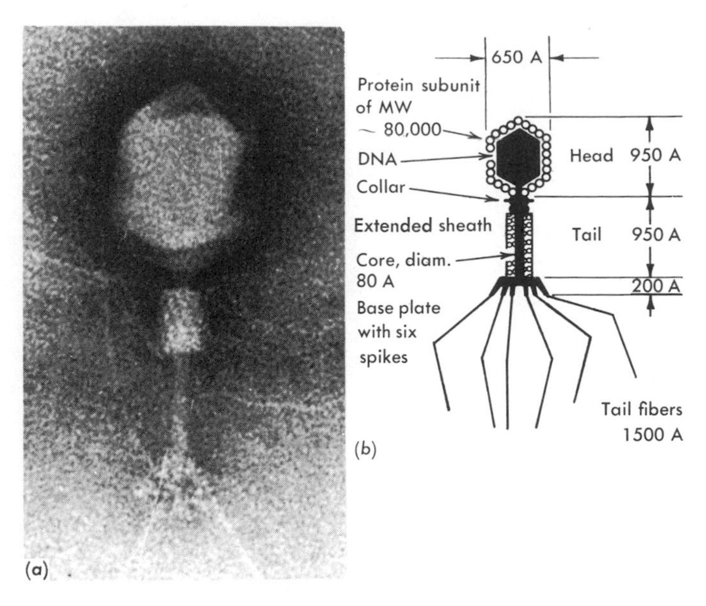

Figure 2A. The structure of the T-even (2, 4, and 6) phage particle. **(a)** An electron micrograph of T2. (Reproduced from R. W. Horne et al., *J. Mol. Biol.*, 1, 281, 1959.) **(b)** A schematic drawing showing detailed features revealed by electron microscopy. (Reproduced with permission from J. D. Watson, *Molecular Biology of the Gene*, 2nd Ed., p. 469. W. A. Benjamin, Inc., 1970.)

chromosome, referred to as **attachment sites**. The inserted phage genome, or **prophage**, is passively replicated as part of the bacterial chromosome.

Bacterial strains carrying a prophage are termed **lysogens**. Phage which are capable of both the lytic and lysogenic response are referred to as **temperate** phage, while those which can undergo only the lytic response and cannot become prophage are termed intemperate or **virulent phage**.

Lysogenic Phage—The Lambdoid Phage

At least two major groups of temperate phage can be distinguished on the basis of their ability to recombine with one another. One class includes the well studied phage λ and its relatives, the lambdoid phages. Included in this category are ϕ80, 434, and 21. λ phage particles consist of a double-stranded DNA molecule (MW 31 million daltons) enclosed in an icosahedral protein capsule head about 0.054 microns in diameter. A 0.15 micron-long tube terminating in a fiber constitutes the tail (Figure 2C).

Lambdoid phages recombine with one another but do not share the same immunity specificity. A λ prophage directs the synthesis of a repressor molecule which recognizes λ operators, represses the synthesis of λ proteins, and prevents the autonomous replication of λ DNA. A λ lysogen will thus prevent a second or **superinfecting** phage from entering the lytic cycle. A λ lysogen is therefore **immune** to superinfection by λ, but not by ϕ80 or 434 which are said to be **heteroimmune** with respect to λ.

Prophage enter the lytic cycle following the inactivation of the repressor. This process, termed **induction**, can be initiated by experimental means. Lambdoid phage are induced by exposing lysogens to ultraviolet irradiation. Also, many mutants of λ have been isolated which synthesize a heat-labile repressor, allowing the induction of prophage carrying this mutation by simply heating a culture of a

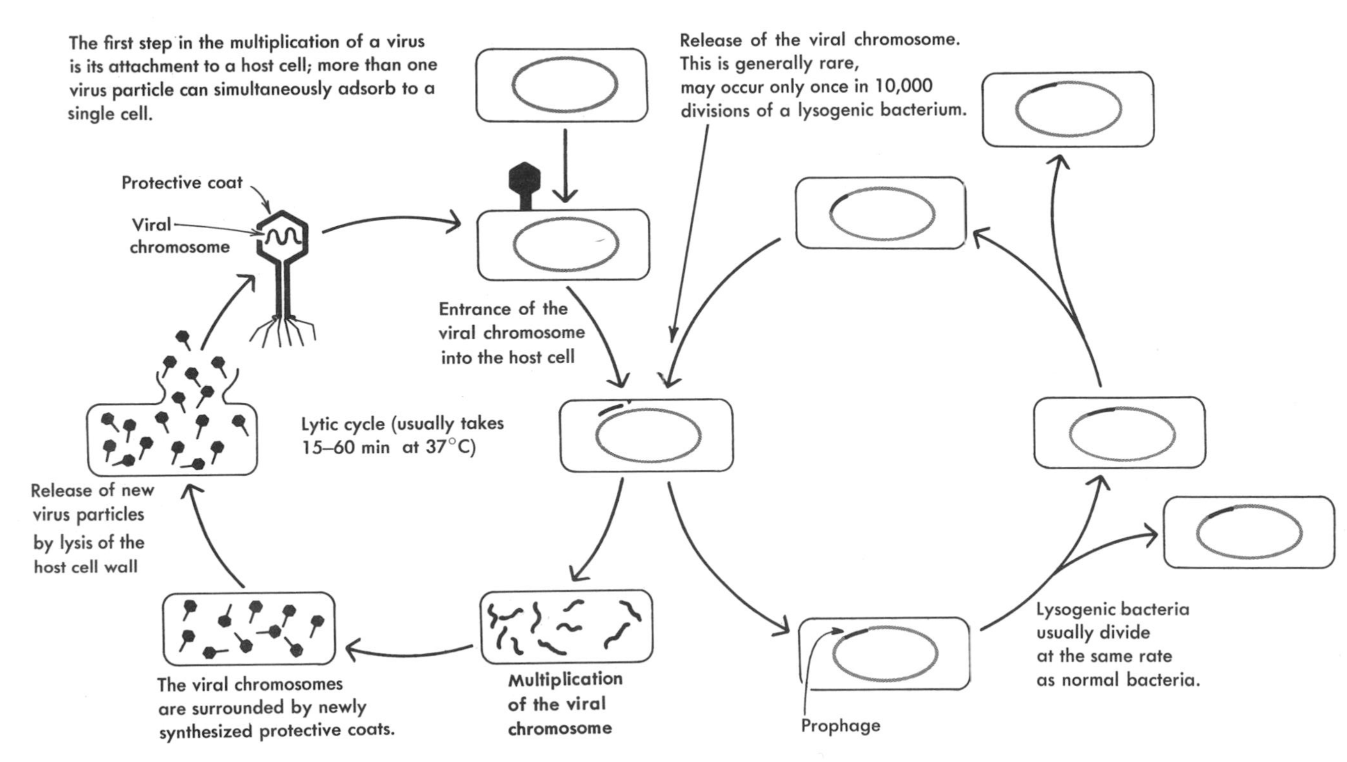

Figure 2B. The life cycle of a lysogenic bacterial virus. After its chromosome enters a host cell, it sometimes immediately multiplies like a lytic virus and at other times becomes transformed into prophage. The lytic phase of its life cycle is identical to the complete life cycle of a lytic (non-lysogenic) virus. Lytic bacterial viruses are so called because their multiplication results in the rupture (lysis) of the bacteria. (Reproduced with permission from J. D. Watson, *Molecular Biology of the Gene*, 2nd Ed., p. 205. W. A. Benjamin, Inc., 1970.)

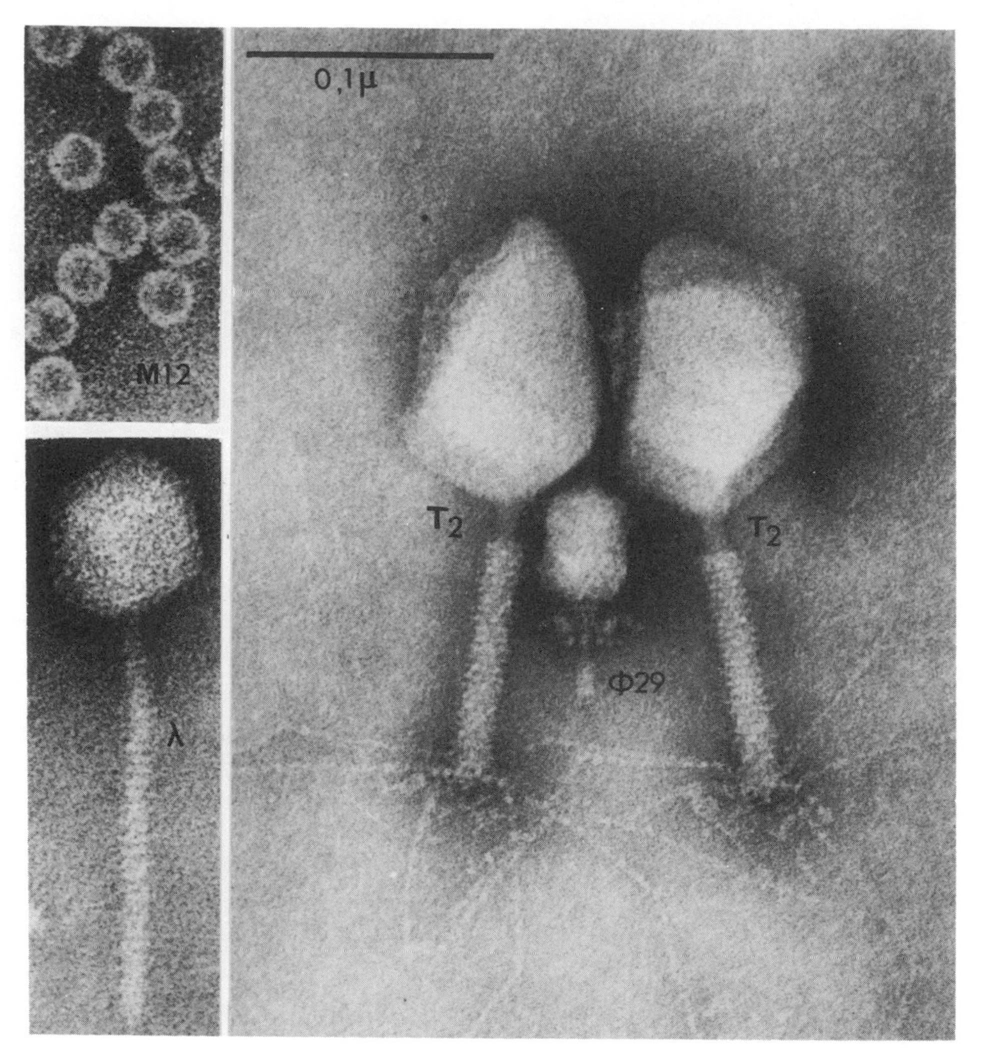

Figure 2C. Bacteriophages M12, ϕ29, λ, and T2 shown at the same magnification (320,000 ×). M12 is a small RNA phage closely related to R17 (Hohn and Hohn, 1970). *B. subtilis* phage ϕ29 has been described by Tsien et al. (1971). Negatively stained preparations by D. L. Anderson (ϕ29 and T2) and F. Eiserling (λ). (Reproduced with permission from Kellenberger and Edgar, 1971.)

lysogen. Following inactivation of the repressor, phage-specific proteins direct the excision of λ from the host chromosome, the autonomous replication of phage DNA, and assembly of virus particles and lysis of the cell wall.

Transducing Phage Lines

The most important practical use of phages such as λ and ϕ80 is the ability of these viruses to incorporate bacterial genes into the phage genome and transduce recipient strains. These abnormal particles, or specialized transducing phage, give rise to stable phage lines which allow DNA enrichment for specific genes. This makes possible the overproduction of specific enzymes after infection or heat induction. The isolation of specialized transducing phage lines is covered in detail in Unit V.

Other Temperate Phage

A second group of temperate phage includes P1 and P2. These phages do not recombine with lambdoid phages and their prophages are not inducible by UV light. Lysates of P1 include defective particles which carry segments of the bacterial chromosome and which can transfer them to other cells. Because of this ability to mediate generalized transduction, P1 is widely used by bacterial geneticists. P1 has no specific attachment site on the *E. coli* chromosome and apparently does not integrate into the host chromosome at all, but is instead passively replicated as an extra-chromosomal (cytoplasmic) element. P2 differs from both P1 and the lambdoid phages in that it has several different attachment sites on the *E. coli* chromosome.

Virulent Phage

Certain bacteriophage cannot integrate into the host chromosome and are not subject to repression. These phage which never become prophage, and consequently form clear plaques on a sensitive host, are termed virulent phage. Virulent mutants of λ and $\phi 80$ which form clear plaques even on an immune host have been isolated (λv; $\phi 80v$).

Among the virulent or intemperate phage are the T-phage (T1, T2, T3, T4, T5, T6, and T7), a group of double-stranded DNA viruses. Also included in this class are a series of single-stranded RNA phages (R17, MS2, f2, M12, fr, and Qβ) and a group of single-stranded DNA phages (fd, f1, and M13). These particular single-stranded viruses specifically infect strains of *E. coli* which carry the F factor (see introduction to Unit II). All of these phages form clear plaques on sensitive strains.

Plaques

When viable bacteria are spread on the surface of a nutrient agar plate, they form an even layer of cells after several hours of incubation. If a single virus particle is introduced, the phage adsorbs to and infects a single cell. After 15–60 minutes the cell lyses, releasing several hundred new virus particles into the medium. These infect, in turn, the neighboring bacteria and after a number of cycles a small clearing or **plaque** is readily visible in the bacterial lawn (Figure 2D). Virulent phage kill virtually all the infected bacteria and form clear plaques. However, a percentage of the cells infected by a temperate phage become lysogenic (and thus immune to superinfection) and survive. The growth of lysogens results in the formation of turbid plaques.

Preparation of Lysates

Lysates, stocks of particular phage lines, are prepared by infecting sensitive bacteria and then permitting additional growth to allow phage production and subsequent cell lysis. Chloroform is added to further disrupt the cell wall. Cell debris is then centrifuged, and the supernatant, which is the lysate, is stored in the cold. After infection cells are either plated in a layer of soft agar on nutrient plates, or else grown with aeration in liquid broth. In the former case enough phage, usually 10^5, are applied to each plate to allow confluent lysis over the surface of the plate. This is a result of the overlapping of plaques. The soft agar layer is scraped into a test tube or centrifuge tube, and then treated as above.

The preparation of lysates in this manner gives high titers for virulent phage,

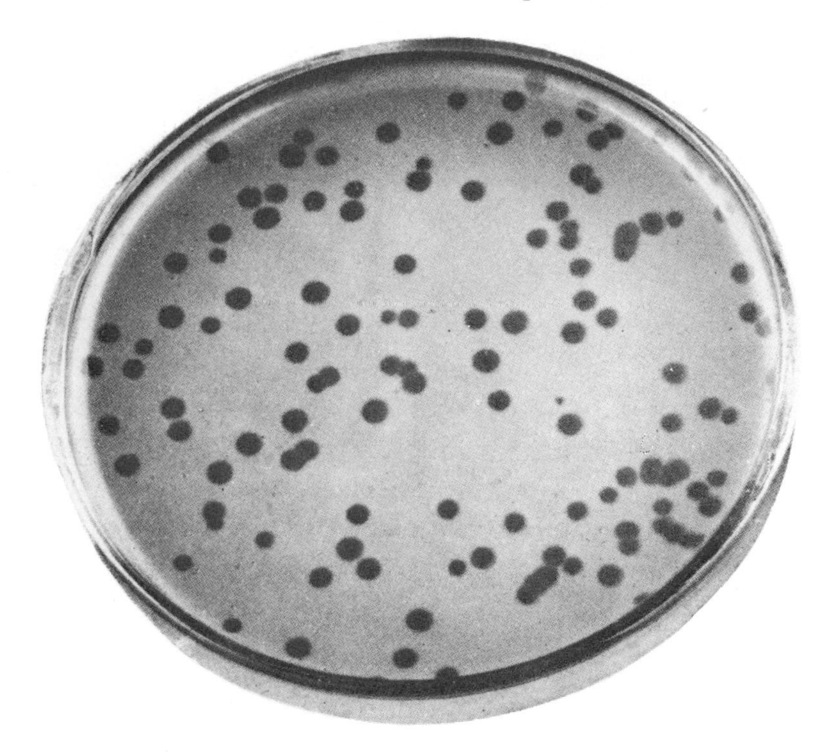

Figure 2D. Photograph of phage T2 plaques on a lawn of *E. coli* bacteria growing in a petri plate. (Reproduced from G. S. Stent, *Molecular Biology of Bacterial Viruses*, p. 41. W. H. Freeman and Co., San Francisco. Copyright © 1963.)

but lower titers for temperate phage which can form lysogens. An effective way of preparing lysates of temperate phage is to induce cultures of lysogens carrying these phage as prophage. Exposure to UV light results in the induction of many prophage, such as λ and $\phi 80$. The exact mechanism is not clear, but it is thought to involve the inactivation of the prophage repressor by accumulation of cell products after irradiation. In addition, certain mutant phage carry a mutation resulting in the production of a heat-labile repressor. Exposing cultures of the appropriate lysogens to high temperature is sufficient to induce the prophage in these strains. This is the most convenient method of preparing lysates.

References

We recommend the following reviews:

CALENDAR, R. 1970. The regulation of phage development. *Ann. Rev. Microbiol. 24:* 241.

HERSHEY, A. D., ED. 1971. *The Bacteriophage Lambda*. Cold Spring Harbor Laboratory.

HOHN, T. and B. HOHN. 1970. Structure and assembly of simple RNA bacteriophages. *Adv. Virus Res. 16:* 43.

KELLENBERGER, E. and R. S. EDGAR. 1971. Structure and assembly of phage particles. *The Bacteriophage Lambda*, p. 271. Cold Spring Harbor Laboratory.

LEVINE, M. 1969. Phage morphogenesis. *Ann. Rev. Genet. 3:* 323.

RADDING, C. M. 1969. The genetic control of phage-induced enzymes. *Ann. Rev. Genet. 3:* 363.

SIGNER, E. R. 1968. Lysogeny. *Ann. Rev. Microbiol. 22:* 451.

SNUSTAD, D. P. and D. S. DEAN. 1971. *Genetics Experiments with Bacterial Viruses*. W. H. Freeman and Co., San Francisco.—Offers detailed descriptions of class laboratory exercises.

STENT, G. S. 1963. *The Molecular Biology of Bacterial Viruses*. W. H. Freeman and Co., San Francisco.—A full discussion of the events involved in phage adsorption and replication.

TSIEN, H. C., E. T. MOSHARRAFA, D. D. HICKMAN, E. W. HAGEN, C. F. SCHACHTELE and D. L. ANDERSON. 1971. Studies with bacteriophage ϕ29 and its infectious DNA. *Informative Molecules in Biological Systems* (L. Ledoux, ed.) North-Holland Publ. Co., Amsterdam.

Method—Plate Lysate of ϕ80*v*

Day 1

Part A

Preadsorb 1–2 × 10^5 phage to 10^8 bacteria by adding 0.1 ml of the appropriate dilution of a ϕ80*v* lysate to 0.2 ml of an exponentially growing culture which is 3–5 × 10^8 cells/ml.* (A *fresh* saturated culture can also be used.) Bacteriophage adsorb to *E. coli* poorly. When preadsorption is carried out with concentrated cells, adsorption is more complete, and the infections more synchronous than if the cells were plated out immediately.

Use a small test tube and incubate at 37° in a waterbath for 10 minutes. Each group should prepare 10 small tubes in this manner. Add 2.5 ml of molten H-top agar (at 45°) to each tube and immediately pour over the surface of an H plate. Alternatively, we can preadsorb in one batch, add agar, and deliver the top layer with a 10-ml pipette. (The H-top agar used for this part of the experiment should be diluted before use by adding 15–20 ml broth to 100 ml H-top agar.)

Allow the agar to harden and then incubate at 37° *face up* for 8–10 hours. (The plates can go overnight, however.) The H plates should be moist and freshly poured (within 36 hours of the experiment).

Part B (8–10 hours later)

Scrape the top agar layer into a large plastic centrifuge tube with a bent glass spreading rod. Add 2 ml of broth per plate. Now add 5 drops of chloroform to each tube and shake vigorously for 30 seconds. Allow to stand for several minutes. Spin down the cell debris in a desk-top centrifuge and save the supernatant. Add 2 drops of chloroform and store in the cold. Care should be taken to use containers which are not sensitive to chloroform!

Part C (titering the lysate)

Make serial dilutions of the lysate as described in Experiment 1. Preadsorb 0.1 ml of different dilutions with 0.1 ml of a fresh overnight of a sensitive strain. Use a 10^{-4}, 10^{-6}, 10^{-7}, and a 10^{-8} dilution. After 10 minutes at 37° plate each preadsorption mixture on H plates with undiluted H-top agar. Allow to dry and incubate overnight at 37° (face down).

Day 2

Count the plaques and compute the number of phage/ml of the original lysate. Remember that you plated 0.1 ml of each dilution. The appearance of 200 plaques on a plate from the 10^{-8} dilution tube thus indicates a titer of 2 × 10^{11} phage/ml in the undiluted lysate.

* The procedure is the same for λ except that it is necessary to preadsorb in 0.01 M $MgSO_4$.

Materials (per group, to prepare 20 ml ϕ80*v* lysate)

Day 1 **Part A**

overnight or exponentially growing culture of a ϕ80-sensitive strain (CSH23)
ϕ80*v* lysate
10 small test tubes, 2 dilution tubes
waterbath at 37°
2 0.1-ml, 2 1-ml, and 3 10-ml test tubes
25 ml of diluted H-top agar, containing 15–20 ml broth per 100 ml soft agar, at 45°
10 freshly poured H plates

Part B

glass spreading rod
20 ml B broth or LB broth
2 plastic centrifuge tubes (30-ml)
desk-top centrifuge
chloroform
screw-cap bottle to store lysate (sterile)

Part C

fresh overnight culture of a ϕ80-sensitive strain
8 0.1-ml, 2 1-ml, and 1 10-ml pipettes
12 ml H-top agar (undiluted) at 45°
4 H plates
5 dilution tubes, 4 small test tubes

Preparation of ϕ80*h* Lysate by UV Induction

In the following procedure an exponentially growing culture of a lysogen (CSH49) is resuspended in 0.1 M $MgSO_4$ and exposed to UV light. After the addition of broth the cells are grown with good aeration for 2.5 hours. Chloroform is added and the lysate is titered for phage. For details on UV irradiation, see Experiment 13.

The UV lamp should be calibrated before the experiment. Killing curves on lysogens and non-lysogens should be done (Experiment 13). In general a proper dose for UV induction of prophage corresponds to approximately 300 ergs/mm^2 or about 50% survival of a non-lysogen. However, the best way to calibrate a lamp for this purpose is to expose cells for increasing periods of time and measure the resulting final phage yields. This can be performed either by the class as an exercise or by the instructor beforehand.

Strain CSH40 (Pro^-Str^r) harbors a ϕ80*h* prophage. The *h* mutation enables ϕ80 to infect *tonB* strains. In addition, this strain contains a ϕ80*h*d*trp* prophage. A lysate containing both ϕ80*h* and ϕ80*h*d*trp* will transduce with high efficiency a Trp^- recipient. Since the particular phage used here contains only the last four *trp* cistrons, only Trp^- recipients with an intact *trpE* gene (first *trp* cistron) can be transduced to Trp^+ (see Experiment 42). In this experiment we will prepare a lysate from CSH49 and titer the plaque-forming phage (ϕ80*h*). The lysate can then be used to map trp^- mutations in Experiment 42.

Method

Day 1

Part A

Strain CSH49 should be grown in the absence of tryptophan until used, to prevent the loss of the *ϕ80hdtrp* phage, which occurs spontaneously at a low frequency. Inoculate 10 ml of LB broth with a colony from a freshly grown minimal plate and aerate at 37° until a density of 2–3 × 10^8 cells/ml is reached. Place in ice for 10 minutes. Spin down the **10 ml** of cells and resuspend in **5 ml** of 0.1 M $MgSO_4$. Transfer to a sterile petri dish. Irradiate with UV light for the predetermined length of time, **in the dark** (absence of direct illumination). Transfer to a dark tube (or an aluminium foil-covered tube) containing 5 ml LB broth and aerate for 2.5 hours at 37°. Good aeration is critical during this time and it is better to use a 100- or 125-ml erlenmeyer flask to shake the cultures, if possible. This should be carried out in the dark (aluminum foil-covered tubes) or in red light.

Part B (2.5 hours later)

Add 5 drops of chloroform and shake vigorously for 30 seconds while securing the tube tightly. Spin down the cell debris and save the supernatant. Add 2 drops of chloroform and store in the cold. Titer the lysate exactly as in the preceding exercise.

Materials

Part A

glucose minimal proline plate with freshly grown colonies of strain CSH49
test tube with 10 ml LB broth
plastic centrifuge tube (20- or 30-ml)
desk-top centrifuge
5 ml 0.1 M $MgSO_4$
glass petri dish, UV lamp
1 5-ml or 1 10-ml pipette
100- or 125-ml flask with 5 ml LB broth

Part B

chloroform, fresh culture of ϕ80-sensitive strain
plastic centrifuge tube (20- or 30-ml)
5 dilution tubes
4 H plates, 10 ml H-top agar
7 0.1-ml, 2 1-ml, and 1 10-ml pipettes

Additional Methods

1. Preparation of a T4 Lysate from a Single Plaque

Plate dilutions of a lysate of the T4 stock to be used, so that 50 to 100 plaques appear. The indicator strain should be a fresh overnight grown in phage broth (see Materials). Plate 0.1 ml of the appropriate dilution of phage together with several drops from the overnight in 2.5 ml of soft agar maintained at 45–47°. Use the media prescribed in the materials section. Allow the soft agar to harden and then transfer the plate to 37° for 5 hours.

Prepare a 25-ml culture of the indicator strain in phage broth at 30° several hours before the next step.

Inoculate 25 ml of an exponentially growing culture of the indicator strain with a single plaque of phage. This is achieved by using a sterile 8-inch length of thin-walled glass tubing. Punch out a single, well isolated plaque from the plate and transfer it to the culture by blowing out the agar plug. Incubate further at 30° with good aeration (Snustad and Dean, 1971).

After the culture clears (usually 8–12 hours for wild-type phage for this volume) add 5–10 drops of chloroform, shake, and centrifuge the debris. Store in a refrigerator. Using this method titers of 10^{11} can be achieved for wild-type phage (Snustad and Dean, 1971).

Media for T4 Phage Lysate*

Phage Broth

per liter:	Bacto nutrient broth	8.0 g
	Bacto peptone	5.0 g
	Sodium chloride	5.0 g
	Glucose	1.0 g

Adjust pH to 7.2–7.4 with NaOH before autoclaving.

Dilution Broth

per liter:	Bacto tryptone	10.0 g
	Sodium chloride	5.0 g

Adjust pH to 7.2–7.4 with NaOH before autoclaving.

Soft Agar

per liter:	Difco minimal agar	6.5 g
	Bacto tryptone	13.0 g
	Sodium chloride	8.0 g
	Sodium citrate (dihydrate)	2.0 g
	Glucose	3.0 g

Bottom agar

Same as soft agar except that 10.0 g agar and 1.3 g glucose are used.

2. Preparation of RNA Phage Lysates

Lysates of RNA phage can be prepared by infecting an exponentially growing culture of male bacteria with a multiplicity of 5–10:1 (RNA phage: bacteria). Rich broth containing 0.005 M Ca^{++} should be used. Aerate vigorously at 37° for 5–6 hours and then add chloroform. Store in the cold. Titers between 10^{10} and 10^{11} phage/ml are obtained by this method.

* Snustad and Dean, 1971.

EXPERIMENT 3

Behavior of Mutants on Indicator Plates

Indicator plates, plates on which different phenotypes are indicated by colony appearance, have a wide use in bacterial genetics. In addition to enabling us to score different phenotypes simply by observing colony color, the proper use of indicator plates makes possible the screening of large populations for a wide variety of mutants. Some indicator plates (EMB, tetrazolium, and MacConkey) make use of the pH differences between colonies which metabolize specific carbohydrates and colonies which do not. Others, such as Xgal, employ specially prepared substrates which release a colored dye when hydrolyzed by the specific enzyme being studied. These substrate dyes are either dissolved in the medium itself, or else sprayed onto the surface of the plate after the colonies have grown.

We describe below some of the indicator plate techniques used today to detect mutant phenotypes in *E. coli*. As an exercise, the appearance of different *lac* mutants is examined on different indicators. Several "unknowns" are provided. Using indicator plate reactions the *lac* genotype of each unknown strain is determined. Recipes for some of these plates are provided in the materials section. In addition, plate methods for detecting auxotrophs have been described by Messer and Vielmetter (1965). A recent review of indicator plate methods for *E. coli* and other bacteria (Hopwood, 1970) provides excellent background reading for this topic.

Fermentation Indicators

A. Tetrazolium (triphenyltetrazolium chloride): *E. coli* growing on rich medium (peptone) reduces the colorless tetrazolium dye to deep red, insoluble formazan. The biological reduction of tetrazolium is inhibited at low pH. Since fermenting bacteria produce acid and lower the pH, these colonies are white or neutral in color. Thus, Lac^+ colonies are white on this medium (when lactose is present) and Lac^- bacteria are deep red (Lederberg, 1948). Intermediate levels of *lac* enzymes often result in pink or light red colonies. Any fermentable sugar can be used instead of lactose, such as galactose, maltose, and arabinose. Rhamnose and xylose give weaker reactions with tetrazolium, and are usually used with EMB medium instead. It is important to avoid overcrowding when using these plates since the indicator reaction does not work if too many colonies are present.

B. EMB (eosin-methylene blue): On this medium colonies which ferment the added carbohydrate are dark purple and often have a green sheen. A negative response to the added sugar results in a white or pink colony. Intermediate levels can be easily distinguished from the full wild-type level (Lederberg, 1947).

C. MacConkey: When lactose is the added sugar, Lac^+ colonies are dark red and Lac^- colonies white. Again, intermediate levels of fermentation produce intermediate or light red color.

D. Bromthymol blue: Buffered minimal medium containing 0.075% yeast extract and 0.02% bromthymol blue together with 0.5% of the carbohydrate has been used to detect fermentation reactions (in this case for β-glucosidase activity) in *E. coli* (Schaefler).

Histochemical Staining

A. Xgal glucose minimal plates (5-bromo-4-chloro-3-indolyl-β-D-galactoside): Plates which contain 40 mg/liter of this galactoside allow detection of various amounts of **constitutive** β-galactosidase inside *E. coli* colonies. The Xgal, which is not an inducer of the *lac* operon, is colorless itself. However, when hydrolyzed by β-galactosidase, this compound releases deep blue 5-bromo-4-chloro-indigo. Constitutive $i^-z^+y^+$ colonies are deep blue, $o^cz^+y^+$ colonies are pale blue, and $i^+z^+y^+$ colonies are almost completely white in the same time period (Davies and Jacob). From the comparison of single colonies it becomes easy to compare constitutive β-galactosidase levels of different strains. If IPTG is included in the medium, then these plates become indicators for maximal levels of β-galactosidase, rather than constitutive levels. Even extremely low levels can be detected after a long period of incubation. Up to 5000 single colonies can be observed on a plate, and i^- colonies can be detected against a background of 100,000 microcolonies.

B. Indicator added to the colonies: In principle, any substrate which gives colored hydrolysis products can be added to the colonies. A compound often used for β-galactosidase is *o*-nitrophenyl-β-D-galactoside (ONPG). Constitutive colonies, which have a high level of β-galactosidase, are able to split this compound and release yellow *o*-nitrophenol (Cohen-Bazire and Jolit). The ONPG concentration has to be high enough (10^{-2} M) so that its entry into the cells is not rate limiting. This can be enhanced by first spraying the cells with toluene. Similarly, *p*-nitrophenyl-phosphate has been used to stain alkaline phosphatase (Garen), and *p*-nitrophenyl-β-glucoside for β-glucosidase (Schaefler and Maas).

C. Coupled indicators; naphthol azo dyes: α-Naphthol compounds when enzymatically cleaved liberate free α-naphthol, which can be coupled with diazonium salts to form insoluble azo-dyes. Large numbers of colonies (as many as 10^5 microcolonies per plate with a stereomicroscope) can be screened, since the dye is not diffusable. Colonies (typically 5000 per plate) are embedded in a double layer of soft agar (see methods for this experiment) and then, for instance, treated with a solution of 6-bromo-2-naphthyl-α-D-galactoside and fast blue RR*. (Usually the colonies are at a diameter of $\frac{1}{2}$mm for this treatment.) This stains for constitutive mutants, with non-constitutive colonies remaining light-colored. However, if constitutive cells are used to start with (for instance i^-) or if inducer (IPTG) is present in the nutrient agar medium, then Lac^+ colonies stain dark and Lac^- colonies remain light (Messer and Melcher). Naphthol azo dyes have also been used to stain alkaline phosphatase mutants. Recipes for these plates are provided in the materials section. Variations of this technique have been used for the detection of bacterial growth requirement mutants (Messer and Vielmetter, 1965).

D. Other techniques: Dyes which react with nucleic acids, such as **Giemsa** or methyl green, have been used to isolate nuclease deficient mutants of *E. coli* (Dürwald and Hoffman-Berling; Wright).

"Giemsa"-staining, the application of an aqueous solution (containing per liter 17.5 ml of Merck's Giemsa solution, 30 mg methylene blue and 30 mg basic fuchsin) to mutagenized colonies imbedded in a double layer of soft agar, has been used to visualize a large collection of different mutants in *E. coli*. Mutant colonies stain dark blue, violet or red, whereas wild type are white (Vielmetter et al.).

Extracellular enzymatic activities can often be detected by plate tests. RNase I mutants of *E. coli* were isolated using this technique. Colonies were gridded onto two duplicate plates and grown overnight. One plate was saved and the other tested by applying soft agar containing EDTA (to effect release of RNase I into the medium) and yeast RNA. After incubation, 1 N HCl was added to cause a white precipitate of polymerized RNA. Around each wild-type colony was a clear halo due to degradation of the RNA by RNase I. Two mutant colonies had no halo and were found to be lacking RNase I (Gesteland).

Special Uses (for tetrazolium, EMB, and MacConkey plates)

A. The dramatic difference between Lac^+ and Lac^- colonies enables us to observe the **segregation** of negative clones from heterogenotes for the *lac* character. When the Lac^+ character is carried on an episome and the host cell is Lac^-, the segregation or loss of the episome itself can be followed easily on indicator plates. Tetrazolium plates are particularly useful for isolating negative mutants (Lac^-), since the Lac^- colony or sector can be seen as a dark red point or section on a light-colored background (the colored dye in these plates is not diffusable). Although overgrowth of colonies interferes with the tetrazolium reaction, in some cases several thousand cells per plate can be viewed; and with the aid of a low magnification viewing scope even more cells can be screened for red sectors.

B. Revertant cells ($Lac^- \rightarrow Lac^+$) can be purified from Lac^- colonies after prolonged incubation. These appear as small microcolonies (**papillae**) growing out of the main colony. After the peptone medium has been exhausted, only those cells which can utilize the carbohydrate will continue to grow. This allows the easy isolation of revertants on tetrazolium, EMB, or MacConkey plates.

* Obtainable from Sigma.

Mutant Colonies on Indicator Plates

The following plates show *E. coli* colonies magnified approximately 50 times. All of the colonies have been plated on nutrient agar in a layer of soft agar. Each stain has been applied after addition of a second layer of soft agar (see text). (Concentrations given are the concentrations added to the double layer of soft agar.) Many of the stained colonies are mutator strains. These have lesions in a locus which result in increased frequencies of spontaneous mutation. Mutator colonies segregate stained sectors at a high rate, regardless of the particular stain applied. These photographs were supplied by G. Hombrecher, K. Reiners, and R. Hohlfeld from the laboratory of W. Vielmetter, Institut Genetik, Universität Köln.

Plate 1. Top: Heterozygous colony stained for *lac*$^+$ *i*$^-$ with Xgal. The white colony is i$^+$.

Center: The yellow colony is a strong R$^-$ (constitutive for alkaline phosphatase) segregant stained with Naphthol-AS-GR-phosphate (1 mg/ml). No azo dye has been added.

Bottom: A mutator colony (see introduction to Unit III) stained with a double stain for constitutive β-galactosidase and also constitutive alkaline phosphatase. The blue Xgal and the yellow naphthol-phosphate technique are as in the top and center figures on this page.

Plate 2. Top and Center: A double stain has been used to show the reactions of different mutants in the *lac* and *pho* (alkaline phosphatase) regions. To stain mutants in the *lac* region, the coupled dyes Bromo-naphthyl-β-D-galactoside (0.2 mg/ml) and Fast Blue RR (1.6 mg/ml) have been used. To mark mutants in the *pho* region, the coupled dyes were Naphthol-AS-MX-phosphate (1 mg/ml) and Fast Blue RR (1.6 mg/ml). Here the colonies are:

Dark Brown: *lac*$^+$, *i*$^-$; *phoR*$^-$, *S*$^-$
Green: *lac*$^-$, *z*$^-$; *phoR*$^-$, *S*$^-$
Red: *lac*$^+$, *i*$^-$; *phoA*$^-$
White: *lac*$^-$, *z*$^-$; *phoA*$^-$

Bottom: Weak R$^-$ constitutive (*phoR*) segregant from a mutator colony. Stain: Naphthol-AS-MX-phosphate (1 mg/ml) coupled with Fast Blue RR (1.6 mg/ml).

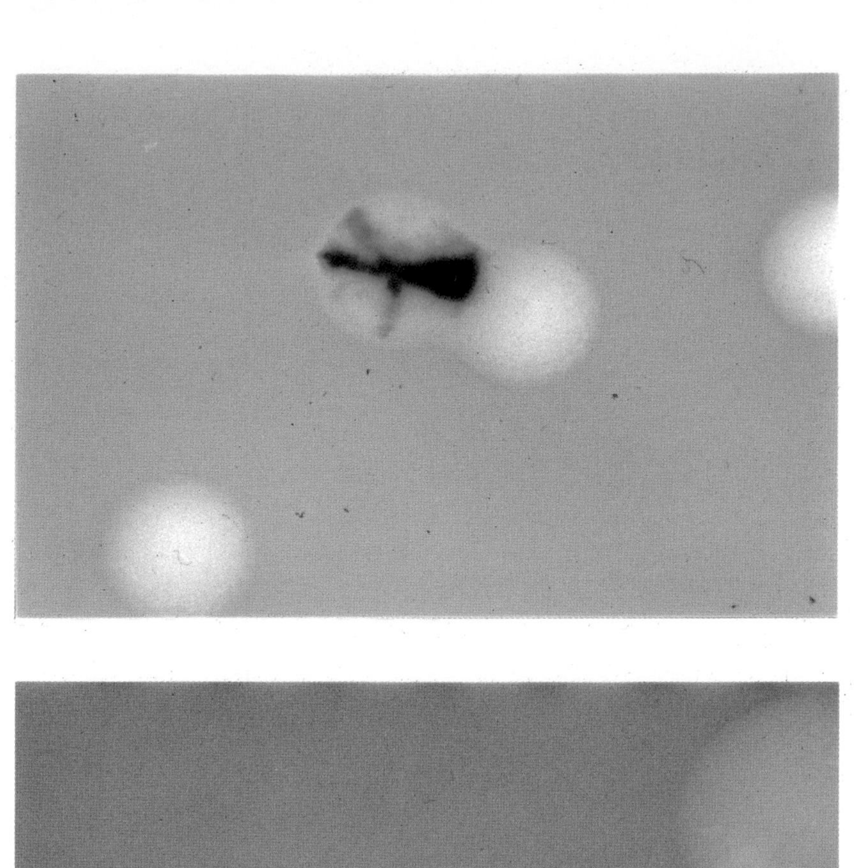

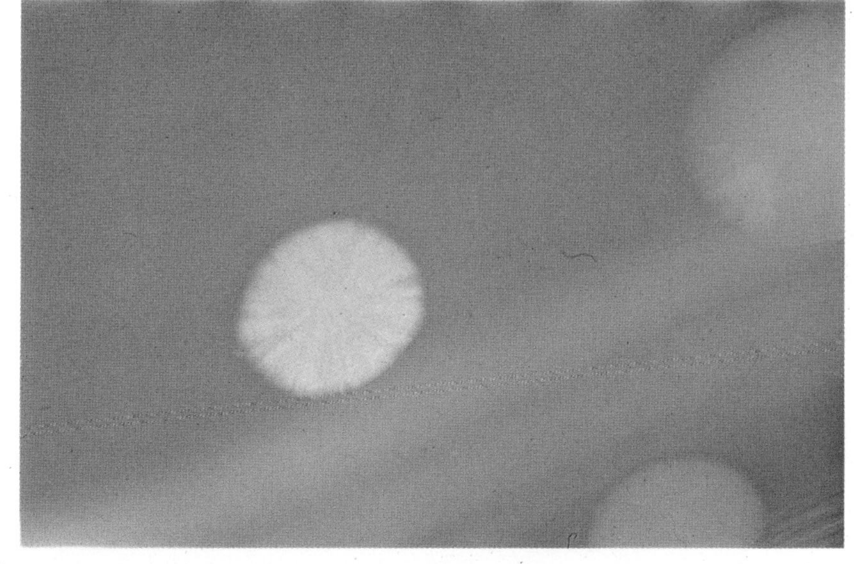

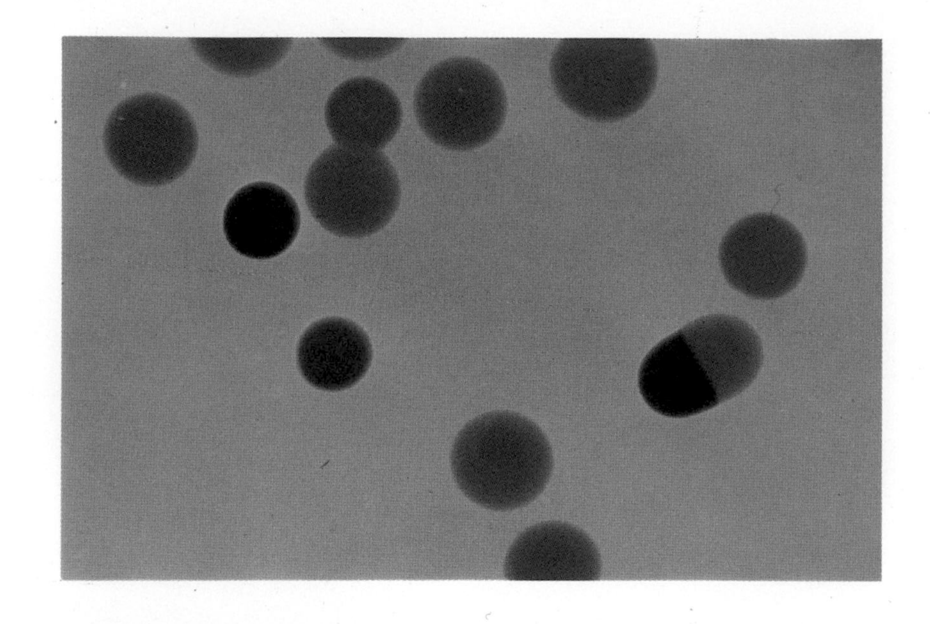

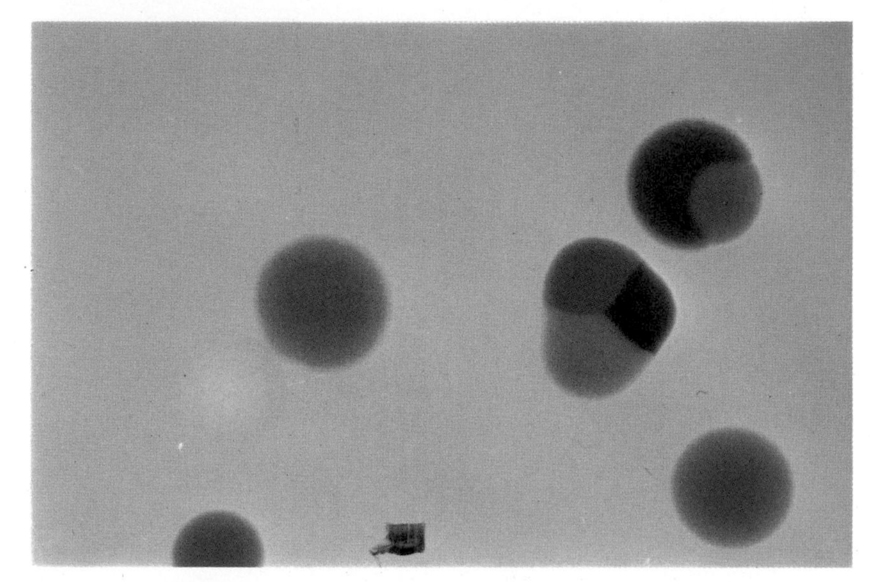

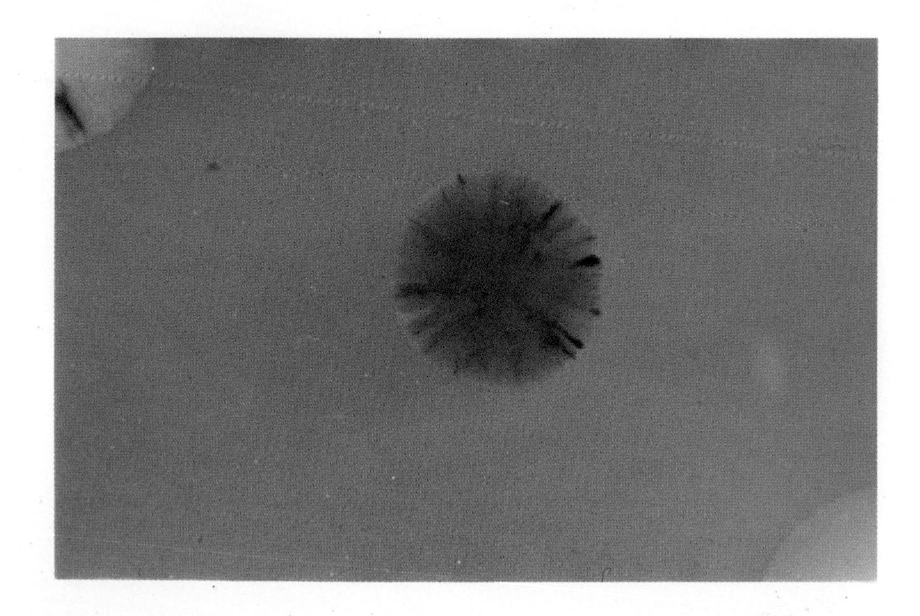

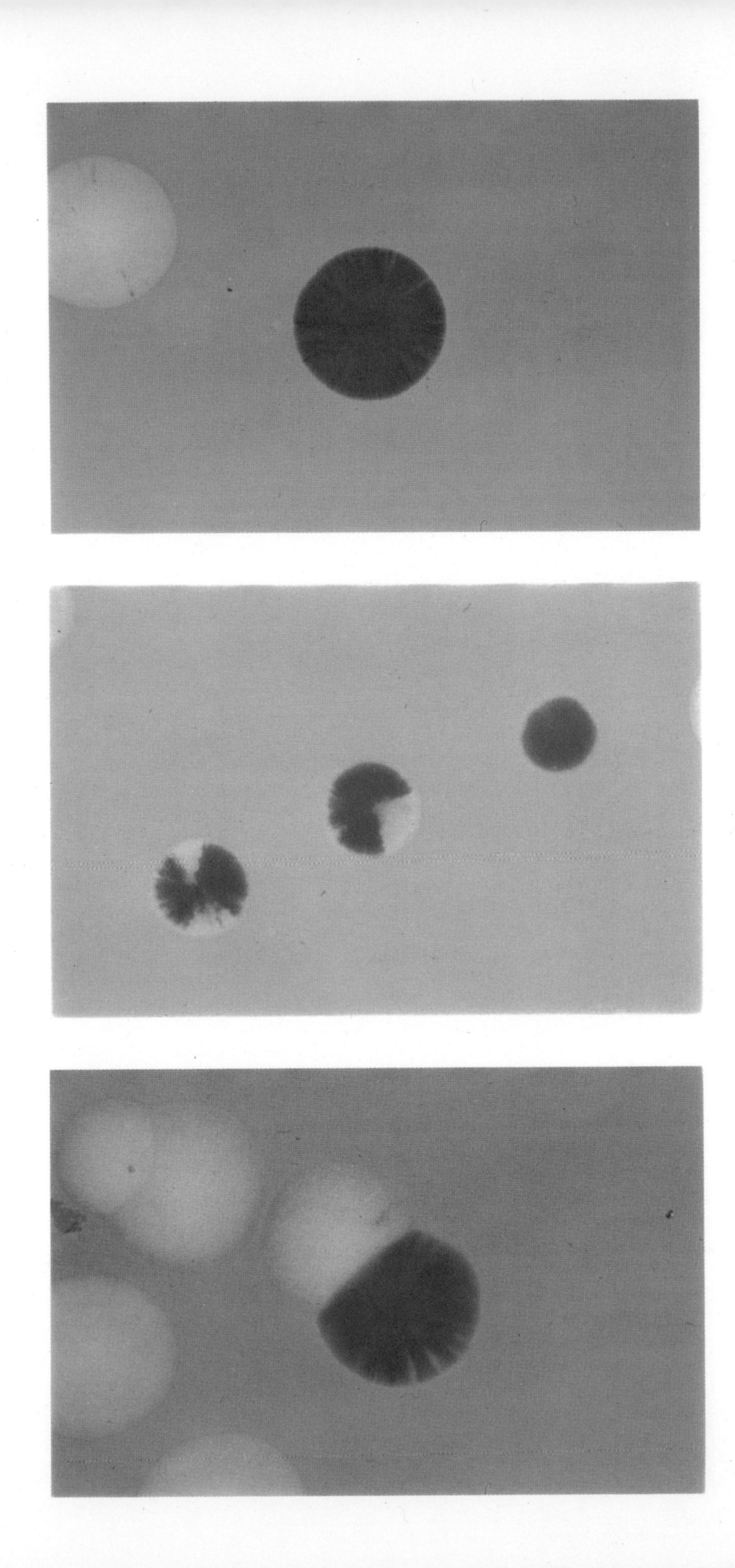

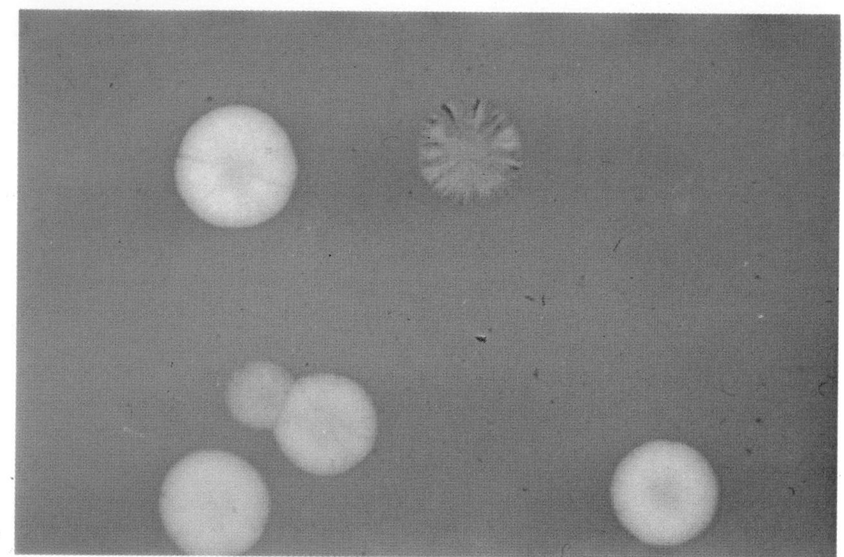

Plate 3. Top: Strong R^- constitutive segregant from a mutator colony. Stain: Naphthol-AS-MX-phosphate (1 mg/ml) coupled with Fast Blue RR (1.6 mg/ml).

Center: Heterozygous colonies stained for the arabinose constitutive phenotype (*araC*) by the method of Lin et al. (see text). Stain: 1% solution of 2,3,5-triphenyltetrazolium chloride.

Bottom: An i^- segregant from a mutator colony throwing out z^- mutants. The i^- mutants stain red here and the z^- mutants stain white. Stain: Bromo-naphthyl-β-D-galactoside (0.2 mg/ml) coupled with Fast Blue RR (1.6 mg/ml).

Plate 4. Top: Mutator colony stained for constitutive alkaline phosphatase (*phoR*). Stain: Naphthol-AS-MX-phosphate (1 mg/ml) coupled with Fast Blue RR (1.6 mg/ml).

Center and Bottom: Mutator colonies stained with "Giemsa" stain. The wild-type colonies appear white (see text for recipe).

C. Isolation of carbohydrate constitutives with tetrazolium: After mutagenesis and outgrowth (see Unit III) approximately 1000 cells are spread onto minimal medium buffered with 0.12 M Tris (pH 7.5) and containing 0.5 g Bacto tryptone and 0.25 g yeast extract. After 48 hours incubation at 37° the plates are sprayed with a solution containing 0.5% chloramphenicol and 10% carbohydrate substrate (for instance, lactose if we were isolating *lac* constitutives). The plates are then reincubated at 37° for 30 minutes to allow the accumulation of reducing substances from the dehydrogenation of the substrate. Plates are then sprayed with a solution of 1 M phosphate buffer (pH 7.0) containing 1% triphenyltetrazolium chloride. Constitutive cells which had broken down the substrate during the 30-minute incubation period turn red from the resulting reduction of the tetrazolium to formazan. This procedure is general for the isolation of constitutive mutants for enzymes involved in carbohydrate metabolism (Lin et al.).

References

Cohen-Bazire, G. and M. Jolit. 1953. Isolement par sélection de mutants d'*Escherichia coli* synthétisant spontanément l'amylomaltase et la β-galactosidase. *Ann. Inst. Pasteur 84:* 937.

Davies, J. and F. Jacob. 1968. Genetic mapping of the regulator and operator genes of the *lac* operon. *J. Mol. Biol. 36:* 413.

Durwald, H. and H. Hoffman-Berling. 1968. Endonuclease I-deficient and ribonuclease I-deficient *Escherichia coli* mutants. *J. Mol. Biol. 34:* 331.

Garen, A. 1960. Genetic control of the specificity of the bacterial enzyme, alkaline phosphatase. *Symp. Soc. Gen. Microbiol. 10:* 239.

Gesteland, R. F. 1966. Isolation and characterization of ribonuclease I mutants of *Escherichia coli*. *J. Mol. Biol. 16:* 67.

Hopwood, D. A. 1970. Isolation of mutants. *Methods in Microbiology*, Vol. 3A, p. 363. Academic Press, New York.

Lederberg, J. 1947. Gene recombination and linked segregations in *Escherichia coli*. *Genetics 32:* 505.

Lederberg, J. 1948. Detection of fermentative variants with tetrazolium. *J. Bacteriol. 56:* 695.

Lin, E. C. C., S. A. Lerner and S. E. Jorgensen. 1962. A method for isolating constitutive mutants for carbohydrate-catabolizing enzymes. *Biochim. Biophys. Acta 60:* 422.

Messer, W. and F. Melchers. 1970. Genetic analysis of mutants producing defective β-galactosidase which can be activated by specific antibodies. *Mol. Gen. Genet. 109:* 152.

Messer, W. and W. Vielmetter. 1965. High resolution colony staining for the detection of bacterial growth requirements using napthol azo-dye techniques. *Biochem. Biophys. Res. Commun. 21:* 182.

Schaeffler, S. 1967. Inducible system for the utilization of β-glucosides in *Escherichia coli*. I. Active transport and utilization of β-glucosides. *J. Bacteriol 93:* 254.

Schaeffler, S. and W. K. Maas. 1967. Inducible system for the utilization of β-glucosides in *Escherichia coli*. II. Description of mutant types and genetic analysis. *J. Bacteriol. 93:* 264.

Vielmetter, W., W. Messer and A. Schütte. 1968. Growth direction and segregation of the *E. coli* chromosome. *Cold Spring Harbor Symp. Quant. Biol. 33:* 585.

Wright, M. 1971. Mutants of *Escherichia coli* lacking endonuclease I, ribonuclease I, or ribonuclease II. *J. Bacteriol. 107:* 87.

Strains

Number	Sex	Genotype	Important Properties
CSH23	F'*lac*$^+$ *proA*$^+$,*B*$^+$	Δ(*lacpro*) *supE spc thi*	$i^+z^+y^+$
CSH36	F'*lacI proA*$^+$,*B*$^+$	Δ(*lacpro*) *supE thi*	$i^-z^+y^+$
CSH35	F'*lacI proA*$^+$,*B*$^+$	Δ(*lacpro*) *supE thi*	$i^sz^+y^+$
CSH22	F'*lacZ proA*$^+$,*B*$^+$	*trpR* Δ(*lacpro*) *thi*	$i^+z^-y^+$

Table 3A. Indicator Plate Reactions of Lac

Indicator Plate	Lac⁻		Lac⁺		
	$i^+z^+y^-$	$i^+z^-y^+$	$i^+z^+y^+$	$i^-z^+y^+$	$o^cz^+y^+$
EMB (lactose)	white	white	dark, with green sheen	dark, with green sheen	dark, with green sheen
MacConkey (lactose)	white	white	deep red	deep red	deep red
MacConkey + IPTG (lactose)	red	white	deep red	deep red	deep red
Xgal (glucose)	white	white	white	deep blue	light blue
Xgal + IPTG (glucose)	deep blue	white	deep blue	deep blue	deep blue
Tetrazolium (lactose)	deep red	deep red	white	white	white

Method

The object of this experiment is to demonstrate the use of indicator plates in characterizing *lac* mutants. Six test strains are provided. These are: $i^+z^+y^+$, $i^-z^+y^+$, $i^sz^+y^+$, and $i^+z^-y^+$; the other two strains are unknowns in this experiment. Each of the strains, including the unknowns, will be provided already grown up on non-indicating glucose minimal plates.

Day 1 Pick a single colony of each strain and streak for single colonies onto each of the following indicator plates: lactose tetrazolium, lactose EMB, lactose MacConkey, Xgal glucose, and also LB. Divide each plate into six sectors and pick one colony of each strain onto each plate. Incubate at 37° overnight.

Day 2 Inspect the plates. By comparing the results of the unknown with those of Table 3A and with the strains of known genotype, determine the genotype of the two unknown strains. Pour 1 ml of a 10^{-2} M solution of ONPG over the LB plate and observe which colonies turn yellow quickly.

Materials

Day 1

6 minimal glucose plates, each with single colonies of strains CSH22, 23, 35, 36, and two unknowns
6 round wooden toothpicks
1 each of the following plates: lactose tetrazolium, lactose EMB, lactose MacConkey, LB, and Xgal glucose

Day 2

2 ml ONPG (MW 301) 10^{-2} M solution in H_2O

Preparation of Indicator Plates

1. Xgal glucose plates: These are prepared in the same manner as glucose minimal plates (see Appendix). After autoclaving add 2 ml of a 20 mg/ml solution of Xgal in N,N-dimethylformamide to the salts directly before mixing.

2. Tetrazolium plates: These plates work best with Difco antibiotic medium #2. Dissolve 25.5 grams of this medium in 950 ml distilled water. Add 50 mg of 2,3,5-triphenyltetrazolium chloride and continue to heat until completely dissolved. Autoclave and then add 50 ml of a 20% solution of the desired sugar (in this case lactose). It is extremely important that the tetrazolium be added before autoclaving. In cases where Difco antibiotic medium #2 is not available, or if colonies do not grow well on this medium, the following two media have given satisfactory results:

per 950 ml distilled water:

1. 1.5 g Bacto beef extract, 3 g yeast extract, 6 g peptone, 15 g agar.
2. 23 g Difco nutrient agar, 1 g NaCl.

3. EMB plates: Prepared medium can be purchased from Difco. This comes with and without added sugar (lactose). The latter is preferred. In the case of EMB-lactose medium, dissolve and autoclave in accordance with the included instructions. For EMB base or EMBO, add 50 ml of the desired sugar (20% stock solution) after autoclaving. EMB plates can also be made from the following formula:

930 ml distilled water
10 g Bacto tryptone
1 g yeast extract
5 g NaCl
15 g Difco agar
2 g KH_2PO_4
Autoclave, and add after autoclaving 10 ml each of a sterile solution of 4% eosin yellow and 0.65% methylene blue. These are autoclaved separately. Also add 50 ml of a 20% solution of the desired sugar.

4. MacConkey plates: These are made from prepared media, and can be purchased from Difco. MacConkey medium with lactose added is available and preferable to the base agar. TonB strains (Experiment 42) grow well on this medium, but do not grow as well on reconstructed lactose MacConkey medium made from MacConkey base. Any sugar (50 ml of a 20% solution) can be added to MacConkey base after autoclaving.

5. Naphthol azo dye for β-galactosidase: Plates used for this technique contain 10 g tryptone, 9 g NaCl, and 12 g agar per liter for the determination of constitutive (i^-) colonies. Lactose is added to a final concentration of 0.2% for the differentiation of Lac^+ and Lac^- colonies. Nutrient agar (Difco) can also be used. The bacteria are plated in 1.5 ml of soft agar (4.5 g agar and 9 g NaCl per liter). As soon as this dries the plates are left open in a 37° incubator for about 45 minutes until cracks in the soft agar are visible. Then a second 1.5-ml layer of soft agar is added. The plates are incubated until microcolonies 0.5 mm in diameter appear.

Staining: Naphthyl dyes are dissolved in a few drops of dimethyl-sulfoxide and then 0.08 M Tris buffer, pH 8.0, is added. For β-galactosidase 0.2–0.4 mg/ml of 6-bromo-2-naphthyl-β-D-galactoside is used, and for alkaline phosphatase 1 mg/ml of naphthol-AS-MX-phosphate is used (final concentrations in buffer).

Approximately 8–10 mg of Fast Blue RR is used per plate. Several drops of dimethyl-sulfoxide are added to dissolve the Fast Blue RR in a small test tube and then 5 ml of the respective naphthol staining solution is added. This is immediately poured over the double soft-agar layer.

For soft-agar layer staining with Xgal, 5 mg is used per plate. This is dissolved in DMSO as described above and 5 ml of 0.1 M Tris buffer, pH 7.0, is added. This is then directly poured over the surface of the plate.

The plates are rinsed with 0.9 M NaCl solution after sufficient color has developed (usually 5–20 minutes) as viewed with a stereomicroscope. Colonies or sectors of colonies can be picked with a fine wire needle or with 0.05 mm diameter glass capillaries.

A wide variety of naphthol or naphthyl compounds are available from Sigma Chemical Company, P.O. Box 14508, St. Louis, Mo., 63178.

EXPERIMENT 4

Replica Plating

Replica plating, which was introduced by Lederberg and Lederberg (1952), enables the transfer in one operation of a replica of a master plate containing 50–100 colonies to several different plates, thereby saving hours of work. Usually sterile velvet pads are employed, although filter paper can also be used. A circular block which just fits inside the petri dish is covered with a velvet pad. The velvet is drawn tight and held by a metal ring (Figure 4A). This is then carefully pressed against the surface of the plate to be replicated. The velvet is then immediately imprinted in a similar manner onto the surface of a fresh plate. The same velvet can be used to transfer to several plates, and when extreme care is used as many as 10–15 transfers are possible. Replica plating techniques are widely used in the isolation of auxotrophic mutants (Experiment 34) and in F′ transfers and Hfr crosses (Experiments 5, 25, 26, 35, and 36).

Master plates can be prepared in several ways. We favor gridding a plate by picking colonies with sterile round wooden toothpicks. Graph paper can be used to construct a number grid with fifty sectors. The graph paper is then placed under the plate. It is important to mark the top of each master plate to denote the orientation of the grid. It is also possible to replicate colonies directly from a selection plate, if the colonies are well separated. This allows the transfer of several hundred colonies.

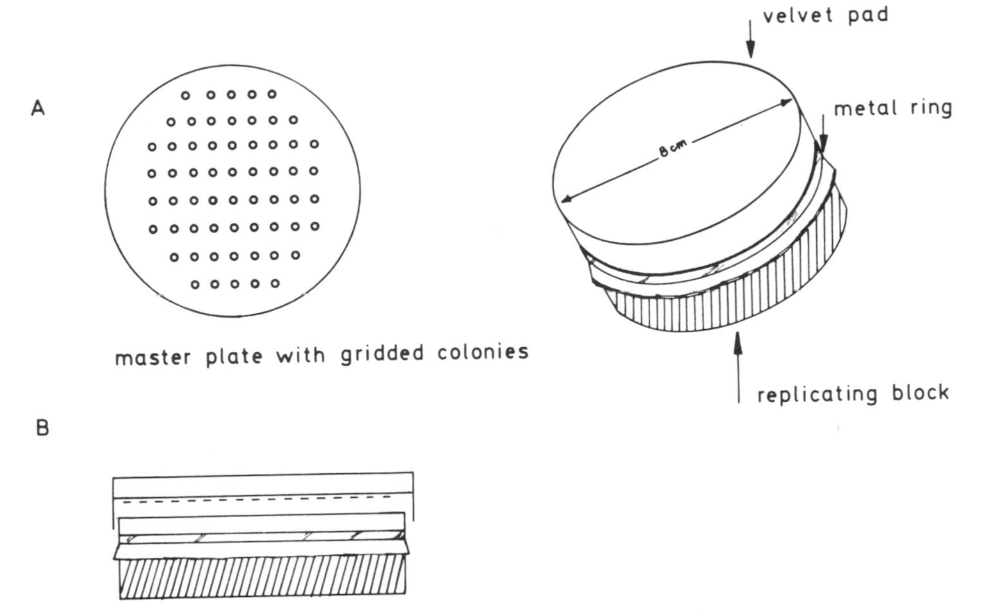

Figure 4A. Replica plating.

The velvets themselves are usually cut into squares or discs about 14 cm in diameter and autoclaved dry. Velvets are washed after each use, or simply brushed before autoclaving. Some investigators wipe velvets with chloroform instead of autoclaving. If an upturned 400-ml aluminum beaker is used as a replicating block, the velvet can be permanently held in place and sterilized by wiping with a chloroform soaked rag after each use. The chloroform is quickly evaporated by warming the inside of the beaker over a bunsen burner (Kemp). Although the bases required to mount the velvets are not always readily available, a simple method for their construction using aluminum automative pistons (8 cm in diameter for a 9–10 cm petri dish) has been described (Adams).

References

Adams, J. N. 1965. Automotive pistons for use as bases in velveteen replication. *J. Bacteriol. 89:* 1627.
Kemp, R. F. O. 1967. Chloroform sterilisation for replica plating. *Microb. Genet. Bull. 26:* 11.
Leberberg, J. and E. M. Lederberg. 1952. Replica plating and indirect selection of bacterial mutants. *J. Bacteriol. 63:* 399.

Strains

Number	Sex	Genotype	Important Properties
CSH61	HfrC	*trpR thi*	Str^s
CSH51	F^-	*ara* Δ(*lacpro*) *strA thi* (φ80d*lac*$^+$)	Str^r

Method

Day 1 Using a graph paper grid, pick colonies of each of the two strains, CSH61 (Str^s) and CSH51 (Str^r), onto alternating sectors of a rich plate. Allow to grow overnight at 37°.

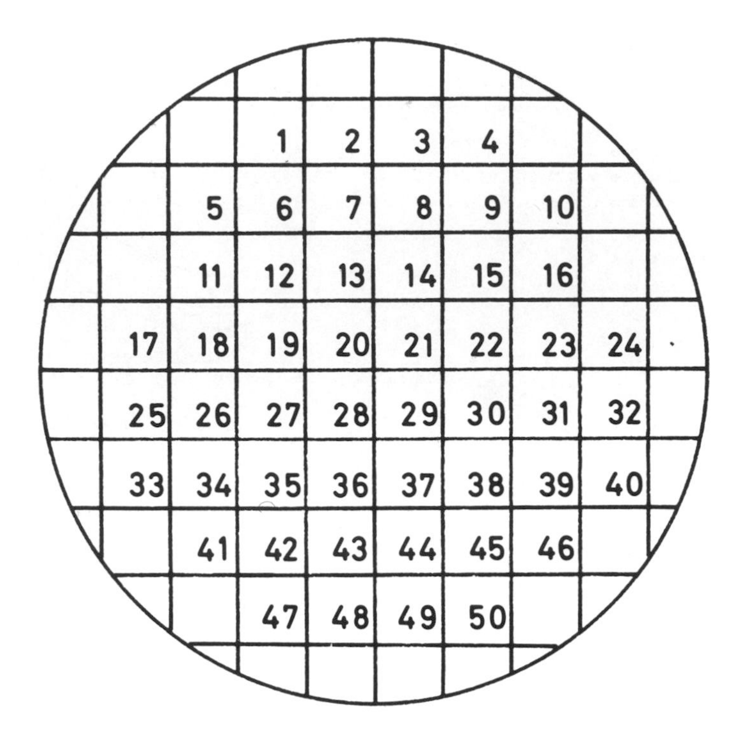

Figure 4B. Grid for 50 colonies.

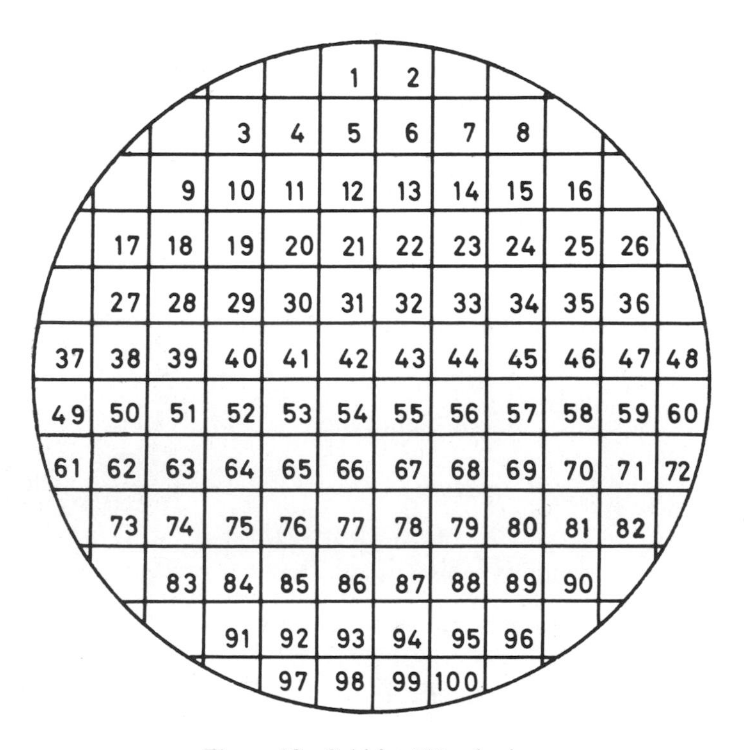

Figure 4C. Grid for 100 colonies.

Day 2 Replicate onto an LB plate with streptomycin, and then onto an LB plate without streptomycin. Incubate at 37° and observe the results the next day and compare with Figure 4D. The instructor will demonstrate the proper technique. Prepare several plates and practice this procedure, for it is used in many experiments in this manual and is invaluable in bacterial genetics research.

Materials

Day 1

2 plates with single colonies of strains CSH61 and CSH51, respectively
1 LB plate
2–10 round sterile wooden toothpicks

Day 2

2 rich (LB) plates, one with and one without streptomycin (100 μg/ml)
replicating block and sterile velvet pad

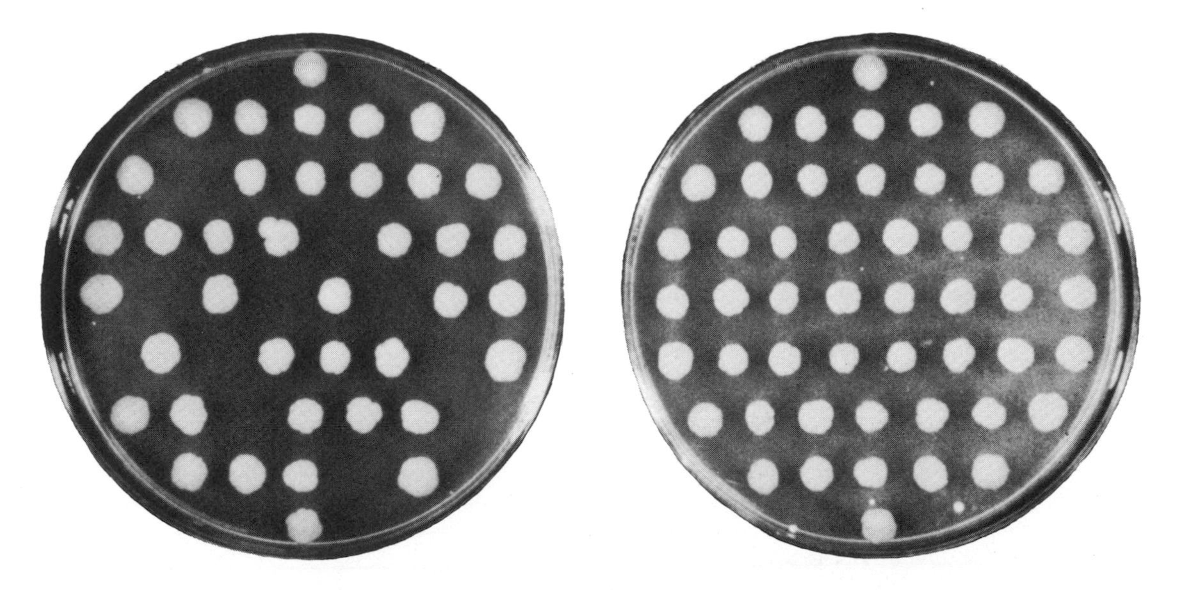

Figure 4D. Fifty single colonies from a mixed population of Str^s and Str^r cells were gridded onto a rich plate and incubated for 10–12 hours at 37°. This master plate was then replicated first onto a rich plate spread with 0.2 ml of a 1% streptomycin solution, and then onto a second plate containing no streptomycin (right side of figure). The plates were grown for 12 hours at 37°. Both Str^s and Str^r cells form colonies on the control plate, while only the Str^r cells appear on the plate on the left. (Photo courtesy of J. Kirschbaum.)

UNIT II

MATINGS BETWEEN MALE AND FEMALE CELLS

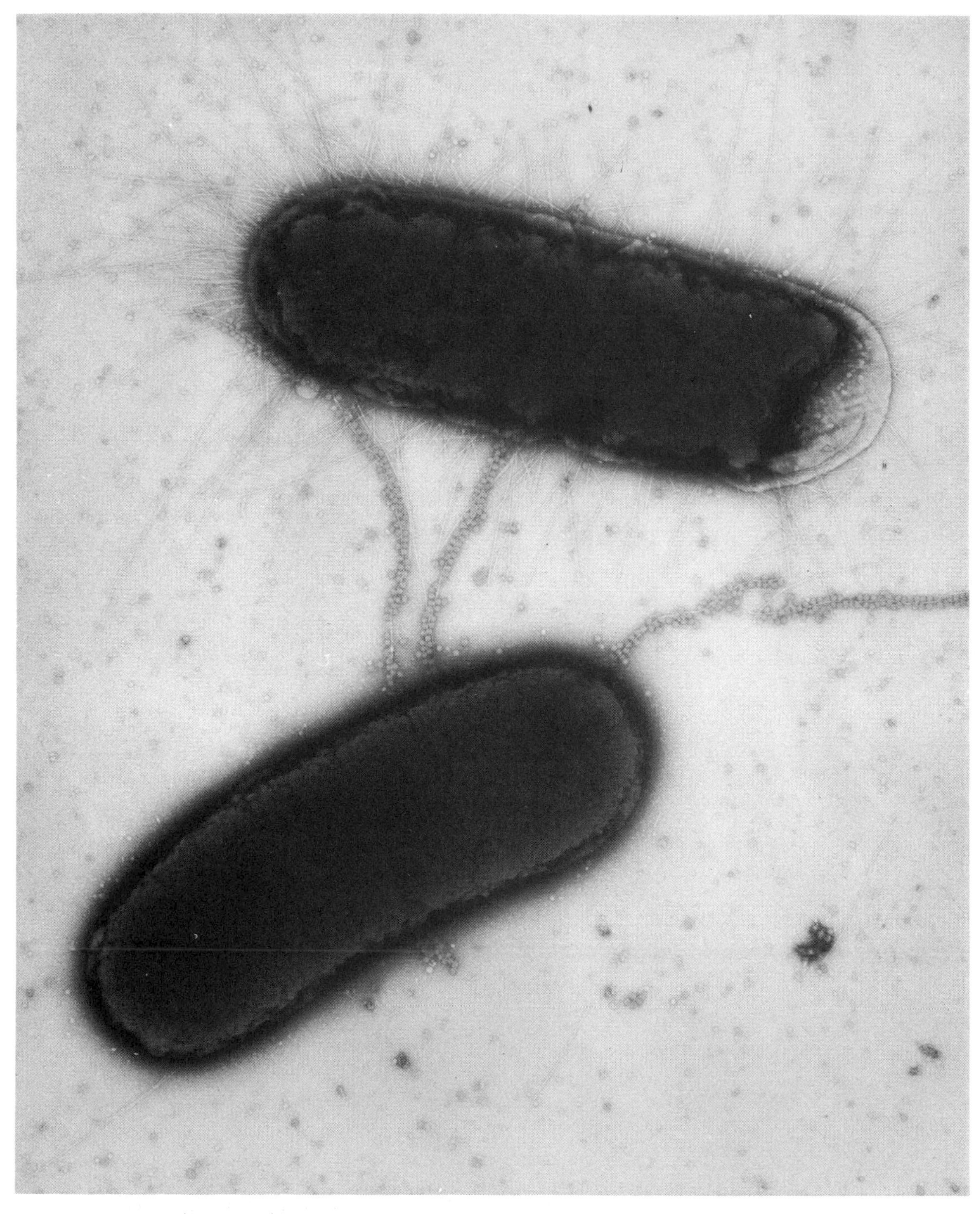

Plate 5. Electron micrograph of a male and female cell of *E. coli* connected by F pili. The male (bottom) has three F pili which are surrounded by MS2 phage that adsorb to the surface. Numerous common fimbriae (see text) can be seen extending out from the female (top). Photograph courtesy of Lucien Caro (× 46,200).

INTRODUCTION TO UNIT II

The essential aspects of bacterial sexuality are derived from the pioneering work of five persons:

Joshua Lederberg, who together with Tatum first employed the concept of selective methods to discover the processes of conjugation and recombination in bacteria;

William Hayes, who defined the unidirectional nature of genetic recombination in bacteria allowing the clear differentiation of donor and recipient "male" and "female" cells, and who, in addition to Cavalli-Sforza and the Lederbergs, discovered the F or fertility factor, the main genetic element responsible for the conjugation properties of bacteria;

L. L. Cavalli-Sforza, who isolated the first Hfr strain and who participated in the original discovery of the F factor;

Elie L. Wollman and Francois Jacob, who fully developed the Hfr × F^- mapping system which forms the basis for genetic mapping in *E. coli*.

Circularity of Bacterial Chromosomes

The first suggestion that the *E. coli* chromosome was circular came from genetic studies (Jacob and Wollman, 1961) which showed the haploid genome to be a closed, continuous linkage group. Soon afterwards autoradiography of *E. coli* directly demonstrated that the physical structure of the chromosome was circular (Cairns, 1963a,b), some 1200 μ in circumference. The replication of this DNA duplex is semi-conservative (Meselson and Stahl, 1958), meaning that each new duplex consists of one old (pre-existing) strand and one newly synthesized strand.

The density transfer experiments of Meselson and Stahl, the autoradiographs of Cairns (1963a,b), early experiments by Bonhoeffer (quoted in Cairns, 1963b) and work by Nagata (1963) and Yoshikawa and Sueoka (1963) all indicated that the bacterial chromosome starts duplication at a single point. More recent experiments provide strong evidence that this replication does start at a single point, at approximately 74 min on the *E. coli* genetic map (Lark, 1966; Donachie and Masters, 1966; Abe and Tomizawa, 1967; Wolf et al., 1968a,b; Helmstetter, 1968; Caro and Berg, 1968; Nagata and Meselson, 1968; Bird and Caro, 1972).

At first it seemed that the replication proceeded in only one direction, but now there is growing evidence for bidirectional replication both in *E. coli* (Masters and Broda, 1971; Bird et al., 1971) and in *B. subtilis* (Wake, 1971). As a general phenomenon bidirectional DNA synthesis is well established for phage λ (Schnös and Inman, 1970), for T7 (Dressler and Wolfson, 1972), and for T4 (Delius et al., 1971).

As a consequence of the closed ring structure of the chromosome all genetic markers can be positioned on a circular genetic map (Jacob and Wollman, 1961; Taylor and Trotter, 1967). The work of Taylor and his collaborators has provided us with a detailed map which charts the position of many markers on the *E. coli* chromosome (Taylor, 1970). This is reprinted in the Appendix.

F^+ and F^- Cells

E. coli strains can be divided into two groups, defined on the basis of pair formation and conjugational mating properties. F^+ or male cells are able to donate chromosomal markers to recipient F^- or female cells, if mixed together under the appropriate conditions.

Properties of F^+ cells. F^+ cells synthesize long, thin, protein filaments termed F pili, which are required for conjugation (see Plate 5). The sex pili (sometimes referred to as sex fimbriae) are distinct from other filamentous protein structures which frequently occur on the surface of bacteria such as *E. coli*. These are termed common fimbriae (Brinton, 1965; Duguid et al., 1966; Hayes, 1968). Only the F pili have been associated with sexual activity. There are a large number of common fimbriae visible per cell, whereas each male cell synthesizes relatively few F pili, perhaps as few as one per F factor copy. These vary in length from 1 to 20 μ (Hayes, 1968).

The F pili also serve as specific adsorption sites for a series of single-stranded DNA and RNA phages (Crawford and Gesteland, 1964; Brinton, 1965). These phages therefore infect only male cells. Cell surface charge alterations are also associated with the F^+ state (Maccacaro and Comolli, 1956) with changes in cell wall antigens being noted in cells carrying the F factor (Ørskov and Ørskov, quoted in Hayes, 1968). These surface components, which are distinct from F pili, prevent F^+ cells from forming mating pairs with other F^+ cells. Only after growth under starvation conditions do F^+ cells lose their surface exclusion properties and conjugate with other males. This change is temporary and normal surface properties are regained upon continued growth in fresh medium. These phenotypic F^- cells are termed phenocopies (Lederberg et al., 1952).

The F factor. All of the properties of F^+ cells are due to the presence of a small circular DNA element termed the F factor (or sex factor) which can exist either as a free circular genome or else integrated into the host chromosome. The F factor has been found to have a molecular weight of approximately 45×10^6 daltons (Freifelder, 1968) compared to 2.9×10^9 daltons for the *E. coli* chromosome. It is one of a general class of elements called episomes. Recently a large number of mutants of F have been isolated (Achtman et al., 1971) which are transfer-deficient (see following sections) and are defective in F pili formation. As of this time ten complementation groups have been defined.

Episomes

Episomes are defined as added genetic elements which can exist in either of two states: the autonomous state in which replication occurs independent of the host chromosome, and the integrated state in which the episome is incorporated into the host chromosome and replicates synchronously with it (Jacob and Wollman, 1961). In addition to the F factor, temperate phages such as λ and $\phi 80$, and certain colicinogeny determinants are true episomes. Other elements such as the drug resistance or R factors are similar, but have not yet been shown to exist in the integrated state (Campbell, 1969).

The R agent (see Watanabe, 1963) carries genes determining resistance to streptomycin, tetracycline, sulfonamide, and chloramphenicol. Transfer of the R factor can occur from one cell to another, and R also mediates chromosomal mobilization (see later section). These factors form pili analogous to those determined by the F factor. Because of their ability to transfer from one cell to another

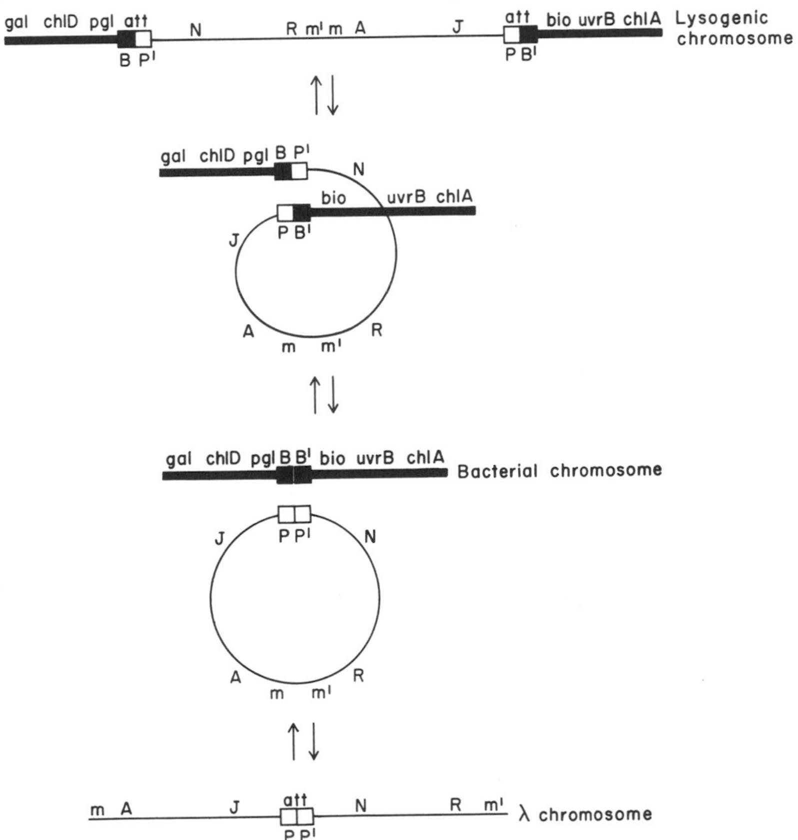

Figure IIA. Insertion of λ into the bacterial chromosome. *A*, *J*, *N*, and *R* are phage genes. *att*P.P′ and *att*B.B′ are the phage and bacterial sites of insertional recombination. *m* and *m′* are the ends of the λ DNA molecule. Bacterial genetic symbols are: *gal*, a cluster of three genes determining enzymes of galactose catabolism; *chlD*, *chlA*, two genes determining chlorate sensitivity; *pgl*, structural gene for phosphogluconolactonase; *bio*, a cluster of at least five genes determining enzymes of biotin biosynthesis; *uvr*B, a gene determining resistance to ultraviolet light. (Reproduced with permission from Campbell, 1971.)

simultaneous resistance to several different antibiotics, the study of R factors has important medical implications.

The Campbell Model

Campbell (1962) first proposed that episomes integrate into the bacterial chromosome by means of a cross-over (breakage and rejoining) between a circular form of the episome and the chromosome (Figure IIA). Genetic studies with phage λ have shown that the order of genetic markers in the non-integrated linear form is a permutation of the order in the integrated form, as predicted by the model (Campbell, 1961; Campbell, 1971).

F Transfer

F^+ cells are able to transfer the F factor to F^- cells at a high frequency after forming conjugation pairs, as depicted in Figure IIB, in contrast to the low frequency of transfer of host chromosomal markers. Whereas every cell in an F^+ population can transfer the sex factor to a female recipient, only a small fraction of the cells in an F^+ population will transfer chromosomal markers. Although several different mechanisms for this latter phenomenon have been proposed (Jacob and Wollman, 1961; Curtiss and Renshaw, quoted in Campbell, 1969), the most widely studied of these involves the stable integration of the F factor into the host chromosome, in a manner outlined by the Campbell model (Figure IIA).

F-mediated Chromosome Transfer

There exist strains (Figure IIC) in which the F factor has stably integrated into the host DNA (Cavalli, quoted in Campbell, 1969; Hayes, 1953). These are termed Hfr strains (high frequency recombination) since every cell now transfers chromosomal markers, and the population as a whole displays a high frequency of transfer relative to an F^+ population.

Once an Hfr has been isolated from an F^+ population and purified, each cell transfers the chromosome in a linear fashion from a fixed starting point or origin, 0, (Figure IID). This occurs because the transfer is mediated by the integrated F factor (Hayes, 1968; Jacob and Wollman, 1961) and begins at a point within the F factor. The F factor is split by this process, F factor genes being the first and last markers to be transferred in an Hfr $\times$ F^- cross. Recipient female cells can be converted to Hfr strains (males) in this type of cross only if the mating proceeds long enough to allow the entire chromosome to be transferred.

Although the F factor does not integrate completely at random into the chromosome, Hfr strains containing F integrated at many sites on the *E. coli* chromosome have been isolated (Jacob and Wollman, 1961). Some of these are pictured in Figure IIE. This type of diagram is used throughout the manual. We usually distinguish F factor DNA from host chromosomal DNA by drawing a wavy line. The F factors in Figures IIB,C, and D are therefore shown in this manner. This convention is also used to depict integrated F factor DNA in Hfr strains.

The origin of transfer of each Hfr strain occurs at a point within the integrated F factor. This point is represented by an arrowhead in the middle of the F factor DNA. The direction of transfer of chromosomal markers is given by the direction of the arrow. Therefore, the Hfr depicted in Figure IIF transfers markers in the order D,E,F . . . . A,B,C. The D,E, and F markers are transferred early after mating, and the A,B, and C markers late. However, the Hfr represented in

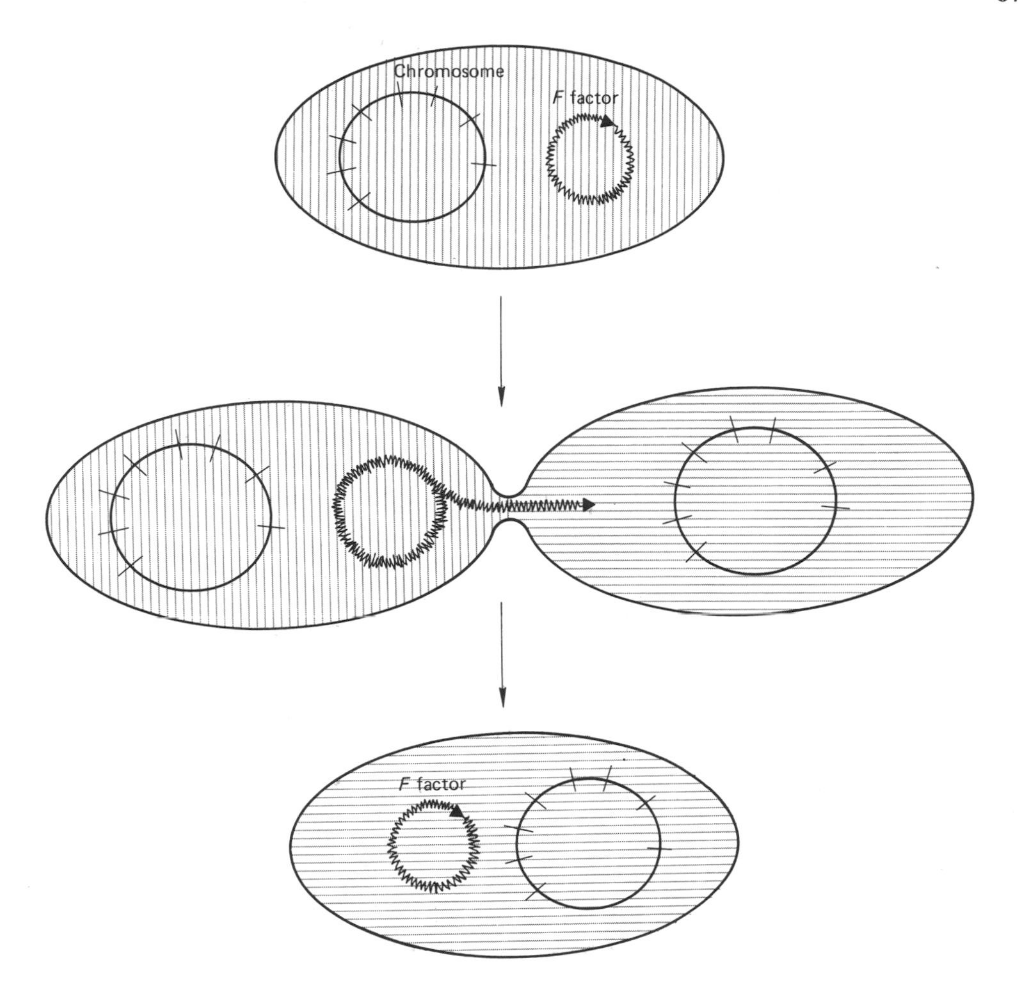

Figure IIB. Transfer of the F factor from a male to a female cell. The donor cell in this mating is shown with vertical shading, and the recipient with horizontal shading. Here the F factor is transferred to the female as a linear element. However, upon entry into the female, a circular form is regenerated. At the same time an F factor is synthesized in the original male strain, with the old one acting as a template. Therefore, male cells do not become cured upon transferring the F factor. (The F factor in these diagrams is not drawn to scale.)

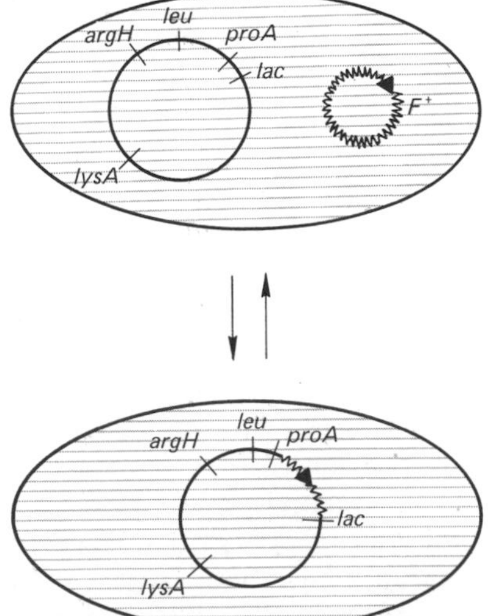

Figure IIC. Integration of F factor into host chromosome.

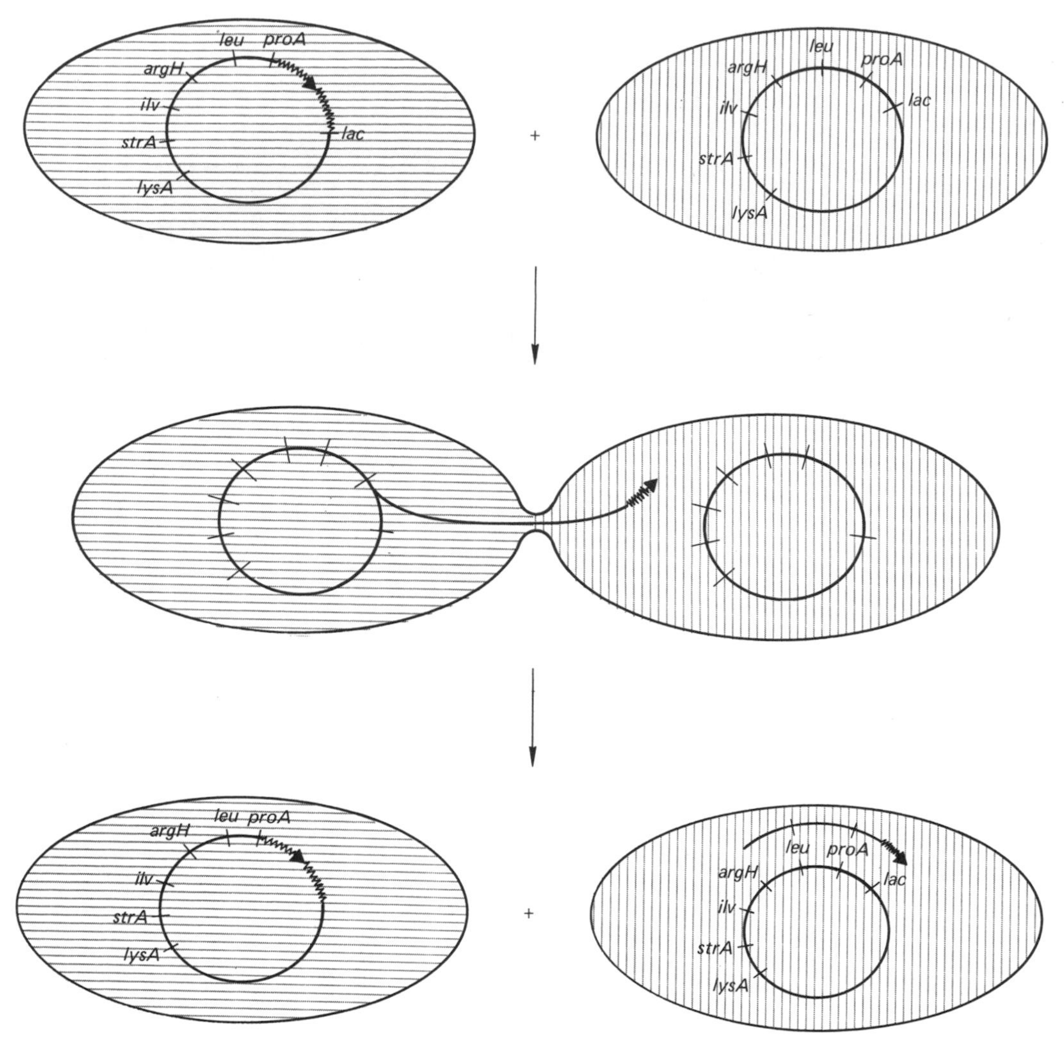

Figure IID. Hfr × F^- mating. An Hfr strain, with the F factor integrated near the *proA* region, is depicted on the left and is shaded horizontally. The F^- strain (right) is shaded vertically.

Figure IIG donates markers in the opposite direction (C,B,A . . . . F,E,D), although from a similar point of origin.

DNA Transfer

After the initial discovery of conjugation in bacteria (Lederberg and Tatum, 1946), the idea that a close cell-cell contact between male and female cells was necessary for DNA transfer was widely accepted, due principally to EM photographs showing close "mating pairs." These showed the formation of a bridge between mating pairs (Anderson et al., 1957). However, the discovery of the relationship between the male (Hfr or F^+) state and F pili suggested that bacterial chromosomes and male-specific phage were transferred through these tubular filaments (Brinton et al., 1964; Brinton, 1965). More recent EM photographs have detected mating pairs connected by a long F pilus. Experimental support for this view has been provided by Ou and Anderson (1970). These authors isolated individual mating pairs of *E. coli* and observed them under the light microscope. After 30 minutes of

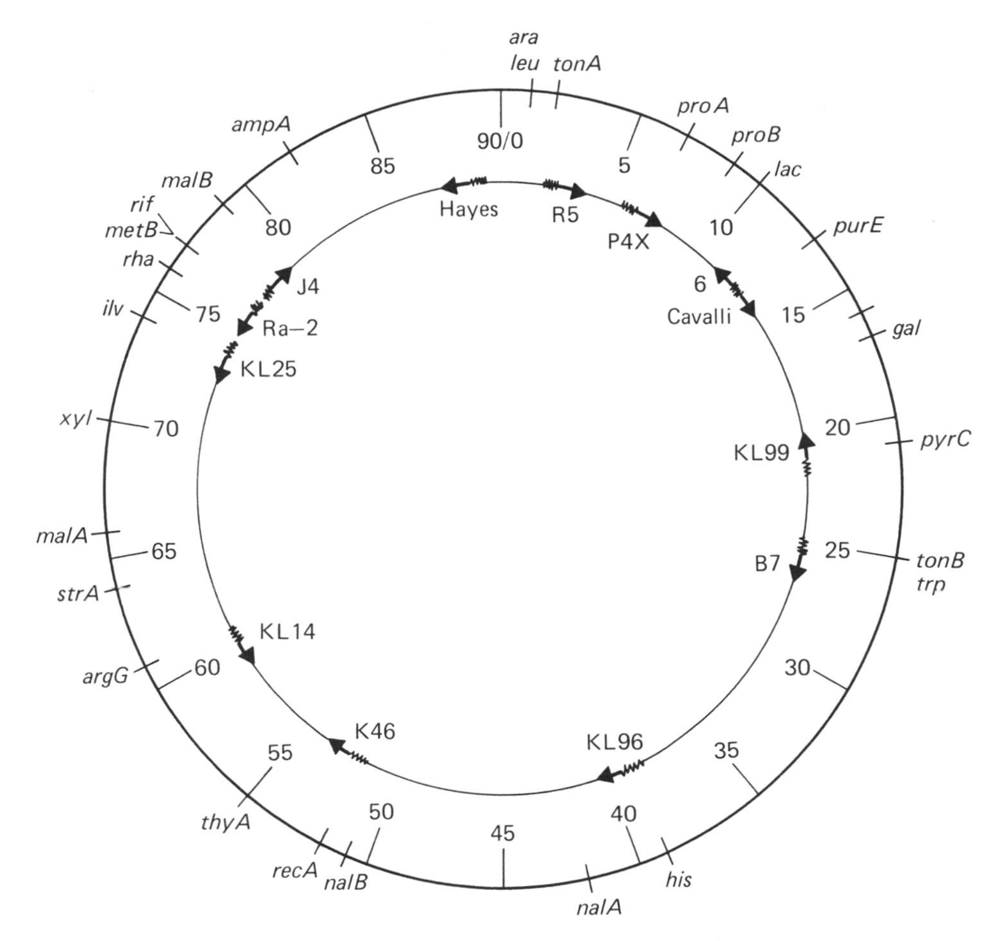

Figure IIE. Location of markers on *E. coli* genetic map.

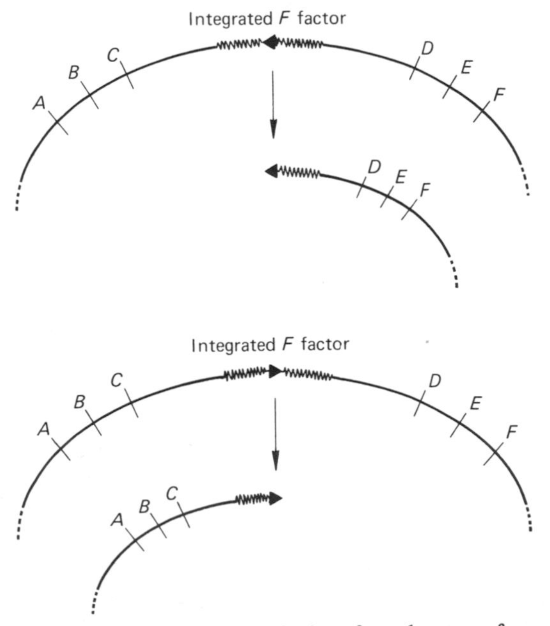

Figures IIF (top) and **G**. Order of marker transfer.

mating, during which time particular pairs were not in cell-cell contact but remained loosely connected (perhaps by invisible F pili), the pairs were separated and plated for single colonies. The finding of recombinants in female cells derived from these loosely connected mating pairs suggests that DNA transfer can occur in cells which are not in close contact, and supports the view that DNA transfer takes place through the F pili. More experiments of this type are necessary to prove this conclusively, however, since the direct visualization of DNA in F pili has not been achieved.

Models which invoke transport of nucleic acid through the F pili are of two types. One has the DNA being actively transported through the pilus, while the second depicts the DNA at the base of a nascent pilus (for instance at the inception of mating) which then grows out towards the female.

Because of the narrow bore of the channel in the F pilus (approximately 30 Å), many scientists find it unlikely that DNA enters through this structure. This is particularly true for single-stranded DNA phages such as fl, fd, and Ml3, which exist in a closed circular form. It is difficult to envision the same simple protein structure actively transporting three different types of nucleic acid in two different directions. These problems led to the suggestion that the F pili serve to connect male and female cells and then retract to bring about a close cell wall-wall interaction (see review by Marvin and Hohn, 1969). After preliminary contact with the F pilus, and subsequent retraction, a more substantial "conjugation bridge" can be constructed. Similarly, RNA phage would be drawn to the base of the pilus, where RNA injection would occur (Marvin and Hohn, 1969).

Another possibility is that the female cell actually ingests the F pili during mating. In fact, experiments by Trenkner, Bonhoeffer, and Gierer (1967) have shown that during infection of sensitive hosts by fd phage, both the DNA and the protein of the virus enter the cell. Labeling experiments indicate that the protein is also incorporated into new virus particles, meaning that at least some intact phage coat protein penetrates the cell during infection. That this mechanism can occur for the filamentous phage indicates that it may also exist for the similarly shaped pili.

F Factors Carrying Bacterial Genes

Not all Hfr strains are equally stable and many Hfr populations contain revertants in which the F factor is no longer integrated in the chromosome, but has returned to the cytoplasmic state. Rare errors can occur during the excision of the F factor from the chromosome which result in the formation of F′ factors carrying bacterial genes, for instance *lac* or *gal* (see Figure IIH). These are called substituted or F′ factors (Adelberg and Burns, 1960). Visualization of F′ DNA can be achieved by electron microscopy (see Figure 37D).

Transfer of F′ chromosomes. If an F′ factor, for instance F′*lac*$^+$, is transferred to a wild-type strain, then a partial diploid (or merodiploid) for the *lac* region is created. Because bacterial recombination enzymes promote recombination between regions of identical sequence ("homology"), F′ factors integrate readily in the host chromosome at these regions of homology (Adelberg and Pittard, 1965). Therefore, in a population of merodiploids of this type, there will be many cells in which the F factor is integrated into the chromosome, and the population will exhibit a much greater transfer of certain chromosomal markers than an F$^+$ population. In these strains the polarity of chromosomal transfer is dependent upon F factor orientation.

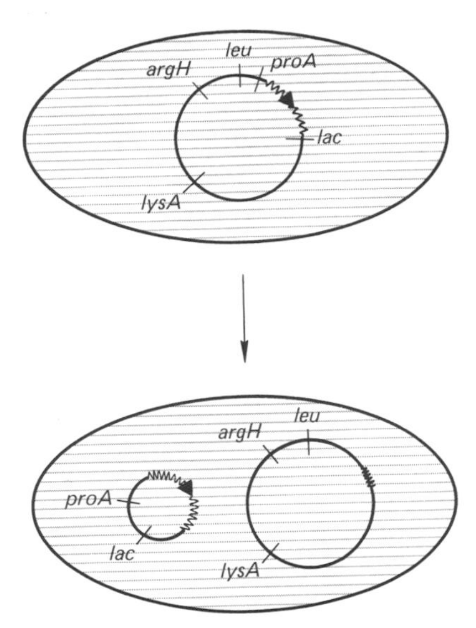

Figure IIH. F′ factor formation. The creation of an F′ factor carrying *proA* and *lac* is depicted here. First an F^+ factor integrates into the chromosome, in this case between *proA* and *lac*, resulting in an Hfr strain (see Figure IIC). Rare cells can be isolated in which the F factor excises in an abnormal manner, incorporating some of the bacterial chromosome. The host chromosome now contains a deletion for the *proA-lac* region. Formation of F′ factors is demonstrated in Experiment 37.

DNA Synthesis During Transfer

After the formation of mating pairs, DNA synthesis is initiated in the male. A single strand of DNA is transferred to the female, where a new complementary strand is synthesized (Gross and Caro, 1966; Rupp and Ihler, 1968). Although the precise mechanism of DNA synthesis during bacterial mating is not known, the process can be easily envisioned in terms of the rolling circle model for DNA replication (Gilbert and Dressler, 1968). The pattern of events is diagrammed in Figure IIJ.

Essentially, the closed circular F factor in the male is nicked at a specific site in one strand (the positive strand). This converts the positive strand into a DNA rod with a 3′-OH terminus and a 5′-P terminus. The 5′ end of the opened positive strand is then peeled back and nucleoside triphosphates are condensed upon the 3′-OH terminus. Continuous elongation of the open positive strand is accompanied by the continued displacement of a single-stranded tail of increasing length. It is this tail which is transferred into the female cell during bacterial mating. Once inside the female the tail is made double-stranded. This is necessary for recombination, the substitution of some of this bacterial DNA for a homologous region of the female chromosome.

Many models have been proposed to explain the mechanism of recombination. These models focus on (a) a breakage and rejoining mechanism (Meselson, 1964, 1967), (b) a hybrid overlap joint formed between the parental strands (Whitehouse, 1963; Holliday, 1964), and (c) breakage of both strands followed by enzymatic digestion exposing single-stranded regions which then pair (Thomas, 1967). These models and others are discussed in detail in a recent review by Signer (1971).

Mapping by Interrupted Matings

Hfr strains can be easily employed to provide mapping data based on the time of entry of markers. If we interrupt the mating pairs at different times by vigorously

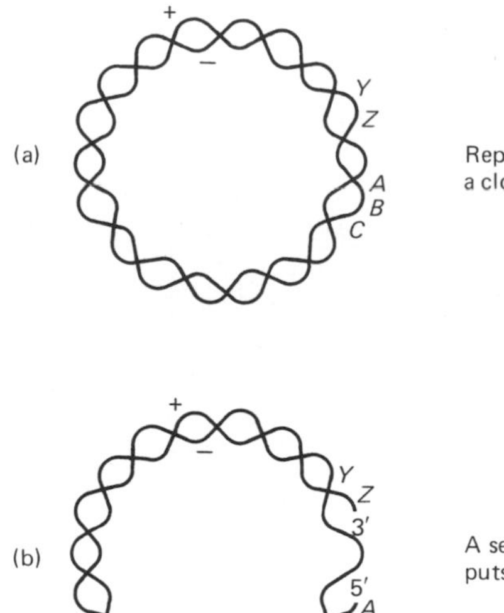

Replication is initiated upon a closed *DNA* circle.

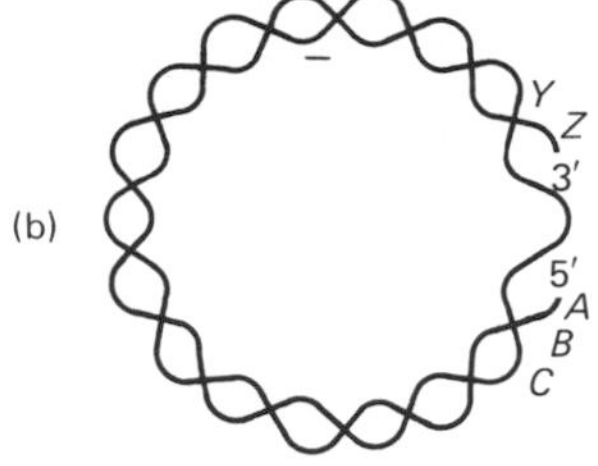

A sequence-recognizing endonuclease puts a nick into the positive strand.

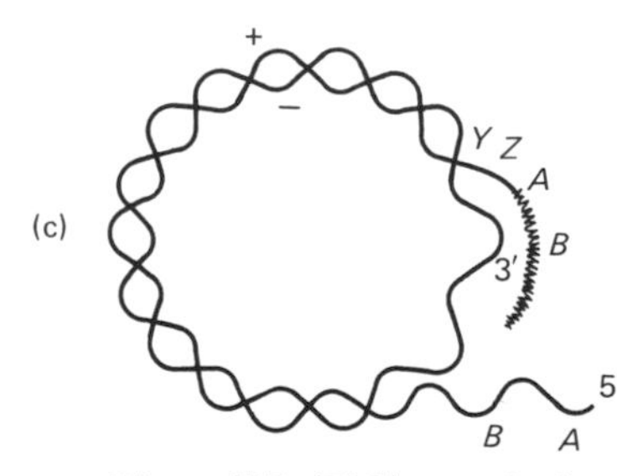

The *DNA* polymerase adds nucleotides onto the 3′ end of the open strand, displacing a tail. The correct nucleotides are chosen by hydrogen bonding to the negative strand template. As new nucleotides are chosen, the positive strand becomes longer than unit length. The 5′ end is then transferred to the female.

Figure IIJ. DNA transfer by a rolling circle mechanism.

agitating the mixture, and then plate aliquots on different selective media to score for the inheritance by the F^- of different Hfr markers, we find that each Hfr transfers chromosomal markers with a definite order and time lag (see Hayes, 1968).

Under optimal conditions the entire *E. coli* chromosome can be transferred in 90 minutes at 37°. The circular genome has been divided into different segments, with distance measured as time intervals, **min** (Jacob and Wollman, 1961; Taylor and Trotter, 1967; Taylor, 1970). The total map covers 90 min (see Figure IIE and Appendix).

Figure IIK shows the results of a mating experiment in which recombinants inheriting different markers from the Hfr donor were selected after disruption of mating pairs at different times. It can be seen that with this Hfr, the Leu^+ marker was transferred early, then the Arg^+ marker, and later the Lys^+ marker. Extrapolation of the very early portion of these curves gives the **time of entry** of these markers, although this cannot be determined too precisely from the data in Figure IIK. Mapping by interrupted mating experiments is demonstrated in Experiment 6; for a full discussion of the factors influencing these curves, see Jacob and Wollman (1961) and also Hayes (1968).

Spontaneous Interruption

Even without mechanically interrupting mating pairs, we find that there is a natural gradient of transmission of markers for every Hfr strain. This is a result

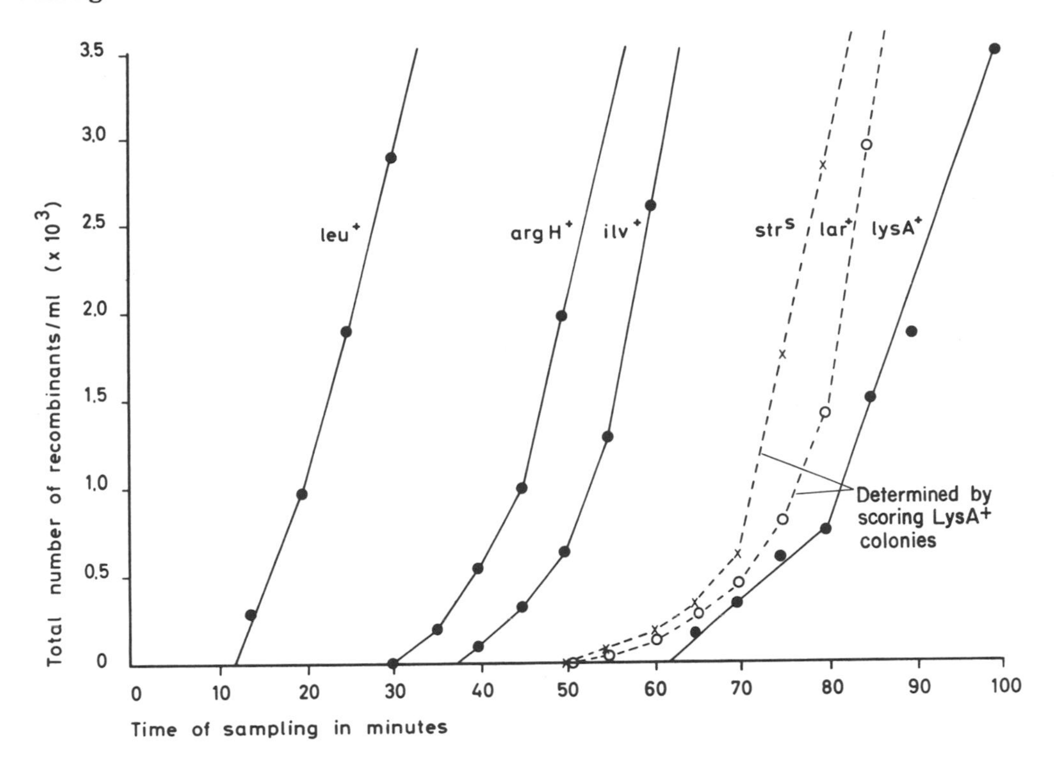

Figure IIK. Interrupted mating curve. (From Kvetkas et al., 1970, *J. Bacteriol.*, 103, 393.)

of the random disruption of mating pairs and subsequent chromosome breakage during transfer, which causes the probability that a marker is transferred to decrease the farther the marker is from the origin. This situation is depicted in Figure IIL, which shows a collection of chromosome fragments present in a recipient population, as might arise from spontaneous interruption of mating pairs. All the DNA ends have the same starting point, determined by the Hfr origin, but different end points, reflecting the random breakage. It is evident from this diagram that recipients from this cross are more likely to inherit the *proA* region from the donor than, for instance, the *argH* region.

Data from a cross in which no mechanical interruption of mating pairs was applied is given in Tabel IIA. In this cross, His$^+$ recombinants are only 13% as frequent as Leu$^+$ recombinants. The right hand portion of this Table shows an analysis of recombinants from this cross. If the **unselected marker** is distal to the **selected marker,** as is the case for all the markers scored in this cross when Leu$^+$ recombinants are examined, then the probability that a marker will be inherited decreases with its distance from the origin (or from the selected marker). However, when the unselected marker is proximal to the selected marker, then the linkage is close to 50%. An exception to this is the case where the proximal marker is within several minutes of the Hfr origin, in which case it is inherited at a reduced frequency. Also, markers very close to the selected marker are inherited at frequencies greater than 50%. The approximate location of markers on the chromosome by use of the gradient of transmission is shown in Experiment 7.

Zygotic Induction

A further demonstration of the linear transfer of the chromosome from a fixed point is achieved by the use of Hfr strains which contain repressible prophage

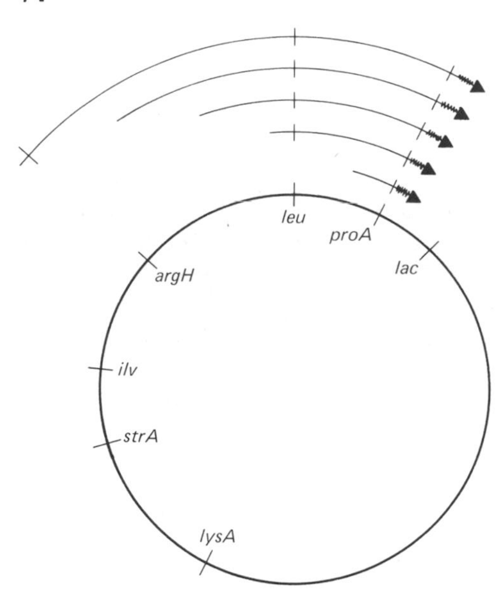

Figure IIL. Gradient of transmission. The collection of temporary partial diploids present in a recipient population after mating with an Hfr with its point of origin between *proA* and *lac*. Random interruption of mating pairs leads to a distribution of cells containing transferred chromosome fragments of differing lengths.

integrated at different sites along the chromosome. After transfer into a sensitive recipient (a cell which is not lysogenic lacks immunity), the prophage becomes induced and replicates. This phenomenon is called "zygotic induction" (Jacob and Wollman, 1957).

The time of appearance of the phage burst is a function of the distance from the Hfr origin to the point of integration of the prophage. This can be seen from Figure IIM which represents three Hfr strains, each carrying a prophage at a different location. Zygotic induction begins approximately 10, 15, and 20 minutes after the inception of mating, respectively, when these three Hfr's are used with a sensitive recipient. Thus, markers transferred **after** the prophage are inherited at a sharply reduced frequency when a non-immune recipient is used (Table IIA; see also Experiment 41A). In Figure IIM for example, the inheritance of *lac* from Hfr(A) is sharply reduced compared to a control mating with an immune recipient. However, the frequency of recombinants inheriting *lac* from Hfr's (B) and (C) is unaffected by the presence of the prophage.

Table IIA. Linkage Data for Uninterrupted Mating

	Relative frequency			Recombinants					
					Unselected marker				
	Gal^+/Leu^+	Trp^+/Leu^+	His^+/Leu^+	Selected marker	Leu^+	Lac^+	Gal^+	Trp^+	His^+
HfrH × F^-	0.59	0.38	0.13	Leu^+	—	58	45	35	14
				Gal^+	62	63	—	52	19
$HfrH(\lambda)^+$ × F^-	0.05	0.13	0.035	Leu^+	—	26	3	6	3
				Gal^+	61	45	—	25	9

Numbers in right half of Table indicate the percent of recombinants inheriting the unselected marker from the donor. (From Jacob and Wollman, 1961.)

bacteria was made (Lederberg and Tatum, 1946a,b) only after medium allowing the selection of rare recombinants was used.

Suppose we cross an Hfr (Leu$^+$) with an F$^-$ (Leu$^-$) and select for Leu$^+$ recombinants by plating the mating mixture on glucose minimal plates without leucine. The F$^-$ parent cannot grow on this leucine-free medium. However, the Hfr parent is Leu$^+$ also. If we simply plated on this medium, then the colonies would be a mixture of both Leu$^+$ recombinants and the parent male strain. It is necessary, therefore, to select against the parent male strain. This is termed *counterselection*. Counterselection is often done using antibiotics or bacteriophage to which the male strain is sensitive.

Selection and counterselection are simply different terms for the same procedure. However, proper usage of this terminology enables us to designate the donor and recipient in the description of each cross. By convention we write that **selection** is performed on the recipient or female strain, and **counterselection** is applied against the donor or male strain. Counterselection against male cells was first extensively used by Hayes and his collaborators (Hayes, 1953; Wollman et al., 1956), who employed streptomycin to prevent the further growth of male donor strains and also T6 phage to eliminate sensitive Hfr cells.

Many of the practical problems which occur in Hfr × F$^-$ crosses or even in F′ transfers to F$^-$ recipients involve the inability to employ selection for recombinant colonies or counterselection against the male parent. These problems and many methods used to overcome them are discussed in Unit IV. We would like to emphasize here, though, the need to employ counterselection in bacterial crosses, since it is often neglected by the beginning student. Experiment 5 presents a simple F′ factor transfer which demonstrates these principles.

Table IIB lists the commonly used counterselective agents, and points out some of their advantages and disadvantages. Many of these markers can also be used as selective markers (a more complete list is given in the introduction to Experiments 30–32). For instance, the *rif*^{r} locus can be crossed from an Hfr to an F$^-$ selecting for rifampicin resistance, provided another marker is used for counterselection. However, many antibiotic resistance alleles are recessive and several generations of growth would be required to allow segregation of Rif^r recombinants. Since it is more convenient to plate mating mixtures out directly, it is preferable to use for selection those nutritional markers which allow direct plating.

Background Growth

The choice of selective and counterselective markers can greatly influence the background growth resulting from a bacterial cross. The higher this growth is, the more difficult it becomes to observe recombinant colonies. Both **leakiness** (residual growth past a nutritional or antibiotic block) and **reversion** contribute to the background. Some markers allow more residual growth than others. In general, vitamin markers are much leakier than amino acid or sugar markers. **Crossfeeding** can also affect background growth. This occurs when one cell excretes a necessary nutrient which is then used by a different cell to satisfy a growth requirement.

Revertants are present in a bacterial population for most markers and limit the resolution of every cross. Markers with low reversion frequencies are therefore the most desirable.

Techniques for Optimizing Mating Frequencies

For Hfr and episome crosses in liquid culture, several procedures are recommended.

1. The preferred temperature for all of these crosses is 37°.

2. The donor should be growing exponentially in rich broth and be at a density of 2–3 × 10^8 cells/ml. Cultures which are at a higher density mate less efficiently. Many investigators find better efficiencies of mating if the F^- cells are also growing exponentially, and we therefore favor also using freshly growing recipient cultures.

The most convenient method for class experiments is to prepare an overnight culture of both the donor and recipient the night before, and to dilute the donor 1:40 and recipient 1:20 into fresh broth on the morning of the experiment. For

Table IIC. Plates Used for Bacterial Mating Experiments

Plate type	Carbon source	Selected marker	Supplements added							
A	Glucose	argG	—	ilv	met	leu	ade	trp	his	str
B	Glucose	met	arg	ilv	—	leu	ade	trp	his	str
C	Glucose	leu	arg	ilv	met	—	ade	trp	his	str
D	Lactose	lac	arg	ilv	met	leu	ade	trp	his	str
E	Glucose	purE	arg	ilv	met	leu	—	trp	his	str
F	Galactose	gal	arg	ilv	met	leu	ade	trp	his	str
G	Glucose	trp	arg	ilv	met	leu	ade	—	his	str
H	Glucose	his	arg	ilv	met	leu	ade	trp	—	str
I	Glucose	control	arg	ilv	met	leu	ade	trp	his	str

maximum efficiency per Hfr cell (which is desired for interrupted mating experiments), at least 10 F^- cells should be used per Hfr cell. In this case both donor and recipient cultures should be growing exponentially and at a density of 2–3 × 10^8 cells/ml before mixing.

3. Aeration during mating is desirable. It can be achieved by gentle shaking (vigorous agitation disrupts mating pairs) or by giving the mixture a large surface to volume ratio. Volumes of 5 ml in a 100-ml erlenmeyer flask, 10 ml in a 250-ml flask, or 20 ml in a 500-ml flask are satisfactory. Also, volumes of up to several milliliters can be used in a test tube which is then placed on the outside ring of a 30–33 rpm rotor drum at 37° (Clowes and Hayes, 1968). For episome transfers, required in many experiments in this manual, we find the latter method most convenient.

Although some investigators prefer to grow donors unaerated (Curtiss et al., 1969), we aerate all donor cultures, since we find no significant difference for the strains used with this manual. A possible exception to this is in Experiment 8. An "unaerated" culture for class experiments can be grown to 1 × 10^8 cells/ml with aeration, and then left at 37° without aeration for 60 minutes.

Spot Tests on Solid Surfaces

Instead of using liquid matings, it is also possible to do Hfr crosses or F′ transfers on solid agar plates. This is convenient for carrying out matings on a large scale, since this only requires "spotting" a drop of both donor and recipient onto a selection plate (see Experiment 5). Cross-streaking and replica plating techniques allow one to perform several hundred crosses with little effort.

Experimental Design

The experiments in this unit are designed to demonstrate many of the techniques used for bacterial crosses and for mapping markers on the chromosome. Hfr strains with different points of origin and directions of transfer are used. These are described in the strain list and depicted in Figure IIC. In addition, the multiply marked recipient, CSH57 (Ilv^-, Met^-, Ara^-, Leu^-, Lac^-, Pur^-, Gal^-, Trp^-, His^-, Arg^-, Str^r) is employed. With this set of strains a series of related experiments can be performed. Table IIC lists the selective markers used in these experiments and also tabulates the supplements required for each type of medium, when CSH57 is used.

Instability of Many Hfr and F′ Strains

Unfortunately, many Hfr strains are unstable for the Hfr character and tend to accumulate a substantial proportion of F^+ revertants. When using an Hfr after recovery from storage, it is advisable to:

A) Streak for single colonies;

B) Grid a master plate with freshly growing colonies (Experiment 4) and incubate at 37° for 6–8 hours;

C) Replicate onto a selective plate onto which several drops of an F^- culture have been spread. Sometimes it is more convenient to spread about 1 ml of the F^- culture on the surface of the plate, and then to pour it off after a few minutes. This achieves a more even spreading of F^- cells over the surface of the plate. Use an F^- which allows selection for a marker donated early by the Hfr, and counterselect with streptomycin if possible.

Recombinants will show confluent growth on these areas of the selection plate onto which Hfr's have been replicated, but only a few colonies will appear where an F^+ strain has been replicated. In this way, even if the stab contains only a few percent Hfr's, these can be isolated and used (from the master plate).

F′ factors are also frequently lost from their hosts upon storage. F′ factors should be kept in strains, whenever possible, which are lacking intact genes that are present on the F′ factor. This is particularly important for large F′ factors, which tend to lose bacterial genes and give rise to smaller episomes. Thus, if an F′arg^+ is carried in an arg^- host which is maintained on medium lacking arginine, segregants which have lost the F′ factor will not grow. If the arg^- strain is $recA^-$, then the chromosomal arg^- marker can never be converted to arg^+. The best strain in which to keep this F′ factor is therefore an arg^-recA^- strain.

References

ABE, M. and J. TOMIZAWA. 1967. Replication of the *E. coli* K12 chromosome. *Proc. Nat. Acad. Sci. 58:* 1911.

ACHTMAN, M., N. WILLETTS and A. J. CLARK. 1971. Beginning a genetic analysis of conjugational transfer determined by the F factor in *E. coli* by isolation and characterization of transfer-deficient mutants. *J. Bacteriol. 106:* 529.

ADELBERG, E. A. and S. N. BURNS. 1960. Genetic variation in the sex factor of *E. coli*. *J. Bacteriol. 79:* 321.

ADELBERG, E. A., and J. PITTARD. 1965. Chromosome transfer in bacterial conjugation. *Bacteriol. Rev. 29:* 161.

ANDERSON, T. F., E. L. WOLLMAN and F. JACOB. 1957. Sur les processus de conjugaison et de recombinaison chez *E. coli*. III. Aspects morphologiques en microscopie électronic. *Ann. Inst. Pasteur 93:* 450.

BERG, C. M. and R. CURTISS, III. 1967. Transposition derivatives of an Hfr strain of *E. coli* K12. *Genetics 56:* 503.

BIRD, R. and L. CARO. 1972. (Personal communication.)

BIRD, R., J. LOUARN and L. CARO. 1971. (Personal communication.)

BIRD, R. and K. G. LARK. 1968. Initiation and termination of DNA replication after amino acid starvation of *E. coli* $15T^-$. *Cold Spring Harbor Symp. Quant. Biol. 33:* 799.

BRINTON, C. C., P. GEMSKI and J. CARNAHAN. 1964. A new type of bacterial pilus genetically controlled by the fertility factor of *E. coli* K12. *Proc. Nat. Acad. Sci. 52:* 776.

BRINTON, C. C. 1965. The structure, formation, synthesis and genetic control of bacterial pili and a molecular model for DNA and RNA transport in gram negative bacteria. *Trans. N.Y. Acad. Sci. 27:* 1003.

CAIRNS, J. 1963a. The bacterial chromosome and its manner of replication as seen by autoradiography. *J. Mol. Biol. 6:* 208.

CAIRNS, J. 1963b. The chromosome of *E. coli*. *Cold Spring Harbor Symp. Quant. Biol. 28:* 43.

CAMPBELL, A. 1962. Episomes. *Adv. Genet. 11:* 101.

CAMPBELL, A. 1969. *Episomes*. Harper and Row, New York.

CAMPBELL, A. 1971. Genetic structure, p. 13. In *The Bacteriophage Lambda* (A. Hershey, ed.), Cold Spring Harbor Laboratory.

CARO, L. G. and C. M. BERG. 1968. Chromosome replication in some strains of *E. coli* K12. *Cold Spring Harbor Symp. Quant. Biol. 33:* 559.

CLARK, A. J. and A. D. MARGULIES. 1965. Isolation and characterization of recombination-deficient mutants of *E. coli* K12. *Proc. Nat. Acad. Sci. 53:* 451.

CLOWES, R. C. and W. HAYES. 1968. *Experiments in Microbial Genetics*. John Wiley and Sons, New York.

CRAWFORD, E. M. and R. F. GESTELAND. 1964. The adsorption of bacteriophage R-17. *Virology 22:* 165.

CURTISS, R., III, L. G. CARO, D. P. ALLISON and D. R. STALLIONS. 1969. Early stages of conjugation in *E. coli*. *J. Bacteriol. 100:* 1091.

DELIUS, H., C. HOWE and A. W. KOZINSKI. 1971. Structure of the replicating DNA from bacteriophage T4. *Proc. Nat. Acad. Sci. 68:* 3049.

DONACHIE, W. D. and M. MASTERS. 1966. Evidence for polarity of chromosome replication in F^- strains of *E. coli*. *Genet. Res. 8:* 119.

DRESSLER, D. and J. WOLFSON. Personal communication.

DUGUID, J. P., E. S. ANDERSON and I. CAMPBELL. 1966. Fimbriae and their adhesive properties in Salmonella. *J. Path. Bact. 92:* 107.

FREIFELDER, D. 1968. Studies with *E. coli* sex factors. *Cold Spring Harbor Symp. Quant. Biol. 33:* 425.

GILBERT, W., and D. DRESSLER. 1968. DNA replication: The rolling circle model. *Cold Spring Harbor Symp. Quant. Biol. 33:* 473.

GROSS, J. D. and L. CARO. 1966. DNA transfer in bacterial conjugation. *J. Mol. Biol. 16:* 269.

HAYES, W. 1953. The mechanism of genetic recombination in *E. coli*. *Cold Spring Harbor Symp. Quant. Biol. 18:* 75.

HAYES, W. 1968. *The Genetics of Bacteria and Their Viruses*, 2nd. Ed. John Wiley and Sons, New York.

HELMSTETTER, C. E. 1968. Origin and sequence of chromosomal replication in *E. coli* B/r. *J. Bacteriol. 95:* 1634.

HOLLIDAY, R. 1964. A mechanism for gene conversion in fungi. *Genet. Res. 5:* 282.

JACOB, F. and E. L. WOLLMAN. 1957. Genetic aspects of lysogeny. In *The Chemical Basis of Heredity* (W. D. McElroy and B. H. Glass, eds.), p. 468. Johns Hopkins Press, Baltimore.

JACOB, F. and E. L. WOLLMAN. 1961. *Sexuality and the Genetics of Bacteria*. Academic Press, New York.

LARK, K. G. 1966. Regulation of chromosome replication and segregation in bacteria. *Bacteriol. Rev. 30:* 3.

LEDERBERG, J., L. L. CAVALLI and E. M. LEDERBERG. 1952. Sex compatibility in *E. coli*. *Genetics 37:* 720.

LEDERBERG, J. and E. L. TATUM. 1946a. Novel genotypes in mixed cultures of biochemical mutants of bacteria. *Cold Spring Harbor Symp. Quant. Biol. 11:* 113.

LEDERBERG, J. and E. L. TATUM. 1946b. Gene recombination in *E. coli*. *Nature 158:* 558.

MACCACARO, G. A. and R. COMOLLI. 1956. Surface properties correlated with sex compatibility in *E. coli. J. Gen. Microbiol. 15:* 121.

MARVIN, D. A. and B. HOHN. 1969. Filamentous bacterial viruses. *Bacteriol. Rev. 33:* 172.

MASTERS, M. and P. BRODA. 1971. Evidence for the bidirectional replication of the *E. coli* chromosome. *Nature New Biol. 232:* 137.

MESELSON, M. 1967. The molecular basis of recombination, p. 81. In *Heritage from Mendel* (R. A. Brink, ed.). Univ. of Wisconsin Press, Madison.

MESELSON, M., 1964. On the mechanism of genetic recombination between DNA molecules. *J. Mol. Biol. 9:* 734.

MESELSON, M. and F. W. STAHL. 1958. The replication of DNA in *E. coli. Proc. Nat. Acad. Sci. 44:* 671.

NAGATA, T. 1963. The molecular synchrony and sequential replication of DNA in *E. coli. Proc. Nat. Acad. Sci. 49:* 551.

NAGATA, T. and M. MESELSON. 1968. Periodic replication of DNA in steadily growing *E. coli:* The localized origin of replication. *Cold Spring Harbor Symp. Quant. Biol. 33:* 553.

OU, J. T. and T. F. ANDERSON. 1970. Role of pili in bacterial conjugation. *J. Bacteriol. 102:* 648.

RUPP, D. and G. IHLER. 1968. Strand selection during bacterial mating. *Cold Spring Harbor Symp. Quant. Biol. 33:* 647.

SCHNÖS, M. and R. B. INMAN. 1970. The position of branch points in replicating lambda DNA. *J. Mol. Biol. 51:* 61.

SIGNER, E. R. 1971. General recombination, p. 139. In *The Bacteriophage Lambda* (A. Hershey, ed.). Cold Spring Harbor Laboratory.

TAYLOR, A. L. 1970. Current linkage map of *Escherichia coli. Bacteriol Rev. 34:* 155.

TAYLOR, A. L. and C. D. TROTTER. 1967. Revised linkage map of *Escherichia coli. Bacteriol. Rev. 31:* 332.

THOMAS, C. A., JR., 1967. The recombination of DNA molecules, p. 162. In *The neurosciences: A Study Program* (G. C. Quarton, T. Melnechuk, and F. O. Schmitt, eds.) Rockefeller Univ. Press, New York.

TRENKNER, E., F. BONHOEFFER and A. GIERER. 1967. The fate of the protein component of bacteriophage fd during infection. *Biochem. Biophys. Res. Commun. 28:* 932.

WAKE, R. G. Personal communication.

WATANABE, T. 1963. Infectious heredity of multiple drug resistance in bacteria. *Bacteriol Rev. 27:* 87.

WHITEHOUSE, H. L. K. 1963. A theory of crossing over by means of hybrid DNA. *Nature 199:* 1034.

WITKIN, E. M. 1969. Ultraviolet-induced mutation and DNA repair. *Ann. Rev. Microbiol. 53:* 487.

WOLF, B., A. NEWMAN and D. A. GLASER. 1968a. On the origin and direction of replication of the *E. coli* chromosome. *J. Mol. Biol. 32:* 611.

WOLF, B., M. L. PATO, C. B. WARD and D. A. GLASER. 1968b. On the origin and direction of replication of the *E. coli* chromosome. *Cold Spring Harbor Symp. Quant. Biol. 33:* 575.

WOLLMAN, E. L., F. JACOB and W. HAYES. 1956. Conjugation and genetic recombination in *Escherichia coli* K-12. *Cold Spring Harbor Symp. Quant. Biol. 21:* 141.

WOLLMAN, E. L. and F. JACOB. 1957. Sur les processus de conjugaison et de recombinaison chez *Escherichia coli.* II. La localisation chromosomique du prophage λ et les conséquences génétique de l'induction zygotique. *Ann. Inst. Pasteur 93:* 323.

YOSHIKAWA, H. and N. SUEOKA. 1963. Sequential replication of *Bacillus subtilis* chromosome. I. Comparison of marker frequencies in exponential and stationary growth phases. *Proc. Nat. Acad. Sci. 49:* 559.

EXPERIMENT 5

Episome Transfers: Direct Selection

In the cross of strain CSH23 (Str^s) with strain CSH50 ($Lac^- Pro^- Str^r$) we wish to transfer an $F'lac^+pro^+$ from the CSH23 background to the CSH50 strain. This could be done in two ways. In one a mating mixture could be plated on glucose minimal plates with streptomycin. All of the CSH23 cells are killed by the streptomycin. Since the recipient cannot grow in the absence of proline, only those CSH50 cells which have received the $F'lac^+pro^+$ from CSH23 will form colonies on this selection plate. Thus, the selection is for Pro^+Str^r (see Figure 5A). Alternatively, we could use lactose minimal streptomycin plates and select for $Lac^+Pro^+Str^r$. In either case there is a direct selection for the desired diploid.

In cases where only the desired recipient (or recombinant in similar situations) grows, but neither of the original parent strains, we are not limited to doing liquid matings. Two additional techniques which are often used in this case are spot matings and replica matings.

Spot matings: Instead of the actual mating occurring in liquid broth, the F′ transfer takes place on the surface of an agar plate. For the cross described above, a glucose minimal plate containing streptomycin is marked into three sections. Onto one a drop of an exponential culture of the donor (CSH23) is applied with a

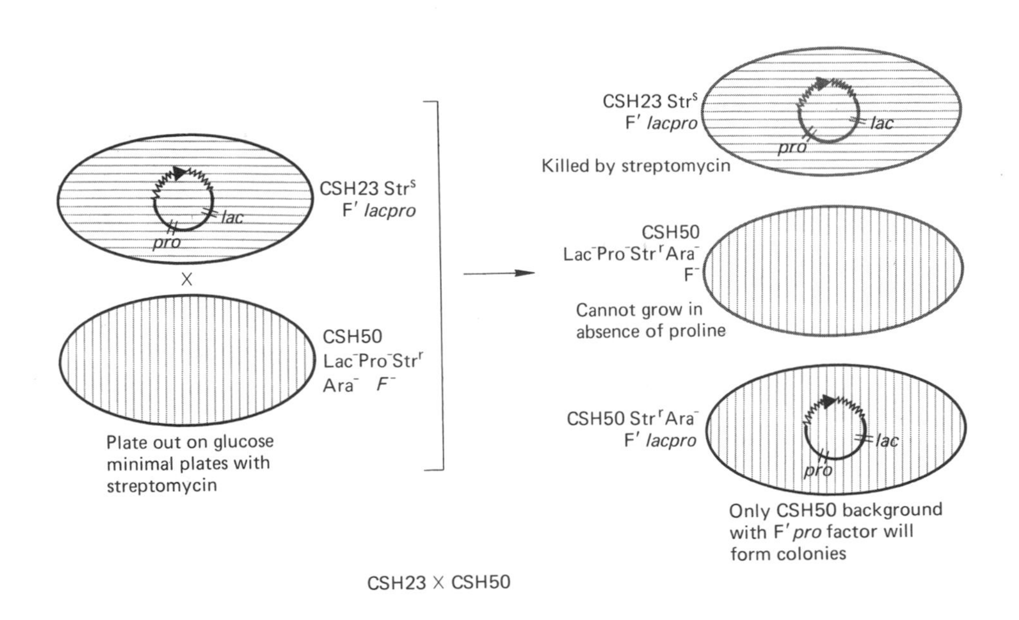

Figure 5A. Transfer of an F'*lacpro* factor from a Str^s donor to a Str^r recipient. Cells of the CSH50 (Str^r) background are shaded vertically, and those of the CSH23 (Str^s) background are shaded horizontally. In this type of diagram the cells which do not form colonies are lighter than those which do on the selection plate.

pasteur pipette, and a drop of the recipient onto a second. Both donor and recipient are spotted together on the third sector, the donor directly on top of the recipient as soon as the recipient spot is dry. After the spots are dry, the plate is incubated at 37° overnight. Only in the space in which the donor and recipient were applied together should there be confluent growth.

Using a toothpick we can pick from the center of the spot and repurify by streaking for single colonies. For best results the agar plate should be as dry as possible. This is usually achieved by drying the plates partially open at 37° for 2 hours. The advantage of this method is that it can be used to prepare as many as eight diploids per plate. It is best to use streptomycin for counterselection. Nalidixic acid should be avoided for plate matings, since it rapidly inhibits DNA synthesis and prevents transfer. The broth applied to a spot can sometimes interfere with the scoring of recombinants, particularly if a nutritional marker is used for counterselection instead of a drug resistance marker.

Replica matings: An efficient way to mate a large number of strains is to replicate a grid of up to 50 donor strains onto a lawn of an F^- strain. Master plates are prepared in advance by gridding freshly growing colonies onto rich plates and incubating for 6–8 hours. From a fresh overnight (or exponentially growing culture) of the recipient 6 drops are spread evenly over the surface of the selection plate. As soon as this dries, the master plate is replicated onto the selection plate (see Experiment 4). The advantage of this method is that 50 colonies can be tested for their donor properties in one step.

Strains

Number	Sex	Genotype	Important Properties
CSH23	F′*lac*$^+$*proA*$^+$,*B*$^+$	Δ(*lacpro*) *supE spc thi*	Str^s
CSH50	F$^-$	*ara* Δ(*lacpro*) *strA thi*	Lac$^-$Pro$^-$Str^r

Method—Cross of CSH23 × CSH50

Day 1 Subculture fresh overnights of both donor and recipient by diluting the donor 1:40 and the recipient 1:20 and growing at 37° until the donor is 2–3 × 10^8 cells/ml. Prepare a mating mixture by mixing 0.5 ml of each together in a test tube. Place on a 30 rpm rotor at 37° for 60 minutes. For alternative ways of preparing a mating mixture see the introduction to this unit. At the end of this time plate out 0.1 ml of a 10^{-3}, 10^{-4}, and 10^{-5} dilution onto glucose minimal plates with streptomycin.

Also plate out 0.1 ml of a 10^{-2} dilution of both donor and recipient alone as a control. Plate 0.1 ml of a 10^{-5} dilution of the recipient onto the selection plate supplemented with proline, to enable the determination of the efficiency of transfer.

Instead of making dilutions, it is also possible to streak a loopful of the mating mixture directly onto the selection plate. In this manner we can streak eight matings per plate.

Do a spot mating as described in the introduction.

Also prepare a master plate and do a replica mating as described in the introduction.

Day 2, 3 Observe the plates and compute the number of diploids generated per ml of mating mixture and also the percentage of cells in the CSH50 population which received the F′ factor in the cross.

Verify that the colonies growing on the selection plate are of the CSH50 background by testing for the Ara marker. Pick several colonies and streak on arabinose indicator plates (or arabinose minimal plates). Recall that CSH23 is Ara$^+$ and CSH50 is Ara$^-$.

Additional exercises: Determine the effect on the efficiency of transfer of different growth conditions of the donor and recipient, and of different temperatures during the mating.

Materials

Day 1 **(for liquid mating)**

fresh overnight cultures of strains CSH23 and CSH50
2 test tubes with 4 ml LB broth each
empty test tube or 100-ml flask
30 rpm rotor drum in a 37° room or a shaking waterbath at 37°
10 0.1-ml, 4 1-ml, and 4 pasteur pipettes
6 dilution tubes
6 glucose minimal streptomycin plates
1 glucose minimal plate with proline and streptomycin

(for replica mating)

freshly grown plate with colonies of CSH23
fresh overnight of CSH50
1 glucose minimal streptomycin plate
velvet with replicating block
10 round wooden toothpicks

Day 2, 3 cultures of CSH23 and CSH50 from Day 1
1 arabinose indicator plate
6 round wooden toothpicks

EXPERIMENT 6

Interrupted Matings and the Time of Marker Entry

Interrupted Mating Apparatus

As described in the introduction to this unit, we can utilize the fact that each Hfr transfers markers linearly from a fixed point of origin to map different markers. In practice, donor and recipient strains are mixed together and at different time intervals the mating pairs are disrupted by agitation. Originally, a Waring blendor was used to separate mating pairs. Although the use of a vortex mixer for this purpose is widespread, it sometimes gives unsatisfactory results; mixtures must be agitated for 30–60 seconds, and even then disruption is not complete. A device first described by Low and Wood (see Appendix) offers the most rapid and efficient agitation to date. When properly fitted into a small motor (such as a saber saw motor), 8–10 seconds is sufficient to bring about complete disruption of mating pairs. We supply this device in the strain kit which accompanies this manual.

Using the strains described in this manual (see strain list and Figure IIE) many different mating experiments are possible. We describe here the procedure for one particular experiment, and outline additional exercises. The object of this experiment is to measure the time of entry of three different markers with two Hfr strains, HfrC and HfrH. We have chosen *lac*, *leu*, and *trp* as the markers. Since *lac* is very close to *pro*, the curves can be compared with Figures 6A and 6B.

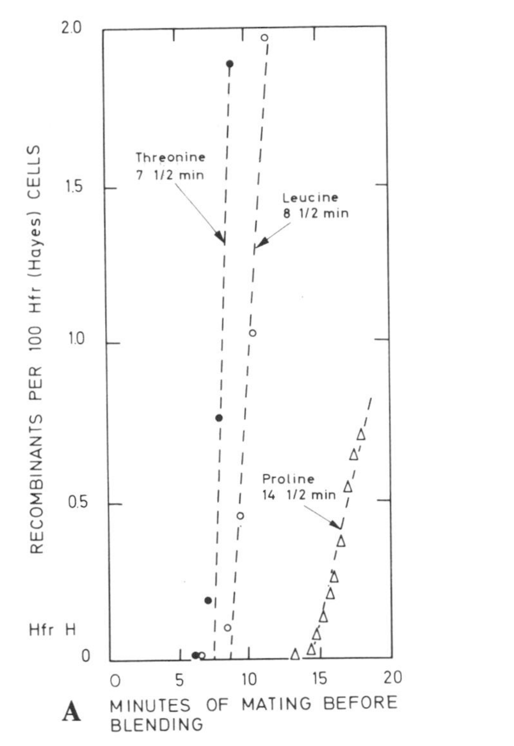

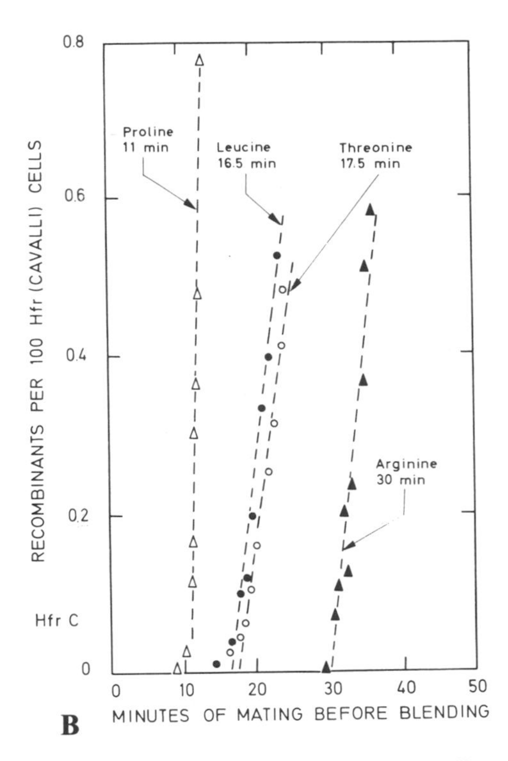

Figure 6A. Results of an interrupted mating experiment using HfrH (Str^s) and an F^- (Thr^- Leu^- Pro^- Str^r) strain. Exponentially growing male and female cultures were mixed in a 1:100 ratio and shaken gently at 37°. Samples for blending were taken at approximately 1 minute intervals, and the number of recombinants of the various types were determined as a function of the time between the start of mating and blending. These numbers were all normalized to the female titer at the time of interruption.

Figure 6B. Results of an interrupted mating experiment using HfrC (Str^s) and an F^- (Thr^- Leu^- Pro^- Str^r) strain. Conditions of the experiment are as in Figure 6A.

Notes on Interrupted Mating Experiments

In determining the time of entry of markers, only the very early portion of the curve is critical. Accurate measurements are easier to obtain for early markers than for late ones, so it is preferable if the marker under study enters within the first 20–30 minutes of transfer. The actual time of entry of a marker, as measured by this technique, is really the earliest time at which recombinants appear above the background. Therefore, the limiting factor for measuring the time of entry is the background number of revertants and recombinants which appear on the zero-minute control (donor and recipient mixed and immediately blended). The procedure we describe for all interrupted matings has been suggested by Drs. B. Low and A. L. Taylor and is the following:

Method A

Donors growing exponentially at 2×10^8 cells/ml are mixed with recipients (at $2–4 \times 10^8$ cells/ml) in a ratio of 1 Hfr:20 F^-. This can be achieved by mixing 0.5 ml of the donor with 10 ml of the recipient in a 125-ml erlenmeyer flask in a 37° waterbath. The flask should be gently shaken throughout the experiment.

Samples are withdrawn by taking 0.05-ml or 0.1-ml aliquots and pipetting directly into 13 × 100 mm test tubes containing 3–4 ml of soft minimal top agar.

The test tube contents are then blended for 10 seconds using the device described in Appendix II and immediately plated on selective medium. With this method no dilutions are required. For doing more careful measurements of early markers, ratios of 1 Hfr:100 F^- cells should be used if 0.05 ml aliquots are to be directly plated.

Figures 6A and 6B show interrupted mating curves done using this procedure. With this method we are plating out aliquots of mating mixtures directly in order to see recombinants at the earliest possible time. Because of the transfer gradient (Experiment 7) the number of recombinants for late markers is much lower than that for early markers, and in order to see a sufficient number of colonies we must plate a large number of cells.

It is necessary to avoid remating on the plate for this type of experiment. If streptomycin or nalidixic acid is used, then remating on the plate does not occur. Therefore, these two antibiotics are highly recommended for use in interrupted mating experiments. Streptomycin is particularly desirable since revertants to Str^r occur at an extremely low frequency. Using nutritional markers to counterselect in this case can lead to difficulties, since the re-formation of mating pairs and subsequent chromosome transfer can still occur after plating.

Method B

An alternative method calls for the dilution of the mating mixture after a brief initial period during which pairs are formed. The dilution prevents the formation of new mating pairs and thus minimizes asynchrony in the initiation of chromosome transfer.

A mating mixture is prepared exactly as in method A, and after 5 minutes carefully diluted 1:500 by adding 0.1 ml to 50 ml of prewarmed broth in a 500-ml flask. Aliquots are withdrawn (usually 0.5 ml or 1 ml are added to 2 ml of saline or buffer) and blended for 10 seconds and then 0.1–0.2 ml are plated in soft agar on selective plates. With this technique, duplicate and even triplicate samples are plated out.

Although we recommend the use of the Low apparatus for interrupted mating experiments, it is not essential for many class exercises. Its most important use is in experiments designed to determine the precise time of entry of markers. Demonstrations of different times of entry of markers can be achieved with the use of a vortex, provided these markers are at least several minutes apart and that method B is used.

Reference

Low, B. and T. H. Wood. 1965. A quick and efficient method for interruption of bacterial conjugation. *Genet. Res. 6:* 300.

Strains

Number	Sex	Genotype	Important Properties
CSH61	HfrC	*trpR thi*	Str^s
CSH62	HfrH	*thi*	Str^s
CSH57	F^-	*ara leu lacY purE gal trp his argG malA strA xyl mtl ilv metA* or *B thi*	$Ilv^- Met^- Ara^- Leu^- Lac^- Pur^- Gal^- Trp^- His^- Arg^- Str^r$

Method

Since the following experiment calls for the taking of points every minute, we recommend that two and preferably three students work together. We also suggest that one group use method A with HfrC, and a second group method A with HfrH; and that two additional groups use method B for these crosses.

Day 1

Prepare mating mixtures as described in the introduction (1 Hfr : 20 F^- cells) and shake gently in a waterbath or in a warm room at 37°. Withdraw 0.05-ml aliquots (for method A) and add directly to 13 × 100 mm test tubes containing 3 ml F-top agar. Do not remove the flask from the waterbath. Cover with parafilm and agitate for 10 seconds. Then plate out on selective plates as described in the unit introduction according to the following schedule, and incubate at 37° for 36–48 hours.

Cross of HfrH × CSH57

Time	
0 min	1 sample for each recombinant type (Leu^+, Lac^+, Trp^+)
5–20 min	1 sample alternating every minute for Leu^+ and Lac^+
21–35 min	1 sample alternating every minute for Lac^+ and Trp^+
40, 45, 50, and 60 min	1 sample for Trp^+

Cross of HfrC × CSH57

Time	
0 min	1 sample for each marker
5–20 min	1 sample alternating every minute for Lac^+ and Leu^+
21–25 min	1 sample every minute for Leu^+
30, 45, and 60 min	1 sample for Leu^+ and Trp^+

Controls

Each parent (donor and recipient) should be plated on each type of selective plate just prior to mating. This is a **control for revertants** in the population to Str^r in the case of the donor, and to Leu^+, Lac^+, and Trp^+ in the case of the recipient. For method A, we want to know the number of F^- revertants in 0.05 ml of the culture at the time of mating. Since this is approximately 10^7 cells, do not be alarmed if several colonies appear. This is merely the background, which is rarely zero in bacterial crosses of this kind.

The second control is the **zero-minute control.** This is done for each marker by mixing the donor and recipient together and immediately agitating and plating on the respective media. Ideally, the number of colonies should not be different from that seen without mixing the donor and recipient. Any increase is due to failure to achieve complete separation of mating pairs.

We recommend preparing a schedule of time points, such as shown on the next page, for the cross of HfrH × CSH57. Here space is left for the actual time that a point is taken. It is not imperative that the 6-minute point, for instance, be taken at exactly 6 minutes, so long as the actual time is recorded. Use the same part of each operation (for instance the point at which agitation is begun) as the point for recording the time.

Day 2, 3 Examine the plates after 30 and 48 hours and record the number of colonies for each marker. Construct curves as in Figures 6A and 6B and determine the time of entry of each marker from the extrapolation of the very early portion of the curve (use the first time point which gave an increase in colonies over the background).

Materials (if Method A is used)

Day 1 **(For cross of HfrH × CSH57)**

fresh overnights of CSH57 and CSH62
2 test tubes with 5 ml broth each
100-, 125-, or 250-ml erlenmeyer flask
shaking waterbath at 37°
44 0.1-ml and 2 pasteur pipettes
44 13 × 100 mm test tubes containing 3 ml F-top agar at 45°
15 type G plates (Table IIC)
18 type D plates
11 type C plates
vibratory interrupted mating device. If method B is used (see Introduction), then it is possible to use a vortex mixer.

(For cross of HfrC × CSH57)

fresh overnights of CSH57 and CSH61
2 test tubes with 5 ml broth each
100-, 125-, or 250-ml erlenmeyer flask
shaking waterbath at 37°
35 0.1-ml pipettes and 2 pasteur pipettes
35 13 × 100 mm test tubes with 3 ml F-top agar, each kept at 45°
11 type D plates
18 type C plates
6 type G plates
vibratory interrupted mating device

Schedule of Points

(HfrH × F^- CSH57)

Scheduled time (min)	Actual time	Selected recombinant	Plate type (Table IIC)	No. of recombinants per ml
0	——— ——— ———	Lac^+ Leu^+ Trp^+	D C G	——— ——— ———
5	———	Leu^+	C	———
6	———	Lac^+	D	———
7	———	Leu^+	C	———
8	———	Lac^+	D	———
9	———	Leu^+	C	———
10	———	Lac^+	D	———
11	———	Leu^+	C	———
12	———	Lac^+	D	———
13	———	Leu^+	C	———
14	———	Lac^+	D	———
15	———	Leu^+	C	———
16	———	Lac^+	D	———
17	———	Leu^+	C	———
18	———	Lac^+	D	———
19	———	Leu^+	C	———
20	———	Lac^+	D	———
21	———	Trp^+	G	———
22	———	Lac^+	D	———
23	———	Trp^+	G	———
24	———	Lac^+	D	———

Schedule of Points

(HfrH × F^- CSH57)

Scheduled time (min)	Actual time	Selected recombinant	Plate type (Table IIC)	No. of recombinants per ml
25	———	Trp^+	G	———
26	———	Lac^+	D	———
27	———	Trp^+	G	———
28	———	Lac^+	D	———
29	———	Trp^+	G	———
30	———	Lac^+	D	———
31	———	Trp^+	G	———
32	———	Lac^+	D	———
33	———	Trp^+	G	———
34	———	Lac^+	D	———
35	———	Trp^+	G	———
40	———	Trp^+	G	———
45	———	Trp^+	G	———
50	———	Trp^+	G	———
60	———	Trp^+	G	———

EXPERIMENT 7

Location of Markers on the *E. coli* Chromosome by the Gradient of Transmission

The introduction to this unit describes the natural gradient of transfer which occurs in Hfr × F^- crosses. In practice, this is often used to determine the approximate location of markers on the chromosome. An example of this is shown in Table 7A which depicts the results of a cross involving an Hfr which donated markers in the order *purC-argA-strA*. Both $PurC^+$ and $ArgA^+$ recombinants were selected and scored for other markers. The analysis of this cross, which was done to

Table 7A. Uninterrupted Mating of Hfr B2 and MK 120*

Selected recombinants	Recombinants per 100 Hfr cells	Number scored	Percentage of recombinants scored as			
			$PurC^+$	$ArgA^+$	Lar^+	Str^s
$PurC^+$	6.8	1,126	100.0	63.1	54.0	36.8
$ArgA^+$	8.7	1,112		100.0	64.8	44.1

* From Kvetkas, Krisch, and Zelle, *J. Bacteriol. 103*, 393, 1970.

roughly map the *lar* locus, indicates that the *lar* locus lies between *argA* and *strA*. Since selection is done for a proximal marker, then the further away the unselected marker (i.e. *lar*) is, the fewer the recombinants that appear (Lar^+).

In the following experiment several Hfr strains are used as donors in 90-minute matings with a multiply marked recipient. Each cross is plated out on selective plates, and partially purified recombinants are then scored for unselected markers.

Strains

Number	Sex	Genotype	Important Properties
CSH57	F^-	*ara leu lacY purE gal trp his argG malA strA xyl mtl ilv metA* or *B thi*	$Ilv^- Met^- Ara^- Leu^- Lac^- Pur^- Gal^- Trp^- His^- Arg^- Str^r$
CSH61	HfrC	*trpR thi*	Str^s
CSH62	HfrH	*thi*	Str^s

Method

Day 1 Use each of the two Hfr strains (HfrC, HfrH) to do a 90-minute mating with CSH57. The mating mixture should be prepared as in Experiment 6. Subculture a fresh overnight of the donor 1:50 in rich (LB) broth and allow to grow at 37° until a density of $2–3 \times 10^8$/ml is reached. Mix 0.5 ml of the Hfr with 4 ml of an exponentially growing (between 2 and 5×10^8 cells/ml) culture of the recipient in a 125-ml erlenmeyer flask in a shaking waterbath at 37°.

Shake gently for 90 minutes, and without interrupting the mating plate approximately 0.1 ml of a 10^{-1}, 10^{-2}, and a 10^{-3} dilution onto the selective plates as indicated below. In this case the use of F-top agar is optional. Because we wish to score several hundred colonies for each cross, it is necessary to have as many colonies as possible on the selective plates.

Plate out a 10^{-1} dilution of the donor and recipient from each cross on each selective plate as a control. Incubate at 37° for 36–48 hours.

Hfr	Selected phenotype	Type of plate
C	Leu^+	C
H	Leu^+	C

Day 3 Pick and grid (50 colonies per plate) the recombinants from each cross. Pick onto the same type of selective plate used for the cross. This procedure, which is not a strict purification, enriches greatly for the selected cells.* Incubate at 37°. Grid at least 350 (7 plates of

* When single colonies are not re-isolated, then some recombinant colonies will be impure for unselected markers, and this will affect the apparent percentage of cells inheriting proximal markers. However, in cases where the mating proceeds for 90 minutes before plating, one finds empirically that only 10% of the colonies are affected in this way. Proximal markers should be inherited between 40–60% in most matings of this type.

50 each). This allows the testing of greater than 50 recombinants per marker for 7 markers.

Day 5 Replicate each of the replica plates onto one of the other types of selective plates (plates A, B, D, E, F, G, H for crosses with HfrC and HfrH). If the replication is done carefully, each replica velvet can be used several times for different selective plates. This would allow 100–150 colonies to be scored for each marker. Incubate at 37°.

Day 6 Score the results. Record the percentage of recombinants which received each unselected marker on a tally sheet such as the one shown below. Can you order all of the markers from the results of each individual cross? Do the combined results of the two crosses enable a better ordering? Does the order match that of the map in Figure IIE? This same experiment can be done using any of the Hfr strains depicted in Figure IIE. Some of these strains can be used by part of the class as unknowns.

Hfr × CSH57—Selection for Leu$^+$ recombinants

Unselected phenotype	Plate type	Genetic marker	% of Leu$^+$ recombinants which inherited marker from donor
Arg$^+$	A	*argG*	_____
Met$^+$	B	*met A* or *B*	_____
Lac$^+$	D	*lac*	_____
Ade$^+$	E	*purE*	_____
Gal$^+$	F	*gal*	_____
Trp$^+$	G	*trp*	_____
His$^+$	H	*his*	_____

Materials (per cross)

Day 1

overnight of Hfr (CSH61 or CSH62) and also of the F$^-$ (CSH57)
2 test tubes with 5 ml LB broth
5 dilution tubes containing 1 part LB broth per 9 parts buffer
125 or 250-ml erlenmeyer flask
shaking waterbath at 37°
5 type C plates (Table IIC)
6 0.1-ml, 5 1-ml, 1 5-ml, and 2 pasteur pipettes
incubator at 37°

Day 3

350 round wooden toothpicks
7 type C plates

Day 5

7 replica velvets, replicating block
3 plates each of the following types: A, B, D, E, F, G, H

EXPERIMENT 8

Determination of the Point of Origin of Hfr Strains

In this experiment four Hfr's are tested for donation of different markers. The origins of two of the Hfr's are known, HfrC and HfrH. The other two are unknowns selected from the strain collection. The results of this test should elucidate both the point of origin of each Hfr and also the direction of transfer. The natural gradient of transmission in Hfr × F^- crosses makes it less likely for a late marker to be inherited by the female than an early marker. If a mating mixture of an Hfr and a multiply marked recipient is plated out on different types of selective plates, this gradient can be observed directly. The earlicr a marker is transferred, the greater is the number of recombinants that will appear.

Matings can also be done directly on the plate by cross-streaking. First the recipient is streaked with a loop or a very small pipette across the surface of the plate. As soon as that dries, each donor to be tested is streaked across this line perpendicularly. The use of 100-μl capillary pipettes to make the cross streaks makes it possible to test 10 or 12 Hfr's per recipient streak. These pipettes also make it easier to apply a constant amount of liquid to each plate. The recipient which we will use in this experiment, CSH57, is Str^r; therefore, we use streptomycin to counterselect. Figures IIC and 34A show the location of the markers in strain CSH57 and also the points of origin of the Hfr strains supplied with this manual.

Reference

Berg, C. M. and R. Curtiss, III. 1967. Transposition derivatives of an Hfr strain of *Escherichia coli* K12. *Genetics 56*: 503.—Describes the cross-streaking technique for matings used in this experiment.

Strains

Number	Sex	Genotype	Important Properties
CSH62	HfrH	*thi*	Str^s
CSH61	HfrC	*trpR thi*	Str^s
CSH57	F^-	*ara leu lacY purE gal trp his argG malA strA xyl mtl ilv metA* or *B thi*	$Ilv^-Met^-Ara^-Leu^-$ $Lac^-Pur^-Gal^-Trp^-$ $His^-Arg^-Str^r$

2 Hfr's with origins known only to the instructor (chosen from those pictured in Figure IIE)

Method

Day 1 Subculture an overnight of each of the four Hfr's to be tested, and also subculture a fresh overnight of the recipient, CSH57. The donors should be diluted 1:50 and the recipients 1:20 into rich (LB) broth. Grow until a density of 2–3 × 10^8 cells/ml is reached.

Each plate to be used should be prewarmed at 37°. Streak the recipient with a 100-μl capillary pipette across each of the nine plates described in the materials section. Use about 100 μl per plate. If capillary pipettes are not available, a standard loop can be used. Pasteur pipettes are also sufficient, although the amount delivered is not standardized. Then streak each donor across the recipient streaks. When all of the streaks dry, incubate at 37° for 36–48 hours.

Day 2, 3 Examine the plates. The amount of recombination can be seen directly. Confluent growth occurs for markers donated early, while only a few colonies are present if the marker is transferred late. From the number of recombinants appearing on each of the selective plates, determine the location of the origin of the Hfr's and also the direction of transfer.

For more precise results, repeat the matings, but make a mating mixture and plate out 0.1 ml of a 10^{-2} dilution after mating for 60 minutes at 37° onto each of the selective plates. Interrupted matings, as demonstrated in Experiment 6, can be used to further verify the results.

Materials

Day 1

overnight of each of the 5 strains to be used
5 test tubes with 5 ml LB each
5 pasteur pipettes
5 100-μl capillary pipettes
1 each of the plates labeled A–I in Table IIC

EXPERIMENT 9

Test for the Sex Factor by Phage Sensitivity

A group of single-stranded RNA phages (R17, MS2, f2, M12, fr, QB) and also a series of single-stranded DNA phages (fd, f1, M13) specifically infect strains of *E. coli* which contain the sex factor (Loeb and Zinder, 1961). These phages, often referred to as male phage, adsorb to F pili, the genetic determinants of which are on the F factor (Crawford and Gesteland, 1964). Therefore, we can distinguish Hfr and F^+ bacteria from F^- bacteria by their response to phages such as R17. Mutants unable to adsorb RNA phage (Cuzin, 1962; Silverman et al., 1967) and which prevent injection of phage RNA (Silverman et al., 1968) have been isolated.

Testing for phage sensitivity can be done by cross-streaking, spot tests (Figure 9A), or looking directly for plaque formation. Clear results are often obtained by the use of sterile filter-paper strips. These are immersed in a phage lysate and gently placed on a top agar overlay containing the cells to be tested. The strip is carefully removed and the plates incubated overnight (Silverman et al., 1968). Cells should be grown at 37° and be in exponential phase for best results, since these conditions favor F pili formation. Normally, these tests will not work at temperatures below 33–34°. We use an Hfr (CSH61) and an F^- (CSH51) in this exercise.

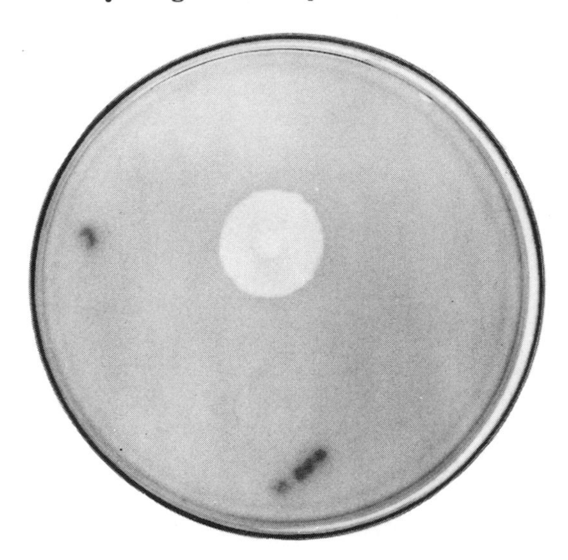

Figure 9A. Spot test with RNA phage. A drop of lysate of R17 was applied to a lawn of an Hfr strain. After overnight incubation at 37° a distinct clearing around the spot can be seen. (Photograph courtesy of J. Kirschbaum and D. Ganem.)

References

CRAWFORD, E. M. and R. F. GESTELAND. 1964. The adsorption of bacteriophage R-17. *Virology 22:* 165.

CUZIN, F. 1962. Mutants défectifs de l'épisome sexuel chez *Escherichia coli* K12. *Compt. Rend. Acad. Sci. 255:* 1149.

HOHN, T. and B. HOHN. 1970. Structure and assembly of simple RNA bacteriophages. *Adv. Virus Res. 16:* 43.

LOEB, T. and N. D. ZINDER. 1961. A bacteriophage containing RNA. *Proc. Nat. Acad. Sci. 47:* 282.

PRATT, D. 1969. Genetics of single-stranded DNA bacteriophages. *Ann. Rev. Genet. 3:* 343.

SILVERMAN, P. M., H. W. MOBACH and R. C. VALENTINE. 1967. Sex hair (F-pili) mutants of *E. coli*. *Biochem. Biophys. Res. Commun. 27:* 412.

SILVERMAN, P. M., S. ROSENTHAL, H. MOBACH and R. C. VALENTINE. 1968. Two new classes of F-pili mutants of *Escherichia coli* resistant to infection by the male specific bacteriophage f_2. *Virology 36:* 142.

STAVIS, R. L. and J. T. AUGUST. 1970. The biochemistry of RNA bacteriophage replication. *Ann. Rev. Biochem. 39:* 527.

VALENTINE, R. C., R. WARD and M. STRAND. 1969. The replication cycle of RNA bacteriophages. *Adv. Virus Res. 15:* 1.

ZINDER, N. D. 1965. RNA phages. *Ann. Rev. Microbiol. 19:* 455.

Strains

Number	Sex	Genotype	Important Properties
CSH61	HfrC	*trpR thi*	male
CSH51	F^-	*ara* Δ(*lacpro*) *strA thi* (φ80d*lac*$^+$)	female

Method

Day 1

Subculture fresh overnights of CSH61 (HfrC) or any male strain, and also CSH51 (F^-) or any female strain. Dilute 1:20 into LB broth and grow at 37° until a density of 2–5 × 10^8 cells/ml is reached.

Cross-streaking: Streak with a loop or with a pasteur or capillary pipette a concentrated lysate of male phage across a rich plate. As soon as this dries streak with a loop each bacterial culture to be tested across the RNA phage streak. Incubate at 37°. After 12–15 hours there should be a clearing on the Hfr cross streak but not on the F^- cross streak. Rich plates (LB) give best results for this test. The advantage of cross-streaking techniques is that it allows the easy screening of a large number of strains. Sometimes, however, it gives ambiguous results, in which case the next two techniques are employed.

Spot tests: A lawn of the culture to be tested is plated in soft agar onto a rich plate. Use H-top agar and H plates, or R-top agar and R plates. Place several drops of the culture to be tested in a small wasserman test tube and add 2.5 ml of top agar. Quickly pour over the surface of a rich plate and allow to dry. Spot a drop of the phage lysate on the top agar layer. As soon as this dries incubate at 37°. Male strains should show a large clearing with faint growth in the center of the spot. Female strains should be unaffected and show confluent growth in the spot. Paper strip tests, as described in the Introduction, give similar results.

Single plaques: Make serial dilutions of the phage lysate. Mix 0.1 ml of each dilution with 3 drops of the culture to be tested. Preadsorb for 10 minutes at 37° in a waterbath, and then plate out in top agar. Use R-top agar and R plates or H-top agar and H plates. By observing the number of single plaques we can determine the relative plating efficiencies of the male phage on different strains.

Materials

Day 1

fresh overnight cultures of CSH51 and CSH61
2 test tubes with 5 ml LB broth
3 pasteur pipettes
lysate of RNA phage stock
1 LB plate, H plate, or R plate

EXPERIMENT 10

Production of F^- Phenocopies

In any population of Hfr or F^+ strains, a small percentage (usually 0.1–1%) of the cells have lost both F pili and also F-determined surface components and temporarily behave as recipients. These phenotypic recipients are referred to as F^- phenocopies. The percentage of phenocopies can be increased by growing the cells under conditions in which the formation of new F pili and cell surface components is inhibited. Continued aeration of saturated cultures at low temperature for 24–48 hours has been shown to increase the percentage of F^- phenocopies, as has starvation in buffered saline in the absence of a required amino acid (Curtiss et al.).

In the following experiment an HfrH is used to donate an early marker, val^r to an HfrP4X strain, CSH70. The "recipient" (CSH70) is grown under different conditions to increase the number of F^- phenocopies in the population. Table 10A tabulates the results of a sample experiment done using these same strains. For a complete discussion of optimal conditions for increasing the recipient ability of Hfr and F^+ strains, see Curtiss et al.

Table 10A

HfrP4X	No. of Val^r recombinants per ml ($\times 10^3$)
Grown exponentially at 37°	8
Aerated for 36 hours at 25°	30
Resuspended in minimal A + glucose for 5 hours	20
Resuspended in minimal A for 5 hours	120
F^- control (CSH57) grown exponentially	1000

0.5 ml of donor (HfrH Val^r Pro^-) was mixed with 0.5 ml of recipient and after 40 minutes, the matings were interrupted and plated on glucose minimal plates with valine, arginine, and methionine.

Reference

CURTISS, R., III, L. G. CARO, D. P. ALLISON and D. R. STALLIONS. 1969. Early stages of conjugation in *Escherichia coli*. *J. Bacteriol. 100:* 1091.

Strains

Number	Sex	Genotype	Important Properties
CSH7	F^-	*lacY strA thi*	Val^s Pro^+
CSH63	HfrH	*val*^{r} Δ(*lacpro*) *thi*	Val^r Pro^-
CSH70	HfrP4X	*argE metB thi*	Val^s Pro^+ Arg^- Met^-

Method

Day 1

Subculture an overnight of CSH63 and when a density of 2×10^8 cells/ml is reached, prepare four identical mating mixtures with:
(A) an exponentially growing culture of CSH7;
(B) an exponentially growing culture of CSH70;
(C) a culture of CSH70 aerated for 36 hours at room temperature, and diluted 1:10 in buffer;
(D) a culture of CSH70 which was first grown up overnight, washed, and resuspended in 1 × A minimal medium with 10^{-3} M $MgSO_4$ and aerated for 5 hours at 37°, and diluted 1:10 in buffer.

Mix 0.5 ml donor (CSH63) with 0.5 ml of each recipient in test tubes and place on a 30 rpm rotor at 37°. Interrupt the matings after 40 minutes and plate 0.1 ml of a 10^{-2} dilution of each mating mixture (and also a 10^{-2} dilution of each parent strain) on glucose minimal plates with valine (50 μg/ml), arginine, and methionine.

In order to titer the viable recipients, plate 0.1 ml of a 10^{-5} dilution of each mixture onto glucose minimal plates with arginine and methionine. (For the mixture containing the CSH70 grown for 36 hours, plate also 0.1 ml of a 10^{-6} dilution.)

Day 2, 3

Observe the plates and count the Val^r recombinants from each cross. Compute the number of recombinants per ml and also per viable recipient (CSH7 or CSH70), and determine which treatment is best for increasing the recipient ability of an Hfr strain. Compared to the F^- strain, how efficient a recipient did the Hfr become? What would you do to increase this proportion?

Materials

Day 1

exponentially growing cultures of CSH7, CSH70, and CSH63
culture of CSH70 aerated for 5 hours in 1 × A medium with 10^{-3} M $MgSO_4$ (no glucose) after overnight growth
culture of CSH70 aerated for 36 hours at room temperature
2 small centrifuge tubes, desk-top centrifuge
10 ml 1 × A minimal medium containing 10^{-3} M $MgSO_4$
24 0.1-ml, 11 1-ml, and 1 pasteur pipettes
5 test tubes, 1 with 5 ml LB broth
18 dilution tubes
7 glucose minimal plates with valine (50 μg/ml L-valine), arginine, and methionine
5 glucose minimal plates with arginine and methionine

EXPERIMENT 11

Curing of Episomes from *E. coli* Strains with Acridine Orange

The replication of episomes can be selectively inhibited by acridine orange (Hirota). Growth of strains carrying F or F′ factors in acridine orange therefore results in loss ("curing") of the sex factor. Concentrations of acridine orange of 50–75 μg/ml are widely used for this purpose. However, different strains are sensitive to different amounts of this compound. For instance, *recA* strains usually do not grow in concentrations much over 30 μg/ml. As a general rule each strain to be cured should be first tested with different concentrations of acridine orange (25, 50, 75, 100, and 125 μg/ml). The highest concentration which still allows growth should be employed.

In the following exercise the strain CSH23, F′ *lacpro*/Δ(*lacpro*), is grown overnight in the presence and absence of acridine orange. Single colonies are then observed on lactose indicator plates and the curing of the episome compared with and without acridine orange treatment. Cured cells are tested for the *pro* marker to verify the loss of the episome.

An effective plate method of curing has been developed recently. The acridine is added directly to the indicator plate and cured colonies are observed without prior growth in acridine orange. The recipe calls for the addition to the medium (for lactose tetrazolium plates in this case) after autoclaving of 2.5 ml of a 10 mg/ml solution of acridine in sterile water, and also 3 ml of 2 N NaOH to 1 liter of final volume. Colonies have a plating efficiency of $\frac{1}{3}$ to $\frac{1}{2}$ the normal efficiency on these plates. Difco antibiotic medium #2 should be used (Fan).

References

FAN, D. P. 1969. Deletions in limited homology recombination in *Escherichia coli*. *Genetics 61:* 351.
HIROTA, Y. 1956. Artificial elimination of the F factor in *Bact. coli* K12. *Nature 178:* 92.
HIROTA, Y. 1960. The effect of acridine dyes on mating type factors in *Escherichia coli*. *Proc. Nat. Acad. Sci. 46:* 57.

Strains

Number	Sex	Genotype	Important Properties
CSH23	F′*lac*$^+$ *proA*$^+$,*B*$^+$	Δ(*lacpro*) *supE spc thi*	Lac$^+$

Method

Day 1

Curing an episome from Lac$^+$ strains: Inoculate 1000–2000 cells (approximately 0.1 ml of a 10^{-5} dilution of a fresh overnight) into 5 ml LB broth, pH 7.6, containing 75 μg/ml acridine orange. The pH of the broth is a critical factor. Grow overnight in the dark to saturation. In a similar manner prepare a control culture without acridine orange.

Day 2

Plate dilutions of the overnight cultures on lactose indicator plates, or streak directly for single colonies. Lactose tetrazolium plates are recommended, but EMB lactose or lactose MacConkey plates can also be used. If the cultures are fully grown, plate 0.1 ml of both a 10^{-4} and a 10^{-5} dilution.

Day 3

Observe the indicator plates. There should be a high degree of curing of the *lac* episome in the culture grown up in the presence of acridine orange. This should be evidenced by a high proportion of Lac$^-$ colonies with respect to the control. Test several Lac$^-$ colonies for Pro$^-$ by streaking on glucose minimal plates using a Lac$^+$ colony from CSH23 as a control.

Materials

Day 1

fresh overnight of CSH23
test tube with 5 ml LB broth containing 75 μg/ml acridine orange; acridine orange is prepared directly before use by dissolving 7.5 mg/ml in sterile water
test tube with 5 ml LB broth
3 dilution tubes
3 0.1-ml pipettes (or 3 pasteur pipettes) and 1 1-ml pipette

Day 2 4 lactose tetrazolium indicator plates
3 dilution tubes
4 0.1-ml and 1 1-ml pipettes

Day 3 1 glucose minimal plate
colonies of CSH23

EXPERIMENT 12

Recombination Deficient Strains: Influence on Episome Transfer and Recombination

Rec$^-$ strains are mutants of *E. coli* with an impaired ability to carry out general recombination. These strains act as recipients for episomes and DNA fragments in conjugational matings, but they produce recombinants at a much reduced frequency. Three *rec* genes (*recA*, *recB*, *recC*) have been identified. All *rec* mutants are deficient in repair functions and are therefore more sensitive to UV than wild type. (It should be noted that *recA* mutants grow more slowly than wild type.) Verification of the Rec$^-$ character of a strain is usually done in one of two ways:

(A) Streak alongside a wild-type control on a rich plate and irradiate with UV light. Incubate at 37°. For *recA* strains use 200 ergs and examine the plates after 10–15 hours. For *recB* and *recC* strains use 500–600 ergs and examine the plates after 6–10 hours. There should be a significant difference in the growth of the two streaks.

(B) Rec$^-$ strains do not form colonies on plates containing methylmethane sulfonate (MMS), while wild-type Rec$^+$ strains do. This is a convenient method, although the plates must be used within 36 hours of being poured.

In this experiment we use CSH51 (Pro^-) and a *recA* derivative as recipients. Crosses with both Hfr strains and F′ donors attempt to demonstrate that the *recA* marker has no effect on the ability of an F^- strain to receive an F′ lac^+pro^+, but seriously lowers the number of Lac^+Pro^+ recombinants in an Hfr cross. Later experiments (28 and 29) demonstrate the effect of the *recA* allele on episome–chromosome recombination.

References

CLARK, A. J. and A. D. MARGULIES. 1965. Isolation and characterization of recombination-deficient mutants of *Escherichia coli* K12. *Proc. Nat. Acad. Sci. 53:* 451.

LOW, B. 1968. Formation of merodiploids in matings with a class of Rec^- recipient strains of *Escherichia coli* K12. *Proc. Nat. Acad. Sci. 60:* 160.

WITKIN, E. 1969. Ultraviolet-induced mutation and DNA repair. *Ann. Rev. Microbiol. 53:* 487.

Strains

Number	Sex	Genotype	Important Properties
CSH23	F′lac^+proA^+,B^+	Δ(*lacpro*) *supE spc thi*	Str^s
CSH51	F^-	*ara* Δ(*lacpro*) *strA thi*	Pro^-Str^r
CSH52	F^-	*ara* Δ(*lacpro*) *strA recA thi*	$Pro^-Str^r Rec^-$
CSH61	HfrC	*trpR thi*	Str^s

Method

In the following exercise CSH61 (HfrC) is used to donate the *lacpro* region to CSH51 (Pro^-Rec^+) and to CSH52 (Pro^-Rec^-). In a second cross, CSH23 is employed as a donor of an F′lac^+pro^+. Both donor strains are Str^s and the recipients Str^r. The selection is for Pro^+Str^r. Mating mixtures are interrupted after 45 minutes and then plated on glucose minimal streptomycin plates. Colonies are counted after 36 hours.

Day 1

Subculture overnights of each of the four strains by diluting 1:40 and allow to grow for approximately 2 hours. When the donors are 2–4 × 10^8 cells/ml, prepare four mating mixtures (CSH23 × CSH51, CSH23 × CSH52, CSH61 × CSH51, and CSH61 × CSH52) by mixing 2 ml of each donor with 1 ml of each recipient in a 125-ml flask, and shake gently at 37° for 45 minutes. (See Unit II Introduction for alternative methods of preparing mating mixtures.)

At the end of this time withdraw 0.1 ml from each of the mating mixtures and pipette into 10 ml dilution buffer. This is a 10^{-2} dilution of the mating mixture. After vortexing vigorously for 30 seconds, dilute this further to make a 10^{-3}, 10^{-4}, and 10^{-5} dilution.

Plate 0.1 ml of a 10^{-2}, 10^{-3}, and 10^{-4} dilution onto the selective plates (glucose minimal with streptomycin), and 0.1 ml of a 10^{-5}

dilution onto a glucose minimal plate with streptomycin and proline (to titer the F^- cells). Incubate all plates at 37° for 36–48 hours.

Day 3 Count the colonies on each plate and compute the number of Pro^+Str^r colonies present per ml, and the percentage of recipients which became Pro^+ in each cross. Determine the effect of the *recA* marker on each type of cross.

Materials

Day 1

overnights of strains CSH23, CSH61, CSH51, CSH52
4 test tubes with 4 ml LB broth each
4 125-ml flasks
shaking waterbath at 37°
16 dilution tubes
24 0.1-ml, 12 1-ml, 2 5-ml, and 4 pasteur pipettes
12 glucose minimal plates with streptomycin
4 glucose minimal plates with proline and streptomycin
vortex

UNIT III

MUTAGENESIS AND THE ISOLATION OF MUTANTS

INTRODUCTION TO UNIT III

Mutant analysis facilitates our study of cellular processes on the molecular level in many important ways.

1. Genetic markers enable us to map the structural gene for a protein under study. This knowledge makes possible the isolation of F′ factors (which are useful for complementation studies) and specialized transducing phage carrying this gene (Unit V). Knowing the location of a marker also simplifies the task of strain construction, the preparation of strains containing different combinations of mutations.

2. With the proper set of mutations, we can help define the enzymatic steps in a biochemical pathway, as was done for the nine steps in histidine biosynthesis in *Salmonella* (Hartman et al., 1960; Loper et al., 1964).

3. It makes possible the characterization of new biosynthetic and regulatory systems, such as the indispensible CAP-cyclic AMP regulatory system for catabolite operons (Unit IX). Another example is the isolation of mutants defective in the Kornberg polymerase (De Lucia and Cairns, 1969), which allows the identification of the true components of the DNA replication system in *E. coli*.

4. On a finer scale, the determination of the detailed structure of operons is in part dependent on the isolation and mapping of the proper mutations. This was the case for the *lac* operon, in which the promoter, operator, and structural genes were clearly delineated (Miller et al., 1968).

5. Protein structure work is aided by studying mutationally altered proteins. Allosteric proteins, altered in one but not both effector sites are an illustration, for example the i^s repressor (Willson et al., 1964).

6. We can select or screen for special classes of mutations. Nonsense mutations

allow us to use different suppressors to insert specific amino acids at the mutated site in these altered proteins. These are also important for polarity studies, which help to determine gene order in an operon (Experiments 22–24). Deletion mutants are useful for fine structure mapping and operon fusion studies (Experiments 21 and 42). Frameshift mutations, in addition to causing polarity, are useful for sequencing altered proteins to deduce mRNA sequences (Streisinger et al., 1966). Temperature-sensitive mutations enable us to isolate strains which lack essential functions at high temperature (Experiment 34).

7. Technical problems are often alleviated by the use of the proper mutant. To facilitate protein purification, mutant strains which overproduce certain proteins can be isolated (Müller–Hill et al., 1968). Strains synthesizing a specific protein with a higher substrate affinity have also been isolated (Gilbert and Müller–Hill, 1966). To reduce RNase degradation of RNA in extracts, RNase$^-$ mutants have been found (Gesteland, 1966), and to eliminate recombination mediated by *E. coli* enzymes, Rec$^-$ strains have been isolated (Experiment 12; Clark and Margulies, 1965).

8. Mutants resistant to certain phage or sensitive to particular drugs must frequently be isolated before certain specific experiments are performed. For instance, *E. coli* is not normally permeable to the RNA polymerase inhibitor streptolydigin. Therefore, streptolydigin-sensitive mutants had to be isolated, before studies of the *in vivo* action of this drug were possible (Schleif, 1969).

Spontaneous Mutations

Mutants occur spontaneously in a bacterial population. However, the observed frequency of a particular mutant in *E. coli* is usually lower than 10^{-5}. Therefore, if we examine colonies from a Lac$^+$ culture of *E. coli*, we would have to examine as many as 100,000 to see a Lac$^-$ colony. In cases where the new phenotype can be selected for, we can detect even very infrequent spontaneous mutants. Often, however, it is not possible to select directly for a particular phenotype, and cells must be screened, either by examining colonies on indicator plates, replica-plating colonies onto different types of media, or even growing up colonies and assaying for a particular activity.

Mutagenic Treatment

Fortunately, we can increase the proportion of mutants in a bacterial population by using "mutagens," chemical or biological agents which induce genetic changes. Some of these (hydroxylamine, nitrosoguanidine, bromouracil, EMS, and 2-aminopurine) induce principally base substitutions, while others (ultraviolet irradiation and nitrous acid) are also efficient deletion mutagens. Frameshifts (both additions and deletions) are induced by acridine compounds, such as ICR 191. Phage Mu-1, on the other hand, causes mutations by integrating into different genes. The mutagenic spectrum of these mutagens has been well characterized, although mostly using *in vitro* phage mutagenesis (Stanier et al., 1970; Drake, 1970).

Mutagenic Compounds

As a result of the work of Freese and collaborators, it is clear that the **base analogs** 5-bromouracil and 2-aminopurine induce **transitions** (the substitution of a purine for a purine or a pyrimidine for a pyrimidine) rather than **transversions** (the substitution of a pyrimidine for a purine or vice-versa) (Freese, 1959a,b,c). This specificity has been verified by direct amino acid analysis of revertants induced by 2-aminopurine in tryptophan synthetase A (Yanofsky et al., 1966), and also in the head protein of phage T4 (Stretton et al., 1966). It is not clear whether these base analogs are misincorporated into the DNA or whether they are first incorporated and then mispair during subsequent replication. It is also not certain whether the rare tautomeric or ionic forms of these molecules play a role in the error process.

Hydroxylamine is specific for the G:C → A:T transition when used to treat phage directly *in vitro*, as a result of its reaction with cytosine (Freese, Bautz–Freese, and Bautz, 1961; Freese, Bautz, and Freese, 1961). However, this compound induces both transitions, G:C → A:T and A:T → G:C, *in vivo*, probably stemming from side reactions with different molecules in the cell (Tessman et al., 1967). Sodium bisulfite shows the same specificity as NH_2OH, but retains this specificity *in vivo*, inducing solely the G:C → A:T transition (Mukai et al., 1970).

Nitrous acid deaminates nucleic acids and results in transitions when used to treat phage *in vitro* (Freese, 1959b; Bautz–Freese and Freese, 1961). However, deletions and possibly transversions are also induced upon *in vivo* use (Schwartz and Beckwith, 1969; Yanofsky et al., 1966).

Acridines such as proflavin and the ICR series of compounds induce frameshifts (Crick et al., 1961; Streisinger et al., 1966; Ames and Whitfield, 1966), probably by intercalating between adjacent base pairs in the DNA double helix and enhancing errors in DNA repair or in recombination (due to unequal crossing over) (Lerman, 1961).

Alkylating agents modify the bases of the DNA itself by adding alkyl groups to the ring nitrogens (Lawly, 1966). Examples of these are EMS (ethylmethane sulfonate), MMS (methylmethane sulfonate) and DES (diethyl sulfonate). After treatment of phage *in vitro*, compounds such as EMS induce primarily transitions (Bautz and Freese, 1960; Freese, 1961), but *in vivo* treatment results in the induction of both transitions and transversions (Yanofsky et al., 1966). Although nitrosoguanidine can effect alkylations, evidence indicates that a decomposition product, diazomethane, is the primary *in vivo* mutagen (Cerdá–Olmedo and Hanawalt, 1968). Nitrosoguanidine induces transitions, transversions, and to a lesser extent small deletions *in vivo* (Whitfield et al., 1966).

Radiation

Ultraviolet radiation induces both base substitutions and deletions in bacteria (Yanofsky et al., 1966; Schwartz and Beckwith, 1969). The weak induction of frameshifts has also been reported (Berger et al., 1966). Several enzyme systems exist which repair photochemical products, such as pyrimidine dimers and hydrates (see reviews by Setlow, 1967, and by Witkin, 1969). Errors in these repair processes may generate many of the observed mutations. Despite the large amount of information concerning the photochemistry of nucleic acids, relatively little is known about the actual mechanism of UV-induced mutagenesis. X-rays have also been used to produce mutations in viruses and bacteria (Brown, 1966; Schwartz and Beckwith, 1969).

Mutator Genes

Mutants have been isolated which exhibit an increase in the spontaneous mutation rate for the entire genome. In some cases the mutations responsible for the increased genetic instability have been mapped and assigned to a "mutator" locus. One such mutation in T4 was shown to be in the structural gene for the T4 DNA polymerase (Speyer, 1965). The best characterized mutator gene in *E. coli*, *mutT*, was first described by Treffers and coworkers and induces the specific transversion A:T → C:G (Treffers et al., 1954; Yanofsky et al., 1966; Cox and Yanofsky, 1967). The mechanism of action of *mutT* is not clear, but it may involve the DNA replication or repair process. The *mutT* locus maps at 1 min.

Mu-1 Phage

Treatment of *E. coli* with lysates of Mu-1 phage results in the induction of mutations at many different locations on the genetic map (Taylor, 1963). Most of the mutations are the result of insertions of the Mu-1 DNA into the host DNA, although deletions are also induced (Bukhari and Zipser, 1972; Boram and Abelson, 1972; Daniell et al., 1972). When Mu-1 DNA is inserted in the middle of a gene, the resulting mutations are non-leaky, polar, and non-reverting (Toussaint, 1969; Bukhari and Zipser, 1972; Daniell et al., 1972). Analysis of Mu-1-induced mutations in the *z* gene of the *lac* operon indicates that Mu-1 integration occurs at many non-specific sites (Bukhari and Zipser, 1972; Daniell et al., 1972). Apparently Mu-1 can integrate into the *E. coli* chromosome at random, although the existence of some favored sites has not been completely ruled out.

Hot Spots

The spectrum of mutations within a gene can be characterized by fine-structure mapping. Rather than mapping at random sites, point mutations tend to fall into a pattern which differs for each mutagen. This was first demonstrated by Benzer (1961) who carried out a classic study of the distribution of spontaneous point mutations in the rII cistrons of phage T4.

Benzer mapped 1609 mutants and found 250 different locations or sites. However, half of the mutations mapped at one of two sites. Sites with a higher than normal distribution of mutations are termed "hot spots." Studies such as this lead to the conclusion that the base sequence surrounding a given codon affects the mutation rate. In accordance with this are the findings of a 3-fold variation of the spontaneous UAG → UAC or UAU reversion and a 9-fold variation in the UGA → UGG reversion induced by 2-aminopurine in the rII system (Stretton et al., 1966). The hydroxylamine-induced transitions from CAG and UGG to UAG and from CAA to UAA also vary over a wide range from specific sites within the rII region (Brenner et al., 1965).

Experimental Design

The purpose of the following exercises is to demonstrate many of these methods of mutagenesis by presenting a series of related experiments involving the isolation and characterization of *E. coli* mutants. Although we use the *lac* operon here, all the procedures are general for *E. coli* mutagenesis.

Two different *E. coli* strains are treated with five different mutagens (ultraviolet light, nitrosoguanidine, ICR 191, 2-aminopurine, and nitrous acid). Cultures are then plated on indicator plates to detect Lac^- mutants and also i^- and o^c mutants. We prefer using lactose tetrazolium plates for the detection of Lac^- colonies, but it is sufficient to use any of the indicator plates described in Experiment 3. In

Table IIIA. Mutagens Used for Bacteria

Mutagen	Apparent *in vivo* specificity	Additional advantages	Disadvantages
2-AP (2-aminopurine)	transitions A:T ⇆ G:C		in some cases a weak mutagen
5-BU (5-bromouracil)	transitions A:T ⇆ G:C		must depress normal thymine incorporation; weak mutagen
NH_2OH (hydroxylamine)	transitions A:T ⇆ G:C		in some cases a weak mutagen
Sodium bisulfite	specific transition G:C → A:T		weak mutagen
Mutator gene (*mutT*)	specific transversion A:T → C:G	no treatment required	genetic construction required
EMS (ethylmethane sulfonate)	mostly G:C → A:T transitions; other substitutions at low rates	powerful mutagen	dangerous to handle
NG (nitrosoguanidine)	mostly G:C → A:T transitions; other substitutions at low rates	very powerful mutagen	dangerous to handle; frequent secondary mutations
ICR 191	frameshifts; small insertions and deletions	powerful mutagen	compound difficult to obtain
Nitrous acid	transitions and probably transversions; deletions		high amount of killing required for good mutagenesis
UV (ultraviolet radiation)	transitions and transversions; deletions; possibly stimulates insertions and chromosomal rearrangements		high amount of killing required for good mutagenesis; certain strains too sensitive
Mu-1 phage	insertions; some deletions	random induction of non-leaky, polar, non-reverting mutations	
Spontaneous (no mutagen)	transitions; transversions; insertions; deletions (frameshifts)	no complications due to secondary mutations	low level of mutants; many siblings in each culture

addition, special strains are provided which enable the selection of spontaneous i^-, o^c, and z^- mutants.

The mutant strains will then be characterized by complementation tests, polarity determinations, reversion patterns, and mapping, and also tested for temperature sensitivity and response to nonsense suppressors.

The strains used throughout this unit are:

CSH51	F^-	*ara* Δ(*lacpro*) *strA thi* (φ80d*lac*$^+$)	
CSH61	HfrC	*trpR thi*	
CSH38	F′*lac*$^+$*pro*$^+$	Δ(*lacpro*) *thi*	i^+p^-
CSH41	F′*lac*$^+$*pro*$^+$	Δ(*lacpro*) *galE thi*	i^-p^-

Strain CSH51 is $F^-Pro^-Str^r$, which allows the easy construction of diploids for complementation tests and mapping with different F′*lacpro* donor strains. CSH38 carries an F′*lacpro* episome, as does CSH41. These strains are used to isolate spontaneous i^- and *lac*$^-$ mutations, respectively. The *lac* regions from the Hfr, CSH61, and from CSH38 and CSH41 can be transferred easily to F^- strains. The mutations from these donors will be tested for response to a nonsense suppressor by crossing the *lac* region into an Su^+ background.

Collecting Mutants

Most of the following procedures call for growing up cells in rich broth after mutagenesis before plating on indicator plates. When this is followed, not all of the mutants found in one culture will be independent. To ensure the isolation of a large number of independent mutants we should (a) prepare a series of cultures using a different single colony to inoculate each one; (b) save only a single mutant from each culture. If we plate directly after mutagenesis, not allowing for cell division, then it is safe to pick many colonies from each original culture. When there is a large induction of mutants over the spontaneous background, as is the case with NG, EMS, and sometimes UV mutagenesis, then there is also little danger of picking "siblings" or mutants from the same original clone, even after overnight growth.

However, it is always necessary to consider each desired mutant phenotype carefully before deciding how soon after mutagenesis to apply the particular selection. This is because several phenomena exist which cause a lag in the expression of some mutations. **Segregation lag** occurs because exponentially growing haploid bacteria often have several identical copies of their nuclei. Segregation patterns are such that if a mutation occurs in one of four nuclei, it would take two generations for a cell with this lesion in all four nuclei to arise. For a recessive marker that means that at least two doublings must occur before the mutant phenotype could be expressed. Many recessive mutations display an additional **phenotypic lag**. For instance phage resistance mutants will not be phenotypically resistant until all of the cell wall receptors are replaced by the synthesis of a new cell wall. This often requires several extra generations, in this case as many as twelve, to allow maximum recovery of all mutants.

Outline of Steps Involved in the Mutagenesis and Characterization of *lac* Mutants

1 Mutagenize both CSH51 and CSH61 (Experiments 13–16)
   A. 2-Aminopurine
   B. Nitrous acid
   C. ICR 191
   D. UV
   E. Nitrosoguanidine

   Plate out cultures on indicator plates at both 37° and 42° (Experiment 18).

2 Select for Lac^-, i^-, and o^c mutants which occur spontaneously, in the special strains described in Experiment 19.

3 Test all Lac^- mutants for growth on lactose, glucose, and melibose (42°). Also test the Lac^- mutants isolated at 42° for growth on lactose at both 42° and 30°.

4 Complementation: Test all mutants isolated in CSH51 for complementation with a series of F′*lacpro* episomes containing i^-, o^c, and y^- mutations (Experiment 20).

5 Fine structure mapping: Map all z^- mutations isolated in CSH51 with a series of episomes which contain deletions ending at various points in the *z* gene (Experiment 21).

6 Reversion: Test all mutants for reversion with NG and ICR 191 (Experiment 17).

7 Test for amber mutants: Test the Lac^- mutants obtained from CSH61 by treatment with NG and 2AP for suppression in the presence of amber suppressors. Also test both the i^-'s and lac^-'s derived on the F′*lacpro* for suppression (Experiment 22).

8 Polarity: Test all z^- mutants for polar effects on *y* (Experiment 24).

References

Ames, B. N. and H. J. Whitfield. 1966. Frameshift mutagenesis in Salmonella. *Cold Spring Harbor Symp. Quant. Biol. 31:* 221.

Bautz, E. and E. Freese. 1960. On the mutagenic effect of alkylating agents. *Proc. Nat. Acad. Sci. 46:* 1585.

Bautz-Freese, E. and E. Freese. 1961. Induction of reverse mutations and cross reactivation of nitrous acid-treated phage T4. *Virology 13:* 19.

Benzer, S. 1961. On the topography of the genetic fine structure. *Proc. Nat. Acad. Sci. 47:* 403.

Benzer, S. and E. Freese. 1958. Induction of specific mutations with 5-bromouracil. *Proc. Nat. Acad. Sci. 44:* 112.

Berger, H., W. J. Brammar and C. Yanofsky. 1966. Spontaneous and ICR 191-A-induced frameshifts in the A gene of *E. coli* tryptophan synthetase. *J. Bacteriol. 96:* 1672.

Boram, W. and J. Abelson. 1972. *J. Mol. Biol.* In press.

Brenner, S., A. O. W. Stretton and S. Kaplan. 1965. Genetic code: the "nonsense" triplets for chain termination and their suppression. *Nature 206:* 994.

Brown, D. F. 1966. X-ray-induced mutations in extracellular phages. *Nature 212:* 1595.

Bukhari, A. and D. Zipser. 1972. Random insertion of Mu-1 DNA within a single gene. *Nature:* In press.

Cerdá-Olmedo, E. and P. C. Hanawalt. 1968. Diazomethane as the active agent in nitrosoguanidine mutagenesis and lethality. *Mol. Gen. Genet. 101:* 191.

Clark, A. J. and A. D. Margulies. 1965. Isolation and characterization of recombination-deficient mutants of *Escherichia coli* K12. *Proc. Nat. Acad. Sci. 53:* 451.

Cox, E. C. and C. Yanofsky. 1967. Altered base ratios in the DNA of an *E. coli* mutator strain. *Proc. Nat. Acad. Sci. 58:* 1895.

Crick, F. H. C., L. Barnett, S. Brenner and R. J. Watts-Tobin. 1961. General nature of the genetic code for proteins. *Nature 192:* 1227.

DANIELL, E., R. ROBERTS and J. ABELSON. 1972. Mutations in the lactose operon caused by bacteriophage Mu. *J. Mol. Biol.* In press.

DE LUCIA, P. and J. CAIRNS. 1969. Isolation of an *E. coli* strain with a mutation affecting DNA polymerase. *Nature 224:* 1164.

DRAKE, J. 1970. *The Molecular Basis of Mutation.* Holden-Day, San Francisco.

FREESE, E. 1959a. The difference between spontaneous and base-analogue induced mutation of phage T4. *Proc. Nat. Acad. Sci. 45:* 622.

FREESE, E. 1959b. On the molecular explanation of spontaneous and induced mutations. *Brookhaven Symp. Biol. 12:* 63.

FREESE, E. 1959c. The specific mutagenic effect of base analogues on phage T4. *J. Mol. Biol. 1:* 87.

FREESE, E. B. 1961. Transitions and transversions induced by depurinating agents. *Proc. Nat. Acad. Sci. 47:* 540.

FREESE, E., E. BAUTZ and E. B. FREESE. 1961. The chemical and mutagenic specificity of hydroxylamine. *Proc. Nat. Acad. Sci. 47:* 845.

FREESE, E., E. BAUTZ-FREESE and E. BAUTZ. 1961. Hydroxylamine as a mutagenic and inactivity agent. *J. Mol. Biol. 3:* 133.

GESTELAND, R. 1966. Isolation and characterization of ribonuclease I mutants of *Escherichia coli. J. Mol. Biol. 16:* 67.

GILBERT, W. and B. MÜLLER-HILL. 1966. Isolation of the *lac* repressor. *Proc. Nat. Acad. Sci. 56:* 1891.

HARTMAN, P. E., Z. HARTMAN and D. SERMAN. 1960. Complementation mapping by abortive transduction of histidine-requiring Salmonella mutants. *J. Gen. Microbiol. 22:* 354.

LAWLY, P. D. 1966. Effects of some chemical mutagens and carcinogens on nucleic acids. *Prog. Nucleic Acid Res. Mol. Biol. 5:* 89

LERMAN, L. S. 1961. Structural considerations in the interaction of DNA and acridines. *J. Mol. Biol. 3:* 18.

LOPER, J. C., M. GRABNAR, R. C. STAHL, Z. HARTMAN and P. E. HARTMAN. 1964. Genes and proteins involved in histidine biosynthesis in Salmonella. *Brookhaven Symp. Biol. 17:* 15.

MILLER, J. H., K. IPPEN, J. G. SCAIFE and J. R. BECKWITH. 1968. The promoter-operator region of the *lac* operon of *Escherichia coli. J. Mol. Biol. 38:* 413.

MUKAI, F., I. HAWRYLUK and R. SHAPIRO. 1970. The mutagenic specificity of sodium bisulfite. *Biochem. Biophys. Res. Commun. 39:* 983.

MÜLLER-HILL, B., L. CRAPO and W. GILBERT. 1968. Mutants that make more *lac* repressor. *Proc. Nat. Acad. Sci. 59:* 1259.

SCHLEIF, R. 1969. Isolation and characterization of a streptolydigin resistant RNA polymerase. *Nature 223:* 1068.

SCHWARTZ, D. O. and J. R. BECKWITH. 1969. Mutagens which cause deletions in *Escherichia coli. Genetics 61:* 371.

SETLOW, J. K. 1967. The effects of ultraviolet radiation and photoreactivation. *Comprehensive Biochemistry* (M. Florkin and E. H. Stotz, ed.) Vol. 27, p. 157. Elsevier, New York.

SPEYER, J. F. 1965. Mutagenic DNA polymerase. *Biochem. Biophys. Res. Commun. 21:* 6.

STANIER, R. Y., M. DOUDOROFF and E. A. ADELBERG. 1970. *General Microbiology*, 3rd Ed., p. 418. Macmillan and Co.

STREISINGER, G., Y. OKADA, J. EMRICH, J. NEWTON, A. TSUGITA, E. TERZAGHI and M. INOUYE. 1966. Frameshift mutations and the genetic code. *Cold Spring Harbor Symp. Quant. Biol. 31:* 77.

STRETTON, A. O. W., S. KAPLAN and S. BRENNER. 1966. Nonsense codons. *Cold Spring Harbor Symp. Quant. Biol. 31:* 173.

TAYLOR, A. L. 1963. Bacteriophage-induced mutation in *E. coli. Proc. Nat. Acad. Sci. 50:* 1043.

TESSMAN, I., H. ISHIWA and S. KUMAR. 1967. Mutagenic effects of hydroxylamine *in vivo. Science 148:* 504.

TOUSSAINT, A. 1969. Insertion of phage Mu-1 within prophage λ: A new approach for studying the control of the late functions in bacteriophage λ. *Mol. Gen. Genet. 106:* 89.

TREFFERS, H. P., V. SPINELLI and N. O. BELSER. 1954. A factor or mutator gene influencing mutation rates in *E. coli. Proc. Nat. Acad. Sci. 40:* 1064.

WHITFIELD, H. J., R. G. MARTIN and B. N. AMES. 1966. Classification of aminotransferase mutants in the histidine operon. *J. Mol. Biol. 21:* 335.

WILLSON, C., D. PERRIN, M. COHN, F. JACOB and J. MONOD. 1964. Non-inducible mutants of the regulator gene in the "lactose" system of *Escherichia coli. J. Mol. Biol. 8:* 582.

WITKIN, E. M. 1969. Ultraviolet light-induced mutation and DNA repair. *Ann. Rev. Genet. 3:* 525.

YANOFSKY, C., J. ITO and V. HORN. 1966. Amino acid replacements and the genetic code. *Cold Spring Harbor Symp. Quant. Biol. 31:* 151.

EXPERIMENT 13

Ultraviolet Light Mutagenesis

Ultraviolet light (UV) induces a wide spectrum of mutations in *E. coli*. In addition to causing base substitutions, it is an effective deletion mutagen. As a source of UV light a low pressure germicidal lamp is used. Every UV lamp, set at a constant distance from the bacteria (usually about 24 inches), should be calibrated. Therefore, a killing curve is required before the optimal conditions for mutagenesis can be determined. Often, a survival of 0.1–1 % is used for mutagenesis (99–99.9 % killing). Since efficient UV repair mechanisms, some of which are stimulated by visible light, exist in *E. coli*, it is preferable to carry out the treatment in the absence of direct illumination. Also, the cells should be grown in the dark or in red light after exposure to UV light.

Survival Curves

Figure 13A shows a survival curve, done with two different *E. coli* strains, one of which (B/r) is a UV-resistant mutant. Curve (A) for each strain was done by irradiating the bacteria directly on solid medium. Curve (B) was done using bacteria suspended in liquid. This clearly demonstrates how both the conditions of the experiment and strain differences can influence the survival curves. UV irradiation

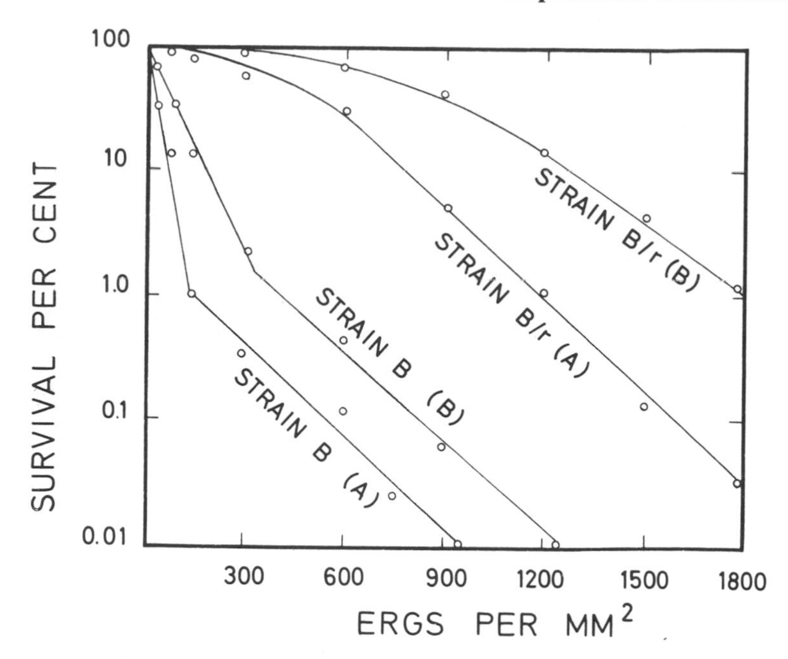

Figure 13A. Survival curve with UV light. (From Witken, 1947).

of complex media produces toxic compounds. Therefore we use 0.1 M $MgSO_4$ to resuspend cells. It is important to keep all conditions, such as density of cells, growth prior to treatment, etc., as standard as possible when doing this type of experiment. Cell densities greater than several times 10^8 bacterial/ml can result in shielding of cells from the irradiation and affect the shape of these curves.

In the following experiment a survival curve is done after UV irradiation. This should be carried out both in the light and in the dark to determine the effect of photoactivated repair mechanisms. The cultures treated with UV in the dark are then grown up overnight and screened for Lac^- and i^- mutants. In order to recover lac mutants, the strain CSH61 (D7011) should be used. CSH51 (X7700) contains a UV-inducible prophage (ϕ80d*lac*) and will thus show a much reduced survival after irradiation. For comparison, one group should do the survival curve with CSH51. It will be seen that different strains show different responses to UV light, and that survival curves are necessary before doing experiments with new strains.

As an additional exercise we recommend using a *recA*$^-$ strain in this experiment to demonstrate the sensitivity of these strains to irradiation. We supply several of these strains in the kit that accompanies this manual (CSH52 and CSH58).

References

DRAKE, J. W. 1970. *The Molecular Basis of Mutation*, p. 161–176. Holden-Day.

HOWARD-FLANDERS, P. 1968. DNA repair. *Ann. Rev. Biochem. 37:* 175.

JAGGER, J. 1961. A small and inexpensive ultraviolet dose-rate meter useful in biological experiments. *Radiation Res. 14:* 394.

MEYNELL, G. G. and E. MEYNELL. 1965. *Theory and Practice in Experimental Bacteriology*, p. 235–237. Cambridge University Press.

RUPERT, C. S. and W. HARM. 1966. *Adv. Rad. Biol. 2:* 1.

SCHWARTZ, D. O. and J. R. BECKWITH. 1969. Mutagens which cause deletions in *Escherichia coli. Genetics 61:* 371.

SETLOW, J. K. and R. B. SETLOW. 1963. Nature of the photoreactive ultraviolet lesion in deoxyribonucleic acid. *Nature 197:* 560.
WITKIN, E. M. 1947. Genetics of resistance to radiation in *Escherichia coli. Genetics 32:* 221.
WITKIN, E. M. 1969. Ultraviolet-induced mutation and DNA repair. *Ann. Rev. Genet. 3:* 525.

Strains

Number	Sex	Genotype	Important Properties
CSH61	HfrC	*trpR thi*	Lac^+ Str^s

Method

Day 1

Subculture an overnight culture of CSH61 into four test tubes containing 5 ml of LB broth. After these have grown to a density of 2–3 × 10^8 cells/ml, spin down and resuspend in 10 ml of 0.1 M $MgSO_4$. Pool all 4 fractions and place in ice for about 5 minutes to prevent further growth. At this point the culture should be titered, by plating a 10^{-5} and a 10^{-6} dilution (0.1 ml) onto LB plates. Now, treat 5-ml aliquots of this with UV for successively longer times. **Be sure the UV lamp has been properly warmed up for at least 30 minutes prior to use.** Carry out all of these operations in the absence of direct illumination! With the lamp set at a distance of 24–30 inches from the bacteria, treat aliquots for 15, 30, 60, 80, 100, 120, and 150 seconds. Exposure to UV light is done in an open glass petri dish. Stir continuously by tilting the plate from side to side. If glass petri dishes are not available, certain plastic petri dishes can be used, although it is harder to spread the bacteria over the surface of the plate. Each irradiated suspension should be titered immediately (0.1 ml of a 10^{-2}, 10^{-3}, 10^{-4}, and 10^{-5} dilution) and then centrifuged and resuspended in 10 ml of LB and grown overnight in foil-covered tubes, along with the untreated control.

Day 2

The overnights from each of the cultures should be plated on lactose tetrazolium plates, along with a control, to look for Lac^- mutants and also on Xgal glucose plates to look for i^- mutants. If the culture is saturated, use a 10^{-4}, 10^{-5}, and a 10^{-6} dilution for the lactose tetrazolium plates and a 10^{-3}, 10^{-4}, and 10^{-5} dilution for the Xgal plates (0.1 ml of each dilution).

Examine the titer plates from the killing curve experiment and plot the survival against time of exposure to UV. A semi-log plot should give a straight line (plotting the log of the surviving fraction against the time of exposure).

Day 3

Examine the indicator plates and compare the Lac^- colonies from each dose. Save one i^- and one Lac^- colony from each culture.

Materials

Day 1 (per killing curve)

overnight of CSH61
4 test tubes with 5 ml LB each
4 small centrifuge tubes (to spin down 5 ml volumes)
ice bucket, 125-ml flask
40 ml 0.1 M $MgSO_4$
UV lamp, stopwatch
7 glass petri dishes
32 dilution tubes
47 0.1-ml, 16 1-ml, 7 5-ml, and 1 10-ml pipettes
30 LB plates

(for mutant screening)

8 test tubes with 10 ml LB broth each, covered with aluminum foil
8 small centrifuge tubes

Day 2 (total requirements for recovering i^- and Lac^- mutants from each dose of one killing curve)

24 lactose tetrazolium plates
24 Xgal glucose plates
56 0.1-ml pipettes (or 56 pasteur pipettes)
16 1-ml pipettes
40 dilution tubes

EXPERIMENT 14

Nitrosoguanidine Mutagenesis

N-methyl-N′-nitro-N-Nitrosoguanidine (NG) is a potent mutagen (and carcinogen) now widely used by bacterial geneticists because it induces a high frequency of mutations at doses which result in little killing. Thus, when Lac^+ cells are treated under conditions in which there is 50% survival, as many as 0.3% of the survivors are Lac^-. In other words, almost one out of three hundred survivors has a mutation somewhere in the *lac* operon. This mutagen induces primarily base substitutions, although small deletions have also been reported. The exact mechanism of its action is not clear, but its mutagenic power stems in part from the generation of diazomethane (Cerdá-Olmedo and Hanawalt). NG induces mutations primarily at the replication point (Cerdá-Olmedo, Hanawalt, and Guerola; Botstein and Jones), which leads to clustering of induced mutations. In fact, recent evidence suggests that the replication region (approximately 2 min on the *E. coli* map) is greater than 200 times more susceptible to mutation by NG than the remainder of the chromosome. Therefore, under the conditions widely used for mutagenesis with NG, it is estimated that as many as 50 additional base changes occur within a map distance of 2 min of an NG-induced mutation (Guerola et al.). The high frequency of double mutations is one disadvantage of NG. However, it is possible to monitor the level of mutagenesis for each experiment.

Procedure for Use

Nitrosoguanidine is dangerous to handle. Solutions should be handled with protective gloves, and never mouth pipetted. Stock solutions are prepared directly before use by dissolving nitrosoguanidine (NG) in the appropriate buffer to give a solution of 1 mg/ml. The buffer used here is a 0.1 M citrate buffer at pH 5.5. Cells are grown up in exponential phase and washed twice in 5 ml of citrate buffer. The NG is added to a final concentration of 50 μg/ml to cells resuspended in 5 ml citrate buffer. This is incubated at 37° in a water bath for 30 minutes. We use 50 μg/ml and 30 minutes here, but different strains are subject to the toxic effects of NG to varying degrees. Therefore, it is advisable to obtain a killing curve on any new strain (loss of viability with increasing dose). One should aim for 50% killing.

After 30 minutes the cells are spun down and washed once in 5 ml phosphate buffer, pH 7.0. This eliminates the NG. The cells can then be transferred to broth and grown up overnight. The grown-up cultures are then plated out on indicator plates to observe Lac$^-$ mutants. The proportion of Lac$^-$ colonies in the mutagenized culture should be much greater than that of the untreated control. If the mutagenesis is insufficient, treat with higher concentrations (100 μg/ml) or for longer time periods.

An alternative procedure is to use exponentially growing cells and employ a short pulse of NG (10 to 40 μg/ml) for 1 to 4 minutes. The NG is subsequently diluted out (Vielmetter, Messer, and Schütte). A method designed to avoid the high frequency of secondary mutations has been developed by Dr. E. Cerdá-Olmedo and coworkers. In this procedure, growing cells are resuspended in broth containing low amounts of NG (1–2 μg/ml) at a density of approximately 5×10^7 cells/ml and aerated at 37° for 4 hours. The cells are then plated out for mutants.

Survival Curves

Survival curves are done with a constant time of exposure and varying doses of NG, or with a constant concentration of NG and different times of exposure. The following exercise includes a procedure for doing the latter type of survival curve. As an alternative to spinning down cells, it is also possible to use a Millipore filter (pore size 0.45 μ) and wash cells directly on the filter. Using this technique Adelberg, Mandel, and Chen have done extensive studies of the effect of NG on survival and mutagenesis under a variety of conditions. They found that killing is much greater when the cells were able to grow during the treatment. Also, high doses of NG result in the induction of a large percentage of auxotrophs. The dose used in this experiment should not yield a significant proportion of auxotrophs. However, each Lac$^-$ mutant should be tested for the ability to grow on glucose minimal agar. The presence of auxotrophs in mutagenized cultures can be minimized by growing cells overnight in minimal media after mutagenesis.

References

ADELBERG, E. A., M. MANDEL and G. C. C. CHEN. 1965. Optimal conditions for mutagenesis by N-methyl-N′-nitro-N-nitrosoguanidine in *Escherichia coli* K12. *Biochem. Biophys. Res. Commun. 18:* 788.

CERDÁ-OLMEDO, E. and P. C. HANAWALT. 1968. Diazomethane as the active agent in nitrosoguanidine mutagenesis and lethality. *Mol. Gen. Genet. 101:* 191.

CERDÁ-OLMEDO, E., P. C. HANAWALT and N. GUEROLA. 1968. Mutagenesis of the replication point

by nitrosoguanidine: Map and pattern of replication of the *Escherichia coli* chromosome. *J. Mol. Biol. 33:* 705.
GUEROLA, N., J. L. INGRAHAM and E. CERDÁ-OLMEDO. 1971. Induction of closely linked multiple mutations by nitrosoguanidine. *Nature 230:* 122.
VIELMETTER, W., W. MESSER and A. SCHÜTTE. 1968. Growth direction and segregation of the *E. coli* chromosome. *Cold Spring Harbor Symp. Quant. Biol. 33:* 585.

Strains

Number	Sex	Genotype	Important Properties
CSH51	F^-	*ara* Δ(*lacpro*) *strA thi* (φ80d*lac*$^+$)	Lac$^+$ Str^r
CSH61	HfrC	*trpR thi*	Lac$^+$ Str^s

Method for NG Killing Curve

Day 1

Subculture an overnight of CSH51 into six text tubes, each with 5 ml of LB (or 30 ml in one 125-ml erlenmeyer flask) and aerate at 37° until a density of 3–5 × 10^8 cells/ml is reached. Spin the cultures down and wash the pellets twice in 5 ml of citrate buffer (or pool all of the samples and spin and wash together). Pool the samples and measure 4.0 ml into each of eight test tubes, after resuspending in 32 ml of citrate buffer (pH 5.5). Place in a 37° water bath. Add 0.2 ml of a 1 mg/ml solution of NG in citrate buffer (pH 5.5) to each tube. This is time 0. The concentration of NG is almost 50 μg/ml in each tube.

At different times withdraw a culture and spin down and wash once in phosphate buffer. Resuspend each pellet finally in 5 ml of phosphate buffer and immediately make serial dilutions and titer on LB plates. Take points at the following times: 0, 5, 10, 20, 30, 45, 60, and 90 min. Plate out 0.1 ml of a 10^{-2}, 10^{-3}, 10^{-4}, and 10^{-5} dilution for the 0, 5, 10, 20, and 30 min points. Plate out 0.1 ml of a 10^{-1}, 10^{-2}, 10^{-3}, and 10^{-4} dilution of the 45, 60, and 90 min points.

Day 2

Determine the viable cell count from the titer plates for each time point. Construct graphs plotting the survival against the time of exposure to NG.

Materials

Day 1

overnight of CSH51
6 test tubes with 5 ml LB broth each
6 small centrifuge tubes or 2 large tubes (to spin down total of 30 ml)
desk-top centrifuge
small flask or large test tube able to hold 40 ml
water bath at 37°

NG and citrate and phosphate buffers are prepared as described in the following section.
NG: 2 ml of a 1 mg/ml stock solution (freshly prepared)
100 ml citrate buffer, pH 5.5
80 ml phosphate buffer, pH 7.0
8 test tubes
8 small centrifuge tubes
32 dilution tubes
32 LB plates
48 0.1-ml, 18 1-ml, and 20 5-ml or 10-ml pipettes

Method for NG Mutagenesis

Day 1

Part 1

Subculture an overnight culture of a Lac^+ strain (CSH51) into LB broth and aerate at 37° until a density of 3–5 × 10^8 cells/ml is reached.

Part 2

Spin down 5 ml of CSH51 and wash the pellet twice in the same volume of citrate buffer. Resuspend in 4 ml of citrate buffer. Add 0.2 ml of a 1 mg/ml solution of NG in citrate buffer. This will bring the final concentration of NG to almost 50 μg/ml. Let stand in a water bath at 37° for 30 minutes. At the end of this time, spin down and wash the pellet in phosphate buffer. Resuspend the washed pellet in 10 ml of broth (LB) and grow up overnight at 37°. Also grow up an unmutagenized portion of the starting overnight as a control.

Day 2

Plate the saturated culture of the mutagenized CSH51 onto lactose tetrazolium plates. Plate 0.1 ml of a 10^{-3}, 10^{-4}, and 10^{-5} dilution. As a control, do the same for an unmutagenized overnight of CSH51. Incubate the plates for 24 hours at 37°.

Day 3

Examine the tetrazolium plates and record the percentage of Lac^- colonies for each culture.

Materials (per culture)

Day 1

0.2 ml freshly prepared solution of NG in citrate buffer (pH 5.5) at 1 mg/ml. Prepare one hour before use, since the NG is difficult to dissolve.
overnight of CSH51
1 tube with 5 ml of LB broth; 20 ml of LB broth
3 empty test tubes; 2 small centrifuge tubes
desk-top centrifuge
37° water bath
15 ml citrate buffer, pH 5.5 (0.1 M)

5 ml phosphate buffer, pH 7.0 (0.1 M)
2 1-ml and 5 5-ml or 10-ml pipettes

Day 2
6 lactose tetrazolium plates
6 pasteur pipettes
6 0.1-ml pipettes and 2 1-ml pipettes
8 dilution tubes

Buffers

Citrate buffer

0.1 M Na citrate, pH 5.5
10.5 g citric acid
4.4 g NaOH
Make up to 500 ml
Adjust to pH 5.5 with 2 N NaOH

Phosphate buffer

0.1 M phosphate, pH 7.0
6.8 g KH_2PO_4
1.16 g NaOH
Make up to 500 ml
Adjust to pH 7.0 with 2 N NaOH

EXPERIMENT 15

Mutagenesis with a Frameshift Mutagen: ICR 191

ICR 191 is one of a series of acridine-based compounds which are highly mutagenic to *E. coli*. This compound (Figure 15A) has been shown to induce primarily the addition or deletion of one or more bases, many of which are frameshifts. (A net change of a multiple of three bases will not result in a frameshift, however.) This is in contrast to 2-aminopurine or nitrosoguanidine which cause principally base substitutions. Although non-alkylated acridines, such as proflavin, are effective mutagens in bacteriophage, they are ineffective in causing mutations in bacteria. However, during conjugation proflavin has a large effect on mutant induction (Sesnowitz-Horn and Adelberg, 1968, 1969).

Since ICR 191 is unstable in solutions at room temperature and in the light, solutions should be stored in the dark in a freezer and thawed directly before use.

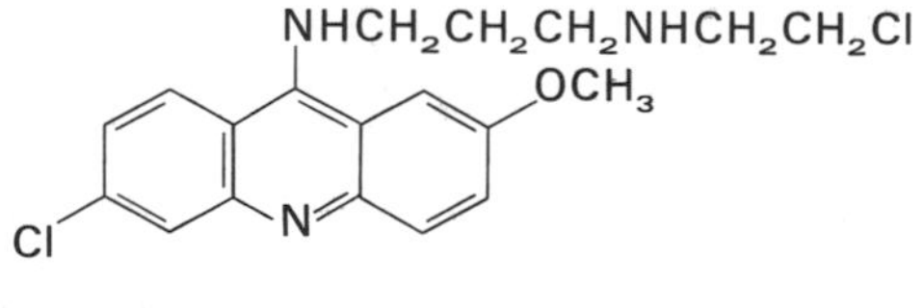

$NHCH_2CH_2CH_2NHCH_2CH_2Cl$

OCH_3

N

Cl

Figure 15A. Structure of ICR 191.

Group #	Concentration of ICR 191 in μg/ml: 0	2.7	4	8	12
I	0/175,000			51/170,000	199/250,000
II	7/170,000			52/290,000	149/125,000
III	0/281,000			28/113,000	34/21,000
IV	1/119,000			16/105,000	26/62,000
V	2/222,000	16/111,000	5/78,000		
VI	0/181,000	2/144,000	10/90,000		

$\frac{\text{\# of i}^- \text{ colonies}}{\text{total \# of colonies}}$

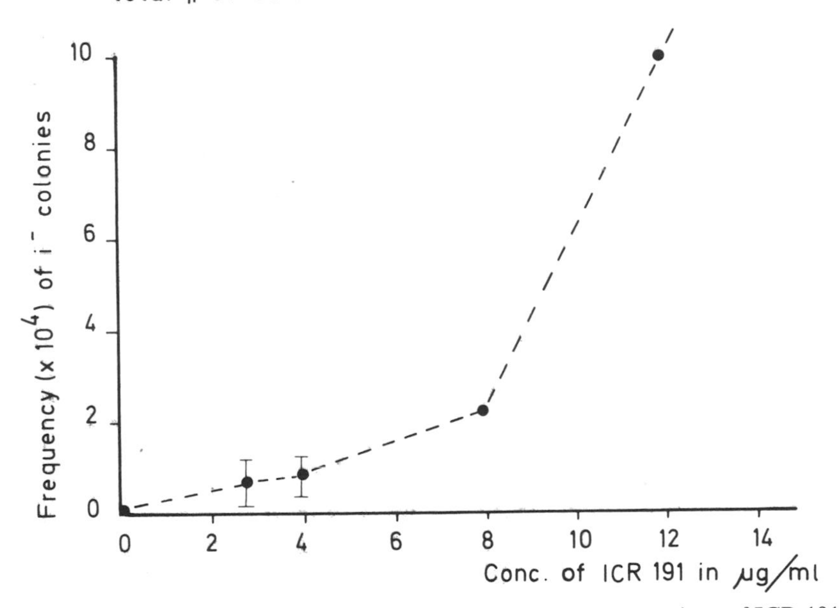

Figure 15B. Induction of i⁻ mutants by increasing concentrations of ICR 191.

A stock solution of 1 mg/ml is recommended. **ICR 191 solutions should not be mouth pipetted.** High concentrations of ICR 191 prevent growth of *E. coli*, and low concentrations are insufficient for good mutagenesis. The best concentration to employ is one which allows slow growth of the bacteria. Although this concentration is often close to 10 μg/ml, it differs with the strain and the conditions. For example, *recA*⁻ strains are more sensitive to ICR 191. Therefore it is advisable to do an "endpoint titration" with ICR 191 first (to determine the amount of growth with different concentrations of the mutagen), and then test several concentrations for their mutagenic effect. Cells should grow 12–15 generations in the presence of the mutagen. It is expected that many of the mutants arising from the same culture are not independent. Therefore, be careful to save only one mutant from each culture for further analysis. This will ensure that we are analyzing an independent collection of mutants. Treatment must be carried out in the dark since acridine compounds can induce mutations photodynamically (most of which are not frameshifts).

Sample Results

Figure 15B is included to show exactly the type of data to expect from a class exercise. This experiment, the induction of i^- mutants with different concentrations of ICR 191, was done in 1970 by a laboratory course consisting of 18 undergraduates at the University of Heidelberg. The students were divided into six groups of three. Each group grew aliquots of the same original culture of an i^+z^+ strain in different concentrations of ICR 191 overnight. Dilutions were then plated on Xgal glucose plates, and i^- colonies were determined the next day by counting the number of deep blue colonies. The table shows the results, the average values of which are plotted in the graph below. The cultures were also plated on lactose tetrazolium plates, and the induction of Lac^- mutants followed by scoring red colonies.

In the following experiment, sensitivity points are determined for both CSH61 and CSH51, after which the cultures are plated out for mutants. Do a sensitivity curve for the *recA*$^-$ derivative of CSH51 as a comparison. Strains deficient in repair functions are extremely sensitive to compounds such as ICR 191. Whereas CSH51 will grow in concentrations of ICR 191 up to 12 μg/ml, the *recA*$^-$ derivative will not grow in concentrations over 2 μg/ml.

References

AMES, B. N. and H. J. WHITFIELD, JR. 1966. Frameshift mutagenesis in Salmonella. *Cold Spring Harbor Symp. Quant. Biol. 31:* 221.

BRAMMAR, W. J., H. BERGER and C. YANOFSKY. 1967. Altered amino acid sequences produced by reversion of frameshift mutants of tryptophan synthetase A gene of *E. coli*. *Proc. Nat. Acad. Sci. 58:* 1499.

NEWTON, A. 1970. Isolation and characterization of frameshift mutations in the *lac* operon. *J. Mol. Biol. 49:* 589.

OESCHGER, N. S. and P. E. HARTMAN. 1970. ICR-induced frameshift mutations in the histidine operon of Salmonella. *J. Bacteriol. 101:* 490.

SESNOWITZ-HORN, S. and E. A. ADELBERG. 1968. Proflavin treatment of *Escherichia coli:* Generation of frameshift mutations. *Cold Spring Harbor Symp. Quant. Biol. 33:* 393.

SESNOWITZ-HORN, S. and E. A. ADELBERG. 1969. Proflavin-induced mutations in the L-arabinose operon of *Escherichia coli*. *J. Mol. Biol. 46:* 1.

Note: Although current supplies of ICR 191 itself are very limited, other substituted acridines (such as those mentioned in Ames and Whitfield) are still available. Their use is exactly as described for this experiment (end-point titration, growth in the dark, etc.).

Strains

Number	Sex	Genotype	Important Properties
CSH51	F^-	*ara* Δ(*lacpro*) *strA thi* (φ80d*lac*$^+$)	Lac^+ Str^r
CSH61	HfrC	*trpR thi*	Lac^+ Str^s

Method

Day 1

Take a fresh overnight culture of the strains to be tested, in this case CSH61 and CSH51, and prepare a 10^{-4} dilution. Pipette 0.1 ml of this into each of several tubes containing 3 ml of enriched A medium and varying concentrations of ICR 191. We suggest using the following concentrations: 0, 2, 4, 6, 8, 10, 12, 14, 16, and 20 μg/ml. Aerate the cultures in the dark at 37° overnight. It is sufficient to cover them with aluminum foil for this purpose. Also do a similar curve with CSH51*recA*$^-$ for comparison. For the *recA*$^-$ strain use the following concentrations: 0.4, 0.8, 1.2, 1.6, 2.0, 3.0, and 5.0 μg/ml. These should be aerated in the dark at the same time as the *recA*$^+$ cultures.

Day 2

Observe the tubes. The cultures with concentrations of ICR 191 that were too low will be indistinguishable from the control. There should be almost no growth with the high concentrations of the mutagen. Some concentration in between will have an intermediate amount of growth. This concentration should be noted for future use for that strain.

Test the mutagenic power of the various concentrations by plating dilutions of the cultures on lactose tetrazolium plates and also on Xgal glucose plates. In addition to the control (no ICR 191), plate the cultures with 2, 8, 12, and 20 μg/ml. Plate 0.1 ml of a 10^{-3}, 10^{-4}, and a 10^{-5} dilution of each, onto both Xgal plates and lactose tetrazolium plates.

Day 3

Record both the number of Lac$^-$ mutants and i$^-$ mutants by counting the red and deep blue colonies on the two types of plates. Also record the total number of cells plated, which can be computed from the highest dilution. Determine the frequency of Lac$^-$ and also of i$^-$ colonies with each concentration of ICR 191. Also repurify and save one deep blue colony from each culture and one red colony (Lac$^-$) for further testing. Mutants which are o^c should appear as medium or light blue colonies, since they are not completely constitutive. Repurify one light blue colony from each culture and save for further testing.

Materials

Day 1 (per strain)

overnight cultures of CSH61, CSH51, and CSH51*recA*$^-$ (CSH52)
2 dilution tubes
10 empty sterile tubes
ICR stock solution of 1 mg/ml in sterile H_2O (just before use make a 50 μg/ml stock solution in enriched A medium). **Handle these solutions carefully, and do not mouth pipette!**
40 ml of enriched A medium. This is minimal A medium supplemented with, per 100 ml: 0.5 ml 1%B1, 2 ml 20% glucose, 0.1 ml 20% $MgSO_4$, and 2 ml LB broth. (If CSH51 or CSH52 is used, proline must be added.)

LB broth contains per liter: 10 g tryptone, 5 g yeast extract, 10 g NaCl
14 0.1-ml, 2 1-ml, and 3 5-ml pipettes
aluminum foil

Day 2

15 Xgal glucose plates (with proline if CSH51 or CSH52 is used)
15 lactose tetrazolium plates
20 dilution tubes
10 0.1-ml, 10 1-ml, and 15 pasteur pipettes

Day 3

1 lactose tetrazolium plate
2 Xgal glucose plates
20 round wooden toothpicks

EXPERIMENT 16

2-Aminopurine and Nitrous Acid Mutagenesis

2-Aminopurine Mutagenesis

2-Aminopurine (2AP) induces primarily transition mutations in *E. coli*. It is not as potent as NG in its action, but this problem is partially overcome by growing bacteria for many generations in its presence. The method is almost identical to that used for ICR 191 mutagenesis. Since each culture will contain many siblings, it is wise to prepare many different cultures and pick only one mutant from each treated culture for analysis.

References

FREESE, E. 1959. The specific mutagenic effect of base analogues on phage T4. *J. Mol. Biol. 1:* 87.
SCHWARTZ, D. O. and J. R. BECKWITH. 1969. Mutagens which cause deletions in *Escherichia coli*. *Genetics 61:* 371.

Strains

Number	Sex	Genotype	Important Properties
CSH51	F^-	*ara* Δ(*lacpro*) *strA thi* (φ80d*lac*$^+$)	Lac$^+$ Str^r
CSH61	HfrC	*trpR thi*	Lac$^+$ Str^s

Method

Day 1 Dilute a fresh overnight culture into 5 ml of LB containing 600 μg/ml of 2AP, so that approximately 100 cells are in the inoculum. Since a typical overnight culture has $1-2 \times 10^9$ cells/ml, this would correspond to 0.1 ml of a 10^{-6} dilution. Grow the 2AP culture, together with a control without 2AP, overnight to saturation.

Day 2 Plate out dilutions of each culture on lactose tetrazolium plates and also Xgal glucose plates. Incubate at 37°.

Day 3, 4 Examine the plates. Pick and purify mutant colonies.

Materials

(2-Aminopurine can be directly dissolved into either water or broth. It need not be autoclaved.)

Day 1

overnight culture of CSH61
3 dilution tubes
4 0.1-ml pipettes
2 test tubes with 5 ml LB broth, one of which contains 600 μg/ml 2AP

Day 2

8 lactose tetrazolium plates
8 Xgal glucose plates
8 dilution tubes
4 0.1-ml and 12 pasteur pipettes

Day 3, 4

2 lactose tetrazolium plates
2 Xgal glucose plates
30 round wooden toothpicks

Nitrous Acid Mutagenesis

Nitrous acid causes both base substitutions and deletions in *E. coli*. It is now widely used to induce deletions.

References

BAUTZ-FREESE, E. and E. FREESE. 1961. Induction of reverse mutations and cross reactivation of nitrous acid-treated phage T4. *Virology 13:* 19.

FREESE, E. 1959. On the molecular explanation of spontaneous and induced mutations. *Brookhaven Symp. Biol. 12:* 63.

KAUDEWITZ, F. 1959. Production of bacterial mutants with nitrous acid. *Nature 183:* 1829.

SCHWARTZ, D. O. and J. R. BECKWITH. 1969. Mutagens which cause deletions in *Escherichia coli*. *Genetics 61:* 371.

TESSMAN, I. 1962. The induction of large deletions by nitrous acid. *J. Mol. Biol. 5:* 442.

Method

Day 1 Nitrous acid should be prepared before use by dissolving sodium nitrite in 0.1 M acetate buffer, pH 4.6, to a final concentration of 0.05 M. The pH of the acetate buffer is a critical factor. Centrifuge a fresh 5 ml overnight of CSH61 and wash in 5 ml of acetate buffer. Resuspend the pellet in 0.3 ml of nitrous acid*. Incubate the mixture at 37° for 10 minutes and then add 5 ml of 1 × A medium (pH 7.0) to stop the reaction. Centrifuge and grow the cells to saturation in 10 ml of LB overnight. Determine the viable cell count immediately before and after treatment with nitrous acid. Survival should be between 0.01% and 0.1%.

Day 2 Plate dilutions of the nitrous acid treated cultures, along with an untreated control, on lactose tetrazolium plates and also on Xgal glucose plates to look for mutants.

Day 3, 4 Examine all plates and determine both the amount of killing with nitrous acid and the extent of mutagenesis. Repeat the procedure using longer times of exposure, if necessary.

Materials

Day 1

5 ml overnight of CSH61 (or CSH51)
sodium nitrite
5 ml 1 × A medium, pH 7.0
15 ml 0.1 M acetate buffer, pH 4.6
0.1 M sodium acetate buffer is prepared as follows:
Solution A contains 11.55 ml concentrated glacial acetic acid made up to 1 liter with distilled water.
Solution B contains 27.29 g sodium acetate · 3 H_2O in 1 liter of distilled water.

* If the cell pellet volumes are large, more nitrous acid should be added (1–2 ml).

Add 25 ml of solution A to 24.5 ml solution B and add 50 ml distilled water. Adjust pH to 4.6 and autoclave.*

pH meter
water bath at 37°
2 small centrifuge tubes
desk-top centrifuge
test tube with 10 ml LB broth
8 dilution tubes
10 0.1-ml, 5 1-ml and 3 5-ml pipettes
6 LB plates

Day 2

8 lactose tetrazolium plates
8 Xgal glucose plates
4 0.1-ml and 12 pasteur pipettes
8 dilution tubes

Other Methods of Mutagenesis

1. EMS (ethylmethane sulfonate, Eastman Kodak Co.): Grow cells to 2–3 × 10^8 cells/ml in glucose minimal medium (1 × A), spin down, wash, and resuspend in $\frac{1}{2}$ the original volume of minimal medium containing 0.2 M Tris, pH 7.5, (and no carbon source). Add 0.03 ml EMS to 2 ml of this suspension and mix vigorously to dissolve. **Do not mouth pipette.** Then aerate at 37° for 1–2 hours. Wash twice. Dilute 10-fold and grow up in broth overnight (Lin et al., 1962). Alternate procedures (Osborn et al., 1967) prescribe a solution of 0.4 ml EMS and 7.6 ml 1 M Tris, pH 7.4, to which 2 ml of resuspended cells are added. Vigorous shaking is then carried out for only 5 minutes at 37°.

2. HA (hydroxylamine hydrochloride, Fisher Scientific Co.): Grow 5 ml of the culture to be treated in minimal medium with aeration at 37° overnight. Centrifuge, wash, and resuspend in 4 ml of 1 M $NH_2OH{\cdot}HCl$, previously adjusted to pH 6.0 with 1 N NaOH. Shake at 37° for 20 minutes. Centrifuge, wash, and either grow in broth overnight or plate directly on selective medium (depending on the type of mutant sought).

3. 5-Bromouracil: Procedures for 5-bromouracil mutagenesis call for a period of thymine starvation during which time 5-bromouracil is incorporated into the DNA. This is achieved either by using a Thy^- strain (these can be constructed in Experiment 31) or else by growing the cells in the presence of inhibitors which specifically interfere with thymine synthesis.

Spin down a saturated overnight culture and wash twice before spreading over the surface of a minimal plate enriched with 20 μg/ml of each amino acid (or with 0.1% casamino acids) and containing 20 μg/ml of 5-bromouracil (this is replaced by thymine in the control). Incubate for 60 minutes at 37° and then recover the bacteria by washing the plates with buffer and plate on indicator or selective plates, or use directly for enrichment procedures (Witkin and Sicurella, 1964).

A simpler technique employs 5-bromodeoxyuridine, which is added at a final concentration of 10^{-2} M to cells which are at a density of 4 × 10^8/ml in supplemented minimal medium. The cells are then incubated at 37° for 65–75 minutes (Osborn et al., 1967). A mutagenesis medium which inhibits thymine synthesis for wild-type cells is minimal medium (1 × A or M9) supplemented with 0.1%

* Recheck pH after autoclaving.

vitamin-free casamino acids, 20 μg/ml tryptophan, 1000 μg/ml sulfanilamide, 25 μg/ml xanthine, 2.5 μg/ml uracil, and 50 μg/ml 5-bromouracil. Cells grown overnight in minimal medium are subcultured into the above medium by diluting 1:25 and then incubated at 37° for 30 minutes without aeration and then for 5 hours with aeration (Clowes and Hayes, 1968).

4. Sodium Bisulfite: Sodium bisulfite induces the specific transition G:C → A:T *in vivo*. Mukai et al. (1970) have used the following protocol: Centrifuge and wash a fresh overnight and resuspend the pellet in the original volume of 0.2 M acetate buffer (pH 5.2) containing 1 M sodium bisulfite. Incubate at 37° for 30 minutes. Centrifuge and wash the cells and plate out for mutants. There is no significant killing with this procedure.

5. Mu-1 Phage: Mu-1 phage induces mutations by inserting into the chromosome, apparently at random (Taylor, 1963). To use this phage, place a drop of a lysate (approximately 10^9 phage/ml) on a lawn of a freshly grown culture of sensitive bacteria on a rich (LB) plate. After 8–10 hours pick cells from the surviving lawn in the center of the spot and (A) streak on indicator plates directly, or (B) resuspend in broth and grow up for mutant selection or penicillin enrichment. This technique gives approximately 1% auxotrophs among the survivors (Bukhari and Zipser, 1972).

References

Bukhari, A. and D. Zipser. 1972. Random insertion of Mu-1 DNA within a single gene. *Nature*. In press.

Clowes, R. C. and W. Hayes. 1968. *Experiments in Microbial Genetics*, p. 79. John Wiley and Sons, New York.

Lin, E. C. C., S. A. Lerner and S. E. Jorgensen. 1962. A method for isolating constitutive mutants for carbohydrate-catabolizing enzymes. *Biochim. Biophys. Acta 60:* 422.

Mukai, F., I. Hawryluk and R. Shapiro. 1970. The mutagenic specificity of sodium bisulfite. *Biochem. Biophys. Res. Commun. 39:* 983.

Osborn, M., S. Person, S. Phillips and F. Funk. 1967. A determination of mutagen specificity in bacteria using nonsense mutants of bacteriophage T4. *J. Mol. Biol. 26:* 437.

Taylor, A. L. 1963. Bacteriophage-induced mutation in *E. coli*. *Proc. Nat. Acad. Sci. 50:* 1043.

Witkin, E. M. and N. A. Sicurella. 1964. Pure clones of lactose-negative mutants obtained in *Escherichia coli* after treatment with 5-bromouracil. *J. Mol. Biol. 8:* 610.

EXPERIMENT 17

Reversion Tests

In order to determine whether the Lac^- mutants we have isolated are point mutants or deletions, it is useful to employ reversion tests. Lac^- point mutations are easily revertible with NG. Deletions are not. The procedure is simply to spread 3 drops of a fresh culture of each Lac^- mutant onto a lactose minimal plate. Place a crystal of NG in the center of the plate.* After 48 hours, a circle of revertants will be around the mutagen, for all *lac* point mutants. There should be no NG-induced revertants at all for deletions. In a similar manner, ICR 191 can be applied to the center of a plate. We recommend spotting one drop of a 500 μg/ml solution of ICR 191 onto the center of the plate, and then incubating in the dark after the spot has dried. ICR 191 should be able to revert many of the frameshift mutations not revertable by NG.

For reversion tests with Lac^- mutants derived from CSH51, proline must be in the lactose minimal plate. It is important to use an excess of proline (200 μg/ml), since *E. coli* can grow slowly with proline as a carbon source, and the background Lac^- cells will therefore use up much of the proline in the plate.

All Lac^- mutations which respond well to NG should be tested, in a similar

* NG should be handled with extreme caution; avoid contact.

manner, for reversion with 2-aminopurine. Place a crystal of 2-aminopurine directly in the center of the plate, onto which a lawn of cells has just been spread. It is also sufficient to spot a drop of a saturated solution of 2-aminopurine. A positive test, indicated by a circle of revertants growing up around the spot, shows that the mutation can be reverted by a transition.

Table 17A. Properties of Frameshift Mutations in the *z* Gene*

Mutation	Induced to revert to Lac$^+$ by: ICR 191	NG
ICR 18	+	0
ICR 27	+	0
ICR 10	+	0
ICR 74	+	0
ICR 22	+	0
ICR 55	+	+
ICR 11	+	+
ICR 38	+	+
ICR 49	0	0
ICR 17	0	0

* From Newton, 1970.

Alternative methods for applying mutagens directly to plates involve the use of small circular filter paper discs. A small circle of filter paper is immersed in a solution of the mutagen, and then carefully placed in the center of the plate. For these experiments, use as controls the Lac$^-$ mutants employed in Experiment 21, since these are nonsense mutants in the *z* gene. Table 17A shows the reversion pattern of several ICR 191-induced mutants in the *z* gene. NG can cause small deletions and thus is able to revert certain frameshifts which involve the addition of several bases.

References

AMES, B. N. and H. J. WHITFIELD. 1966. Frameshift mutagenesis in Salmonella. *Cold Spring Harbor Symp. Quant. Biol. 31:* 221.

BALLBINDER, E. 1962. The fine structure of the loci tryC and tryD of *Salmonella typhimurium.* II. Studies of reversion patterns and the behavior of specific alleles during recombination. *Genetics 47:* 545.

NEWTON, A. 1970. Isolation and characterization of frameshift mutations in the *lac* operon. *J. Mol. Biol. 49:* 589.

OESCHGER, N. S. and P. E. HARTMAN. 1970. ICR-induced frameshift mutations in the histidine operon of Salmonella. *J. Bacteriol. 101:* 490.

YANOFSKY, C. 1963. Amino acid replacements associated with mutation and recombination in the A gene and their relationship to in vitro coding data. *Cold Spring Harbor Symp. Quant. Biol. 28:* 581.

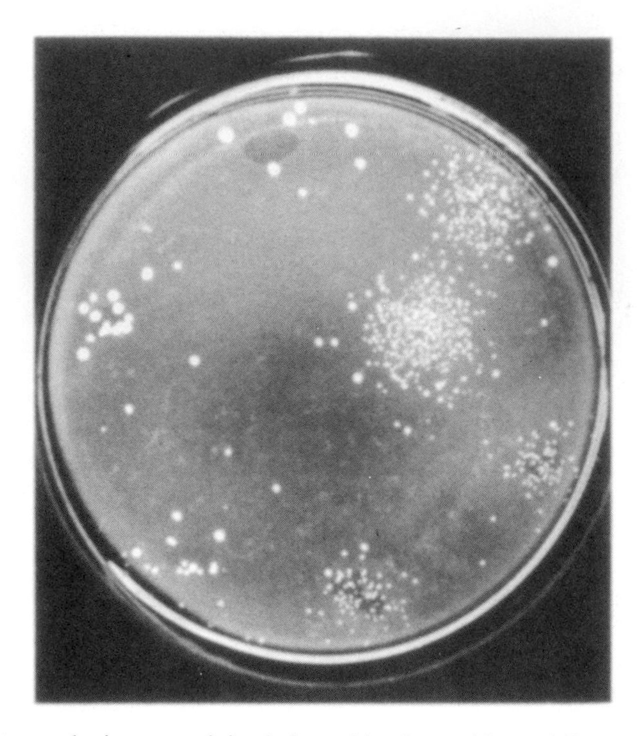

Figure 17A. Revertant colonies on a minimal plate with a lawn of frameshift mutant C207 (0.1 ml of a full grown culture) to which 5 μg of each of the indicated ICR mutagens had been added (as 5 μl of a 1 mg/ml aqueous solution). The plates were incubated 2 days at 37°. ICR 170 is at 12 o'clock and ICR 191, ICR 312, ICR 292, ICR 217, and ICR 171 follow (clockwise). The cluster of colonies slightly off center is from 5 μg of ICR 191 (from a solution that had been stored in the freezer for a month). All compounds were hydrochlorides and dissolved readily in sterile distilled water, and the solutions were checked to show they were essentially sterile. All operations were done in dim light, as the acridines are known to be very light-sensitive.

A trace of histidine (0.2 μmole) is added to the plate so that the background lawn can grow slightly and so that any inhibition of the compounds can be seen. A control plate with no mutagen gave about 15 revertant colonies. (From Ames and Whitfield, 1966.)

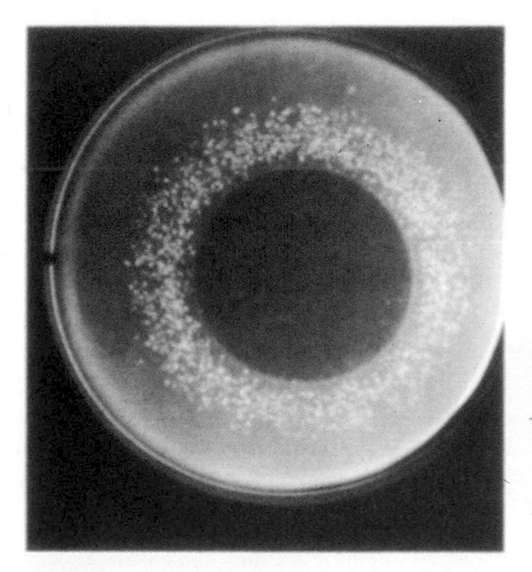

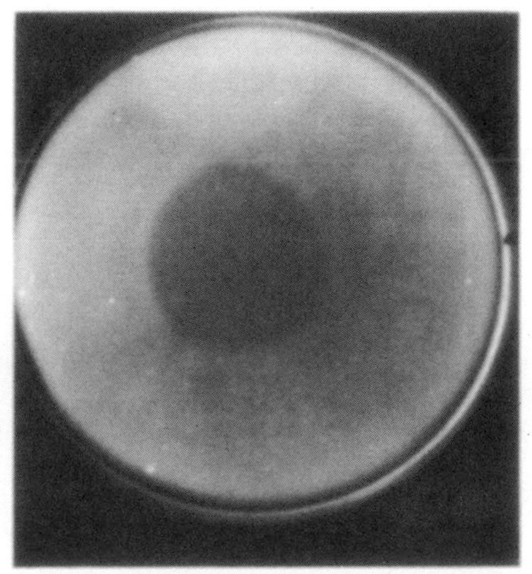

Figure 17B. Action of NG on a missense mutant (left) and a frameshift mutant (right). The method is the same as in Figure 17A except that a small crystal of NG was added to the center of each plate. The zone of inhibition by the NG can be seen. (From Ames and Whitfield, 1966.)

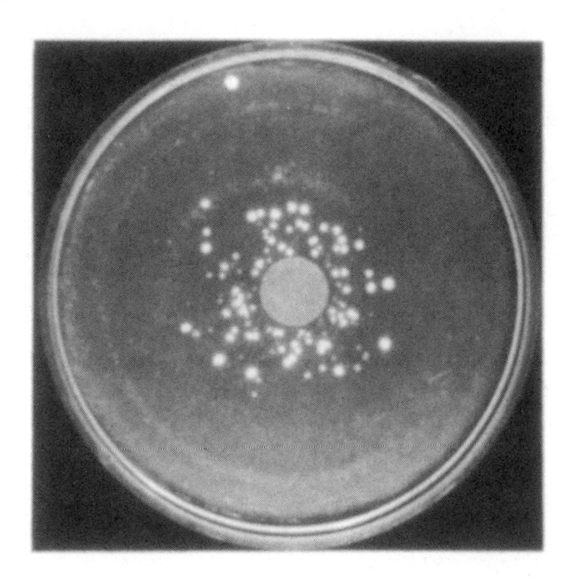

Figure 17C. EMS-induced reversion of a Trp^- mutant, A-46. (From Yanofsky, 1963.)

Materials (per Lac^- mutant)

2 lactose minimal plates (with 200 μg/ml proline only when necessary)
overnight of Lac^- mutant
pasteur pipette
ICR 191 solution (500 μg/ml)
nitrosoguanidine
2-aminopurine

EXPERIMENT 18

Isolation of Temperature-sensitive Mutants in the *lac* Operon

Many enzymes can be altered to a form which is inactive at high temperature, but active at low temperature. Low and high temperatures are often 25–32° and 40–43°, respectively. Likewise, cold-sensitive enzymes can be found. The use of temperature-sensitive mutations is of great value in the study of lethal functions in bacteria. Normally, lesions in a gene for an essential function are fatal to the cell, making the isolation of strains with mutations in such genes impossible by the usual selection procedures. However, mutations which allow growth under one set of conditions but not another (termed "conditional lethal mutations") enable the selection of strains carrying lesions in important genes (Epstein et al., 1963; Kohiyama et al., 1966). Experiment 34 covers the isolation of conditional lethal mutants. In this experiment we isolate temperature-sensitive lac^- and i^- mutants.

References

EPSTEIN, R. H., A. BOLLE, C. M. STEINBERG, E. KELLENBERGER, E. BOY DE LA TOUR, R. CHEVALLEY, R. S. EDGAR, M. SUSMAN, G. H. DENHARDT and A. LIELAUSIS. 1963. Physiological studies of conditional lethal mutants of bacteriophage T4D. *Cold Spring Harbor Symp. Quant. Biol. 28:* 375.

KOHIYAMA, M., D. COUSIN, A. RYTER and F. JACOB. 1966. Mutants thermosibles d'*Escherichia coli* K12. *Ann. Inst. Pasteur 110:* 465.

Strains

Number	Sex	Genotype	Important Properties
CSH51	F^-	*ara* Δ(*lacpro*) *strA thi* (ϕ80d*lac*$^+$)	Lac^+ Str^r
CSH61	HfrC	*trpR thi*	Lac^+ Str^s

Method

The procedure for collecting Lac^- mutations which are temperature sensitive is straightforward. After mutagenesis*, the cultures are plated out on lactose tetrazolium plates at high temperature (42°). Purified Lac^- colonies, which are red on this medium, are tested for growth on lactose at both 30° and 42°. It is advisable, however, to make a master plate with 50 different colonies per plate, and replicate it onto two lactose tetrazolium plates. One is subsequently grown at 30° and the other at 42°. Temperature-sensitive Lac^- mutants appear redder at 42° than at 30°. Then, the candidates from this test are examined for growth on minimal plates at the two temperatures. In a similar manner, i^- mutants can be isolated and screened for temperature sensitivity.

The Lac^- and i^- mutants obtained spontaneously (Experiment 19) should also be tested. Since these are isolated at 42°, they can simply be gridded and replicated as indicated above.

* Cultures should be grown overnight after mutagenesis to allow the segregation of *lac*$^-$ mutations.

EXPERIMENT 19

Isolation of Spontaneous Lac^- and i^- Mutants

We are often interested in obtaining mutants without the aid of a mutagen in order to avoid the complications due to secondary mutations. Spontaneous mutants carrying lesions in a particular gene appear in bacterial populations at frequencies usually between 10^{-5} and 10^{-6}. In previous experiments we mutagenized cells and examined survivors on indicator plates for altered phenotypes. When the percentage of mutants in a culture is greater than 1 per 5000, indicator plate screening is a convenient method. However, to detect mutants which occur at significantly lower frequencies than this, special selections must be employed.

For the practical purpose of designing selection procedures, we can envision mutations as going in either of two directions. A forward mutation results in the acquisition by mutation of the ability to grow on a particular compound or in the absence of an added nutrient (see introduction to Unit II). For instance, many Lac^- strains can acquire by mutation the ability to utilize lactose as the sole carbon source ($Lac^- \rightarrow Lac^+$). Selection for this class of mutant is straightforward, Lac^- cells being plated on lactose minimal medium. A mutation in the other direction results in the loss of this property ($Lac^+ \rightarrow Lac^-$). In order to select for this latter class of mutant, we must find a situation in which the Lac^- cells grow and the Lac^+ cells do not. In practice this usually involves the use of a toxic compound or environment to which the new mutant is now resistant.

Suicide Compounds

Toxic compounds transported by the *lac* permease (*y* gene product) can be used to select against y^+ cells. One example is *o*-nitrophenyl-β-D-thiogalactoside, TONPG (Smith and Sadler, 1971). Cells with a high level of *lac* permease transport this metabolic poison resulting in severe growth inhibition. TONPG is not an inducer of the *lac* operon, and therefore IPTG is included in the medium for y^- selections. In the absence of added inducer only i^-y^+ cells are affected whereas i^+y^+ cells are not. Thus, TONPG can also be used to select against constitutive cells.

Actually, any galactoside which yields a toxic product when cleaved by β-galactosidase can be used to select against lac^+ or z^+ cells. Phenylethylgalactoside and ONPG (ortho-nitrophenylgalactoside) have also been used in this manner, and the application of cinnamylgalactoside and butylgalactoside for this purpose has also been reported.

Additional selections against lac^+ cells can be set up with the aid of genetic manipulations. For instance, strains carrying certain *gal* mutations (*galE* or *galU*) are sensitive to galactose. These cells are also killed by lactose, since galactose is an enzymatic cleavage product of lactose. With the use of $galE^-$ strains a simple lac^- selection is possible. Compounds such as phenylgalactoside, which are also cleaved by β-galactosidase to yield galactose but which do not induce the *lac* enzymes, can be used to select against constitutive cells when used in conjunction with $galE^-$ strains (see discussion below).

Spontaneous Lac⁻ Mutants

Mutants of *E. coli* lacking the enzyme UDPgal 4-epimerase (E in Figure 19A) lyse in the presence of galactose. This is apparently the result of the accumulation of UDPgal and the formation of defective cell walls. Lac⁺ strains generate internal galactose when lactose is hydrolyzed. Therefore lac^+ $galE^-$ strains will not grow on glycerol in the presence of lactose, although lac^- $galE^-$ strains will. This provides us with a convenient selection for Lac⁻ colonies (Malamy). In practice it is found that this medium (0.2% glycerol + 0.02% lactose) selects primarily

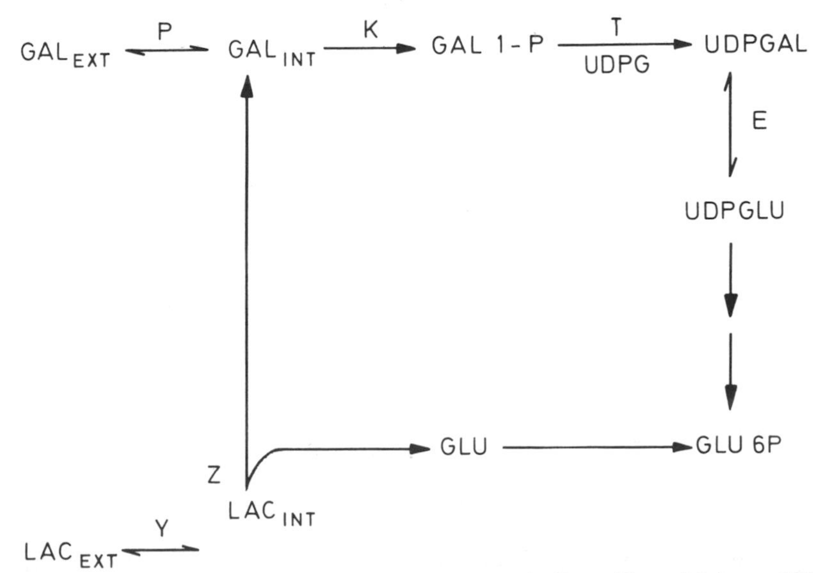

Figure 19A. Pathways of lactose and galactose metabolism. (From Malamy, 1966).

for strains lacking permease activity. Therefore, mutants are usually y^- or z^- polars.

The existence of substrates for β-galactosidase not wholly dependent on the *lac* permease to enter the cell makes possible the direct selection of z^- mutants. Phenylgalactoside is an example. In a *galE*$^-$ background z^- cells but not z^+ cells will grow on glucose medium supplemented with phenylgalactoside. IPTG must be present since phenylgalactoside is not an inducer of the *lac* operon. In the absence of IPTG we can select for i^+ cells using phenylgalactoside. On glucose medium supplemented with this compound (0.03%) i^-z^+ cells will not form colonies whereas i^+z^+ cells will.

GalE$^-$ strains are sensitive to very low levels of galactose. For example, the small contamination of galactose in commercial arabinose preparations prohibits the use of arabinose in medium for *galE*$^-$ strains. Many indicator plate preparations contain contaminants which inhibit the growth of these strains.

Table 19A. Compounds Used for *lac* Selections

Sugar	Selected phenotype					
	i^+	i^-	z^+	z^-	y^+	y^-
Lactose			√ (z^+y^+)		√ (z^+y^+)	√ *
Melibiose		√ **			√	√ *
Phenyl-β-D-galactoside	√ *	√	√ †	√ *†		
o-Nitrophenyl-β-D-thiogalactoside	√					√ †

* in *galE* background
** in p^- background
† + IPTG

References

DAVIES, J. and F. JACOB. 1968. Genetic mapping of the regulator and operator genes of the *lac* operon. *J. Mol. Biol. 36:* 413.

ERON, L., J. R. BECKWITH and F. JACOB. 1970. Deletion of translational start signals in the *lac* operon of *E. coli*. *The Lactose Operon*, p. 353. Cold Spring Harbor Laboratory.

FUKASAWA, T. and H. NIKAIDO. 1961. Galactose-sensitive mutants of Salmonella. II. Bacteriolysis induced by galactose. *Biochim. Biophys. Acta 48:* 470.

MALAMY, M. H. 1966. Frameshift mutations in the lactose operon of *E. coli*. *Cold Spring Harbor Symp. Quant. Biol. 31:* 189.

SMITH, T. F. and J. R. SADLER. 1971. The nature of lactose operator constitutive mutations. *J. Mol. Biol. 59:* 273.

To avoid the necessity of using IPTG in the following exercise, we will use an $i^-z^+galE^-$ strain and select z$^-$ colonies on glucose minimal plates with phenyl-galactoside. To help minimize the effects of phenotypic lag (resulting from the need to dilute out the β-galactosidase in the cell) we have introduced into the *lac* region a promoter mutation, p^-, which lowers the levels of all of the *lac* enzymes 15-fold. This is done merely to gain a higher yield of mutants. However, in a Gal$^+$ background, p^-*lac* strains are still Lac$^+$ so this should not affect further characterization of z$^-$ mutants. The properties of the strain used in this selection are therefore as follows:

Strains

Number	Sex	Genotype	Important Properties
CSH41	F′*lacI,P proA*$^+$,*B*$^+$	Δ(*lacpro*)*galE thi*	i$^-$ GalE$^-$ Str^s

Method

Day 1 Inoculate a single colony into 5 ml broth and aerate overnight at 37°.

Day 2 Plate 0.1 ml of the overnight and also 0.2 ml of a 10^{-1} dilution onto the selective plate, glucose minimal with 0.03% phenyl-galactoside. Incubate the plate for 36–48 hours at 42°. This is done to prevent the growth of mucoid colonies, which do not grow at high temperature. *GalE*$^-$ mutants can overcome galactose sensitivity by reverting to *galE*$^+$, or by the occurence of a secondary *gal*$^-$ mutation in other *gal* genes. The frequency of the latter event is of the same order of magnitude as the frequency of spontaneous z$^-$ mutants. *Gal* revertants can be distinguished from z$^-$ colonies in this experiment by their insensitivity to galactose and by the fact that they still synthesize β-galactosidase (constitutively in this case).

To distinguish z$^-$ colonies from other survivors, resulting colonies can be picked or replicated onto Xgal glucose plates. It is also possible to use Xgal directly in the selective plates. This eliminates the necessity of picking and allows the visualization of a larger number of colonies. It also gives a striking demonstration of all the different phenotypes which occur and survive this selection. If Xgal is used, the cultures should be spun down, washed in buffer and resuspended in the original volume of buffer (1 × A is sufficient). This dilutes out any β-galactosidase in the medium. Plate 0.2 ml of a 10^{-1} dilution of the resuspended overnight at 42°. Also plate a 10^{-5} dilution of the overnight on glucose minimal plates to titer the culture.

Day 4 Examine the plates. If Xgal was in the plate, there should be three types of colonies visible. Deep blue colonies (mutations in the *gal* region) and white colonies (z$^-$ mutants) should predominate. Some colonies will be pale blue, and these are either i$^+$ revertants or leaky z$^-$'s. These can be distinguished by their inducibility by IPTG.

Materials (per starting colony)

Day 1

single colonies of CSH41 growing on a glucose minimal plate in the absence of proline
test tube with 5 ml broth

Day 2

1 glucose minimal plate
2 glucose minimal plates containing 0.03% phenylgalactoside (stock solutions of 1% phenylgalactoside in water are autoclaved for 10 minutes and then 30 ml is added to one liter of medium directly before pouring)
1 dilution tube
1 pasteur pipette, 3 0.1-ml and 2 1-ml pipettes
incubator at 42°
additional: same plates as above, but also containing 40 μg/ml indoxylgalactoside (see Experiment 3)
2 centrifuge tubes to spin down 5 ml of culture
10 ml of phosphate buffer, pH 7.0

Spontaneous i⁻ Mutants

Earlier experiments make use of the Xgal indicator to isolate i^- colonies after mutagenesis. Spontaneous i^-'s are less frequent ($0.3–1.0 \times 10^{-5}$) and to look for them on indicator plates would be tedious. Selections for i^- are based on the requirement for a constitutive level of either β-galactosidase (when phenylgalactoside is used as the only carbon source) or permease (when sugars such as raffinose or melibiose are used as the only carbon source). A procedure for the use of phenylgalactoside to directly select i^- mutants is detailed at the end of this experiment.

At 42° the *lac* permease is required to transport melibiose into the cell. Melibiose itself is a weak **inducer** of the *lac* operon. Thus, the level of *lac* enzymes present when *E. coli* is grown on melibiose is not the maximal level. For wild-type strains this is sufficient, however, to allow growth on melibiose at 42°. If we use a strain with a *lac* promoter mutation (p^-), then the levels are reduced an additional 15-fold. We find that i^+p^- strains do not make sufficient permease in the presence of melibiose to allow growth on this sugar at high temperature. However, i^-p^- strains, not dependent on the **inducing** properties of melibiose, do synthesize sufficient levels of permease to be able to grow well at 42° with melibiose as the only carbon source. Also, $i^+o^cp^-$ strains grow on this sugar.

In summary, we can use an i^+p^- strain easily to select spontaneous i^- and o^c mutants in the *lac* operon. This is done by streaking individual colonies on a section of a melibiose plate and incubating at 42°. A large percentage of the revertant colonies are i^- or o^c. These can be picked first onto Xgal plates as an initial screening procedure to distinguish between these types. The *i* gene in this strain carries a mutation (i^Q) which results in a 10-fold greater synthesis of repressor than normal. This is an advantage in detecting nonsense mutants, since low levels of suppression will be magnified due to the overproduction of the repressor.

Strains

Number	Sex	Genotype	Important Properties
CSH38	F′*lacP proA*$^+$,*B*$^+$	Δ(*lacpro*) *thi*	i$^+$ p$^-$ Str^s

Method

Day 1 Pick a single colony from a freshly grown minimal glucose plate onto a section of a melibiose minimal plate. Pick eight per plate. Streak each colony without flaming the loop a second time. In other words, do not streak for single colonies, but for heavy growth. Incubate at 42° for 36–48 hours. This selection works well at 39–40° and often even at 37°. Check the procedure beforehand.

Day 3 Examine the plates. In the ideal case there should be only a few revertant colonies per section, with only a faint background. Occasionally jackpots of spontaneous mutants are seen. Pick one colony from each section, and streak directly onto Xgal glucose plates for single colonies. Incubate at 37°.

Day 4, 5 Examine the Xgal purification plates. Some Mel$^+$ revertants are at an outside locus. These colonies have the normal basal level of β-galactosidase and are light blue. Some colonies are z$^-$ and are completely white. These are deletions which either fuse the *lac* genes to an outside operon, or else these are *o-z* deletions (see Eron et al.). O^c colonies range in color from deep blue to medium blue, and i$^-$ colonies are deep blue. We recommend picking all deep blue colonies and testing these for nonsense mutants as described in Experiment 22. Attempt to distinguish between i$^-$ and o^c phenotypes by doing complementation tests as described in Experiment 20. This can be done by transferring all of the episomes into CSH51 (i$^+$Pro$^-$) or preferably CSH52 and examining the diploids for constitutivity. Mutants which are i$^-$ should be repressed in this strain.

Materials (per student)

Day 1 glucose minimal plate with single colonies of strain CSH38
8 melibiose minimal plates
64 round wooden toothpicks
incubator at 39–42°

Day 3 8 Xgal glucose minimal plates
64 round wooden toothpicks

Additional Methods

Phenyl-β-D-galactoside selection for constitutive z^+ colonies. Prepare minimal agar plates with 0.7 to 1.0 g per liter of filter-sterilized phenylgalactoside (Sigma). Spread dilutions of a mutagenized culture on these plates and incubate at 42° for 48 hours. If undiluted portions of a culture are used, then the cells should be washed twice in buffer. Some authors recommend the use of Lac^- bacteria to deplete the agar of residual carbon sources and reduce the background. This is accomplished by spreading 10^7 cells from a washed culture of Lac^- bacteria several hours before use. However, we find that at 42° the selection works well without the use of Lac^- bacteria. Incubation at high temperature also reduces interference from mucoid colonies. Phenylgalactoside is reported to have an autocatalytic effect on the induction of β-galactosidase. Previously induced cells maintain high levels of the *lac* enzymes when grown on phenylgalactoside. A complete description of the use of phenylgalactoside plates for selecting different constitutive levels is given in Smith and Sadler (1971).

***o*-Nitrophenyl-β-D-thiogalactoside selection.** This galactoside (Cyclo) is used in minimal medium at 1×10^{-3} M, and selects against y^+ cells. Because this compound is not an inducer, IPTG (at 5×10^{-4} M) is included in the medium for y^- selections. It is reported that the selection works better when succinate or acetate is used as the carbon source (Smith and Sadler, 1971).

EXPERIMENT 20

Complementation Tests of Lac Mutants—Intracistronic Complementation

In practice, complementation tests involve the construction of strains with different pairs of mutations and the examination of these cells for some degree of restoration of wild-type activity. For the study and definition of genes and groups of genes, complementation analysis is an essential tool. This type of analysis requires the existence, at least temporarily, of a duplicate set of genes for the character under investigation. Although many organisms contain diploid sets of chromosomes, *E. coli* does not.* However, a variety of techniques have been used to create partial diploids for various genes. Temporary partial diploids or **"merozygotes"** are produced during interrupted matings. Although these have been used in a number of studies (Pardee et al., 1959; Helling and Weinberg, 1963), they are not widely employed today. Stable partial diploids are of much greater utility. These are constructed by introducing substituted F′ factors (Experiment 37) or by the integration of specialized transducing phage (Experiment 39).

It is important to stress the difference between genetic complementation and genetic recombination. Complementation employs the interaction of gene products, whereas recombination involves the reconstitution of a gene or set of

* During rapid growth in rich medium, *E. coli* can have several copies of the same chromosome. However, nuclear segregation patterns insure the stable inheritance of only a single chromosome.

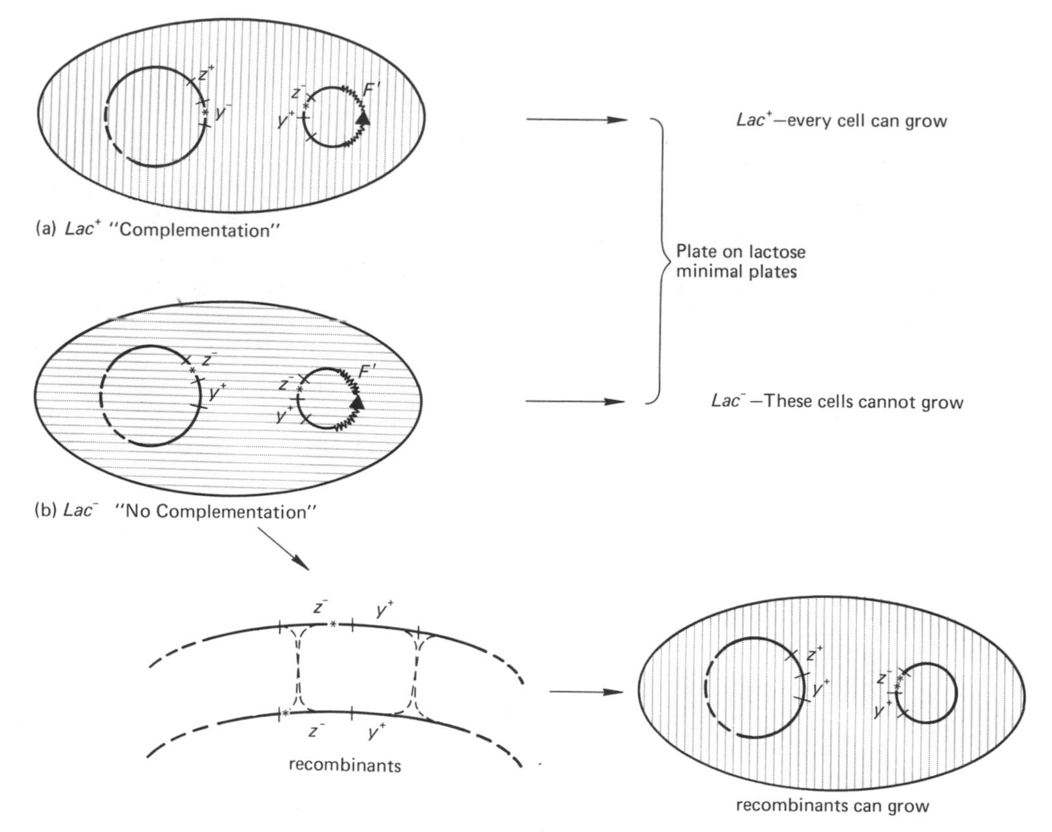

Figure 20A. Complementation *vs* recombination. In **(a)** a diploid for the *lac* region is pictured in which a z^+y^- chromosome is together with a z^-y^+ episome. Here complementation occurs and every cell is Lac$^+$ (indicated by vertical shading). In **(b)** two z^-y^+ elements are together. Even though the z^- mutations are different, there is no complementation, and each cell is Lac$^-$ (indicated by horizontal shading). Rare recombinants can occur, though, giving rise to a Lac$^+$ phenotype.

genes. This is illustrated more clearly by considering the example of the *lac* operon (see Figure 20A).

Certain pairs of Lac$^-$ mutants will complement with each other. Diploids containing one z^+y^- *lac* region and one z^-y^+ *lac* region will be phenotypically Lac$^+$. Only one intact copy of the *z* gene and one wild-type *y* gene are required for the cell to be able to grow with lactose as the only carbon source. This could also occur by recombination, which would result in this case in the formation of a z^+y^+ *lac* region. By constructing merodiploids with different *lac*$^-$ mutations, complementation groups can be defined. Most of the *lac*$^-$ mutations will fall into one of two complementation groups. To a first approximation, z^- mutations should complement only with y^- mutations, and vice versa. However, several phenomena exist which upset this simple picture.

1. Polarity: As discussed in Experiments 22 to 24, some mutations in a gene are "polar." These result in the reduction of the levels of enzymes synthesized by genes distal to the gene in which the mutation occurs. Most polar mutations result in premature polypeptide chain termination. Some are the result of insertions of large amounts of genetic material into the middle of a gene. The effects of polar mutations result in a z^-y^- phenotype. Of course, deletions in the *z* gene extending into or past the *y* gene are also y^-.

2. Outside mutations: At least four different types of mutations which are not in the structural *z* or *y* gene can result in a Lac^- phenotype. Mutations in the *lac* promoter and also i^s mutations have been mentioned in the introduction. Two types of unlinked mutations, involved in the cyclic AMP-regulated positive control of the *lac* operon, are covered in Unit IX. The occurrence of any of these mutations could give a negative result in complementation tests with either *z* or *y* mutations, since they can result in a z^-y^- phenotype.

3. Intracistronic complementation: An extensive review of this topic is provided by Ullmann and Perrin. Complementation between different pairs of point mutations in a gene probably involves interaction of protein subunits. However, certain deletion mutations in *z* also exhibit complementation with other z^- mutations. Deletion mutants which still synthesize the distal part of the β-galactosidase molecule are termed omega donors. These deletions complement with certain z^- mutations which still allow the synthesis of an unaltered proximal part of the β-galactosidase molecule. The strains carrying these mutations are termed omega acceptors. A similar phenomenon involving the proximal (alpha) portion of the β-galactosidase molecule has also been described. Both alpha and omega complementation are due to non-covalent reassociation of peptide fragments.

Negative complementation occurs if the altered form of the enzyme interacts with the wild-type enzyme, probably at the subunit level, and lowers the activity of the wild-type protein. This results in a dominant negative phenotype. Although this phenomenon does not occur frequently amongst z^- mutations, a number of dominant i^- mutations have been characterized. These have been termed i^{-d}. These can be distinguished from an o^c (operator constitutive) phenotype by the use of *cis-trans* tests. A diploid of the type shown in part (a) of Figure 20B is constructed. If the i^- is really a dominant i^- mutation, it can inactivate a wild-type *i* gene which is on a different DNA element. If it can mediate its effect through the cytoplasm or *in trans*, then it is an i^{-d}. If on the other hand, only the genes on the same DNA element, or *in cis*, are constitutive, then the mutation is probably in the operator. In this case the diploid shown in part (b) of the figure would be constitutive, but not that shown in part (a).

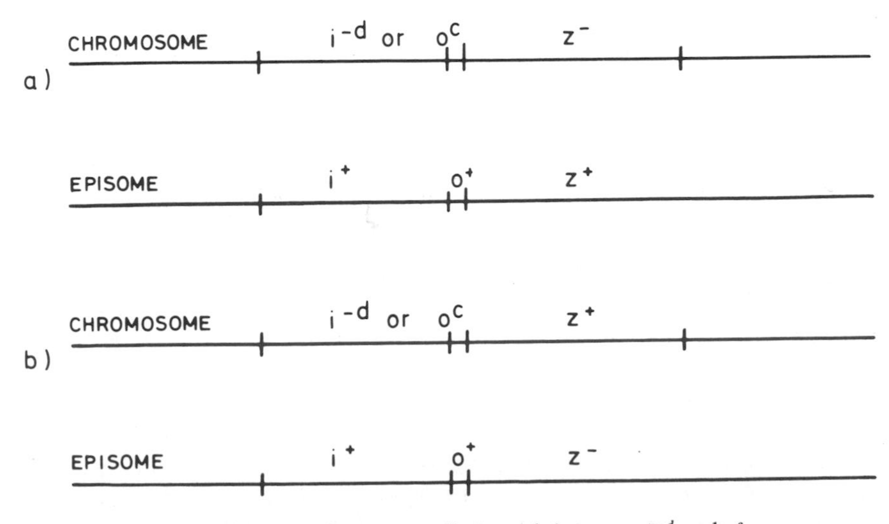

Figure 20B. *Cis-trans* test to distinguish between i^{-d} and o^c.

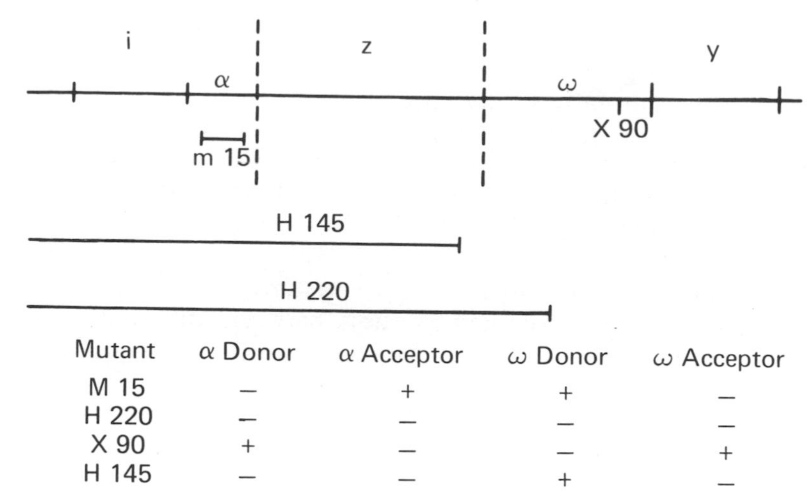

Mutant	α Donor	α Acceptor	ω Donor	ω Acceptor
M 15	–	+	+	–
H 220	–	–	–	–
X 90	+	–	–	+
H 145	–	–	+	–

Figure 20C. Intracistronic complementation.

In summary, it should be recalled that complementation is a test for a particular **phenotype**, in this case Lac$^+$. The Lac$^+$ phenotype depends on the presence of a sufficient level of active β-galactosidase and permease. Any factors which influence either the amounts or activities of these enzymes can have a bearing on the outcome of complementation tests.

In the following exercise, we will test the complementation properties of the *lac*$^-$ mutations in CSH51 by using different *lac* mutations on an F′ *lacpro* episome. All the z^- mutations can also be tested for intracistronic complementation with a set of four different z^- mutations depicted in the accompanying diagram (Figure 20C). These should enable an unambiguous determination of the alpha and omega donor and acceptor properties of any z$^-$ mutant, and furnish a great deal of information about the location of the mutant and what type of mutation it is.

References

FINCHAM, J. R. S. 1966. *Genetic Complementation.* W. A. Benjamin.

HELLING, R. B. and R. WEINBERG. 1963. Complementation studies of arabinose genes in *Escherichia coli. Genetics 48:* 1397.

PARDEE, A. B., F. JACOB and J. MONOD. 1959. The genetic control and cytoplasmic expression of "inducibility" in the synthesis of β-galactosidase by *E. coli. J. Mol. Biol. 1:* 165.

ULLMANN, A., F. JACOB and J. MONOD. 1968. On the subunit structure of wild-type *versus* complemented β-galactosidase of *Escherichia coli. J. Mol. Biol. 32:* 1.

ULLMANN, A. and D. PERRIN. 1970. Complementation in β-galactosidase. *The Lactose Operon,* p. 143. Cold Spring Harbor Laboratory.

Strains

Recipients: Lac$^-$ derivatives of CSH51: F$^-$ Bl$^-$ Pro$^-$ Str^r
Donors: F$^-$ Bl$^-$ Str^s with *lacpro* episomes containing different *lac* mutants

Number	Sex	Genotype	Important Properties
CSH22	F′*lacZ proA*$^+$,*B*$^+$	*trpR* Δ(*lacpro*) *thi*	i$^+$z$^-$(M15) y$^+$
CSH21	F′*lacZ proA*$^+$,*B*$^+$	Δ(*lacpro*) *supE nalA thi*	i$^+$z$^-$(X90) y$^+$
CSH17	F′*lacZ proA*$^+$,*B*$^+$	Δ(*lacpro*) *supE thi*	i$^-$ z$^-$(H145) y$^+$
CSH20	F′*lacZ proA*$^+$,*B*$^+$	Δ(*lacpro*) *supE thi*	i$^-$ z$^-$(H220) y$^+$
CSH36	F′*lacI proA*$^+$,*B*$^+$	Δ(*lacpro*) *supE thi*	i$^-$z$^+$y$^+$
CSH40	F′*lacY proA*$^+$,*B*$^+$	Δ(*lacpro*) *thi*	i$^+$z$^+$y$^-$

Method

Day 1 Construct partial diploids for the *lac* region for each strain to be tested. This can be done easily by crossing F′*lacpro* episomes, containing various *lac* markers, into the Lac^- mutants isolated in CSH51 (which is Pro^-). The selection is for Pro^+. Then the Pro^+ diploids are examined for Lac^+ or Lac^- phenotype. Spot an overnight of each CSH51 derivative to be tested onto a glucose minimal plate with streptomycin. Place a drop of a donor culture (growing exponentially in LB broth at 2×10^8 cells/ml) on top of the CSH51 derivative spot. Allow to dry and incubate overnight at 37°. Each donor is Str^s.

Day 2 After 24 hours, pick from the center of each spot onto a fresh glucose minimal streptomycin plate and purify for single colonies.

Day 3,4 Test individual colonies for growth on lactose and for their appearance on lactose indicator plates. Test two colonies from each diploid. When complementation occurs, practically all colonies from this diploid should be Lac^+. Determine the nature of the particular Lac^- mutation you have isolated, from its behavior in the complementation tests.

Likewise, i^- and o^c mutants can be distinguished from one another by crossing-in, in a similar manner, an episome which is i^+z^-. The diploid should have an i^+ phenotype in the case of an i^-, but an o^c will retain its partially constitutive phenotype in the presence of another *i* gene.

Genotype of diploid: Chromosome (X7700)	Genotype of diploid: Episome F′ lacpro	Phenotype
$i^+ z^+ y^-$	$i^+ z^- y^+$	Lac^+
$i^+ z^- y^+$	$i^+ z^+ y^-$	Lac^+
$i^+ z^- y^+$	$i^+ z^- y^+$	Lac^-
$i^+ z^+ y^-$	$i^+ z^+ y^-$	Lac^-
$i^+ z^- y^-$	$i^+ z^+ y^-$	Lac^-
$i^+ z^- y^-$	$i^+ z^- y^+$	Lac^-
$i^- z^+ y^+$	$i^+ z^- y^-$	i^+
$i^+ o^c z^+ y^+$	$i^+ z^- y^-$	o^c

Figure 20D. Complementation patterns in *lac*. The *lac* genotype of the chromosome in CSH51 (X7700) derivatives and of different F′*lacpro* episomes are shown in the two left-hand columns, and the resulting phenotype of each diploid in the right-hand column (see text).

Materials

Day 1 ***(per strain tested)***

overnight of Lac^- strain
exponential cultures of CSH7, CSH20, CSH21, CSH22, CSH36
1 glucose minimal streptomycin plate
6 pasteur pipettes

Day 2

1 glucose minimal streptomycin plate
6 round wooden toothpicks

Day 3,4

2 lactose minimal plates
2 lactose tetrazolium plates

Additional Experiments

1. Determine the specific activity of the β-galactosidase in the crude extracts of the different heterogenotes which exhibit intracistronic complementation and compare these values with that of the parent strain, CSH51. For methods of IPTG induction and β-galactosidase assay, see Experiment 47. Calculate the efficiency of complementation.

2. Demonstrate that the enzyme from intracistronic complementation is different from that of wild type by doing a heat inactivation curve for β-galactosidase, using the crude extracts from heterogenotes and from CSH51. For details consult

ULLMANN, A., F. JACOB and J. MONOD. 1968. On the subunit structure of wild-type *versus* complemented β-galactosidase of *Escherichia coli*. *J. Mol. Biol. 32:* 1.

ULLMANN, A. and D. PERRIN. 1970. Complementation in β-galactosidase. *The Lactose Operon*, p. 143. Cold Spring Harbor Laboratory.

3. Assay for omega complementation *in vitro*. Mix extracts of donor and acceptor strains and assay for β-galactosidase. For details consult the above references. Demonstrate the kinetics of this type of reaction by doing a time course of omega complementation.

EXPERIMENT 21

Fine Structure Mapping of Lac$^-$ Mutants

The following exercise demonstrates deletion mapping in the *z* gene. We will employ episomes carrying the *lacpro* region and containing deletions of different lengths extending into the *lac* operon. It will be sufficient to use spot tests on lactose minimal streptomycin plates, since the donor strains are sensitive to streptomycin. In the first part of this exercise, ten known nonsense mutations in the *z* gene will be mapped against eight deletions. The positions of both the deletions and the nonsense mutations in the *z* gene are shown in Figure 21A. In addition, each group will be provided with three nonsense mutants in the *z* gene, and also two additional deletions, whose map position is known only to the instructor.

The second part of the exercise will involve the mapping of the z^- mutations isolated in the preceding experiments. By comparing the results of spot tests of each z^- mutant with each episome donor, all of the new mutations can be ordered with respect to one another. Their position relative to previously mapped markers is also revealed. Since the deletions are on *lacpro* episomes, proline need not and should not be present in the lactose selection plates.

References

BECKWITH, J. R. 1970. *Lac:* The genetic system. *The Lactose Operon*, p. 5. Cold Spring Harbor Laboratory.

BENZER, S. 1955. Fine structure of a genetic region in bacteriophage. *Proc. Nat. Acad. Sci. 41:* 344.

BENZER, S. 1959. On the topology of the genetic fine structure. *Proc. Nat. Acad. Sci. 45:* 1607.

BENZER, S. 1961. On the topography of the genetic fine structure. *Proc. Nat. Acad. Sci. 47:* 403.

BENZER, S. 1962. The fine structure of the gene. *Sci. Amer. 206:* 70.

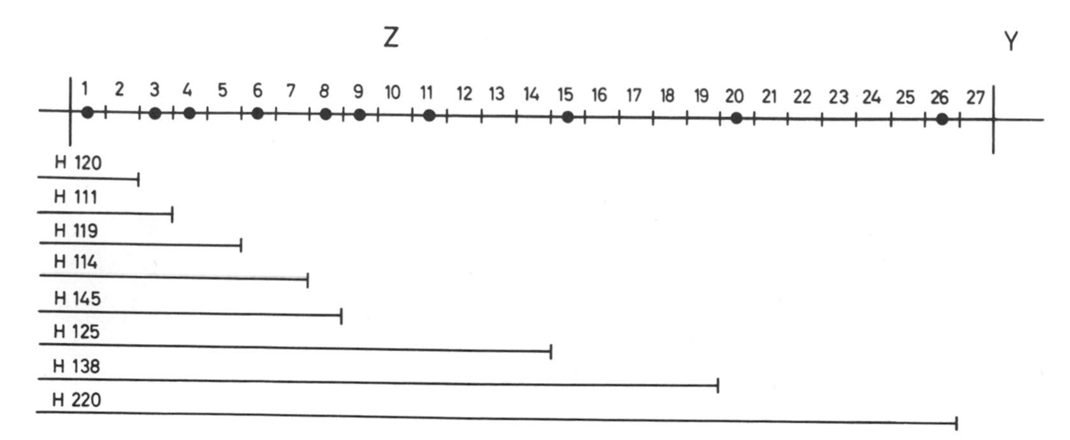

Figure 21A. Map of nonsense mutations in the *z* gene. The *z* gene has been divided into 27 intervals. Some of the deletions used to map this region are depicted above. Each dot represents the site of a nonsense mutation. All intervals have been defined by Zipser et al., *J. Mol. Biol.*, 49, 251, 1970.

Notes on Spot Tests

It is easy to do six spot tests per plate. With care, eight per plate is possible. To map each of the 10 deletions against each of the 13 point mutations would require 26 plates. Use two plates for each recipient, with five sectors on each plate for donors, and one for the recipient alone as a control. If two plates are used to spot each of the donors alone, this makes 28 plates in all. We recommend that several groups work together. The female strains used in part A require tryptophan, which must therefore be present in the selective plates. CSH51, used in part B, does not require tryptophan. It is preferable that the lactose streptomycin-selective plates used in part B do not contain tryptophan (or proline).

Strains

Donors—all of the following type:

Sex	Genotype	Important Properties
F′*lacZ proA*$^+$,*B*$^+$	Δ(*lacpro*) *supE thi*	Str^s, carry deletion on F′ factor ending within *z* gene

Number	Deletion (Figure 21a)
CSH13	H120
CSH14	H111
CSH15	H119
CSH16	H114
CSH17	H145
CSH18	H125
CSH19	H138
CSH20	H220
CSH20a,b	unknowns

Recipients—all of the following type:

Sex	Genotype	Important Properties
F^-	*lacZ trp strA thi*	Str^r, carry nonsense mutation in interval of *z* gene

Number	Interval (Figure 21a)	Number	Interval (Figure 21a)
CSH1	1	CSH8	11
CSH2	3	CSH9	15
CSH3	4	CSH10	20
CSH4	6	CSH11	26
CSH5	8	CSH11a,b,c	unknowns
CSH6	9		

Method

Part A

Day 1 Subculture a fresh overnight of each donor (4 or 5 drops into 5 ml LB) and grow at 37° with aeration to a density of 2–4 × 10^8 cells/ml. (This can be approximated by diluting a fresh overnight 1:20 and allowing to grow for an additional 2.5 hours.) Including the unknowns there are a total of ten donor strains. Do spot matings by placing a drop of each donor on top of a drop of each recipient on sections of a lactose minimal streptomycin plate. This is easier to do if pasteur pipettes are used. The plates should be pre-dried overnight at 37°, and then 2 hours prior to the experiment they should be allowed to sit partially open, face down, at 37°. It is essential that the plates are dry for this experiment. The recipients should be applied first, and as soon as these spots dry the donor spot should be applied. Including unknowns, there are 13 recipients, each containing a different z^- nonsense mutation. Also spot each donor and recipient alone as a control. Incubate at 37°.

Day 2 Examine the spots. The controls should show little or no growth in the spot, as should all of those in which there was no recombination. A positive result is indicated, in most cases, by almost confluent growth of the spot. In a few cases, the deletion end will be close to the mutation, and a low level of recombination will be seen. It is important to observe the plates early enough, since revertants appear after prolonged incubation. Usually 24–30 hours is sufficient to see recombination. The plates should be examined again after 36 and 48 hours. Note from the controls which nonsense mutants revert with the highest frequency.

Construct a map of the *z* gene from these results. Does this map agree with the one provided in the figure? From the results with unknown nonsense and deletion mutants, can you define a new deletion interval?

Part B

Using the same procedure, map the z^- mutants derived from CSH51.

Materials

Day 1

fresh overnights of each of the 10 donors and 13 recipients (It is preferable for the recipients to be grown in simple YT broth, since the amount of rich broth carried over in the spot matings is responsible for the background.)
10 test tubes with 4 ml each of rich (LB) broth
10 pasteur pipettes

For each entire set of spot tests, the following is also required:
23 pasteur pipettes
28 lactose minimal streptomycin plates with tryptophan (pre-dried)

Part B

overnight cultures of the CSH51 z^- strains
lactose minimal plates with streptomycin (no tryptophan)

Additional Exercises

Do any of the mutants, either the ones provided or the new set isolated from CSH51, display omega complementation? Several of the deletions are omega donors, and several of the nonsense mutants are omega acceptors. Complementation can be distinguished from recombination in this type of spot test. Recombination will not yield completely confluent growth of the spot in many cases, whereas complementation will. Attempt to locate the omega boundary by observing which pairs of mutants show this type of complementation.

INTRODUCTION TO EXPERIMENTS 22–24

Chain Termination, Polarity, and Suppression

The three triplets UAG, UGA, and UAA do not code for an amino acid. These signals, termed "nonsense" codons, are involved in chain termination. Two protein factors (R1 and R2) recognize these triplets both *in vivo* and *in vitro* and effect release of the polypeptide chain from the ribosome-mRNA-tRNA complex. Occasionally mutations arise which result in a nonsense codon appearing in the middle of a gene, causing premature chain termination and the production of a protein fragment. Fragments such as these rarely have enzymatic activity. The effect of a nonsense mutation can be reversed or **suppressed** by a second mutation which results in the synthesis of an altered tRNA molecule (Figure 22A). This suppressor tRNA recognizes a nonsense codon and inserts an amino acid at that point in the polypeptide chain (Figure 22B). For instance, Su1 is a suppressor which inserts serine with as high as a 65% efficiency at an amber codon (UAG). It belongs to the amber class of suppressors, which respond only to the codon UAG. The amber suppressor Su2 inserts glutamine, and Su3 inserts tyrosine. Suppressors for UAA (termed ochre) and UGA have also been isolated. All ochre suppressors recognize both amber and ochre codons, UAG and UAA (Table 22A).

The suppressor from Su3 has been isolated and fully sequenced, as has the wild-type tRNA from which it is derived (Goodman et al.). A single base change

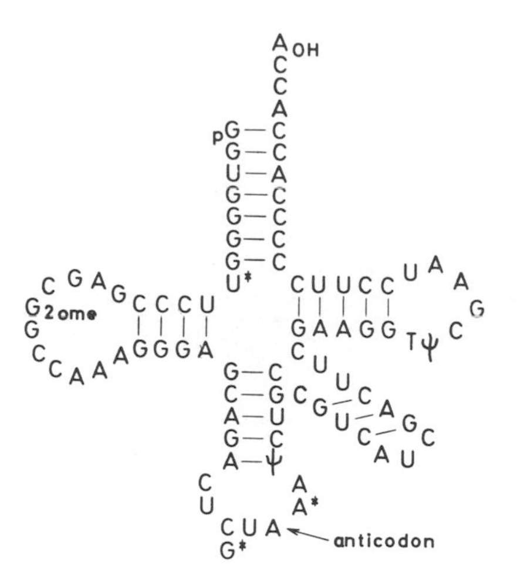

Figure 22A. Su$^+$ tyrosine tRNA of *E. coli*. A single base change (from G to C) in the anticodon has converted a tyrosine tRNA to a suppressor tRNA. (From Goodman et al., 1968.)

from a G to a C (see Figure 22B) has occurred and has converted a tyrosine tRNA into a tyrosine amber suppressor, since this tRNA now recognizes the amber triplet instead of the tyrosine codon. However, this is not always the case, since the sequencing of several UGA suppressors has shown that these molecules are tryptophan tRNA's with a single base change **outside** the anticodon loop (Hirsh). The existence of many well characterized nonsense mutations in the *z* gene of the *lac* operon enables us to isolate nonsense suppressor strains by screening revertants of these mutants. The revertants of amber mutants will include both back mutations at the amber site and also outside suppressors (Experiment 23).

Since different nonsense suppressors insert different amino acids, some amber mutations will not be suppressed by certain amber suppressors. For instance, a glutamine residue inserted by Su2 may not be an adequate amino acid at that site

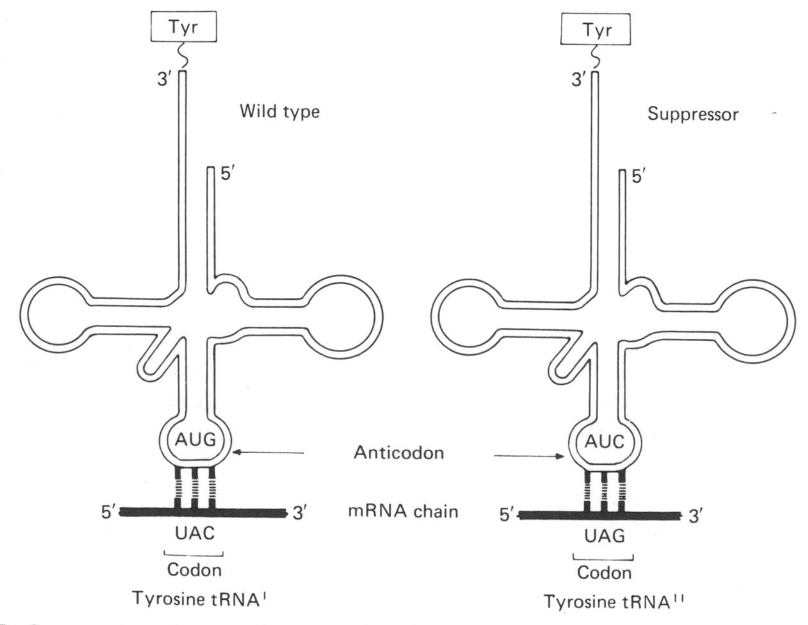

Figure 22B. Suppression of an amber mutation by a suppressor. (Modified from J. D. Watson, *Molecular Biology of the Gene*, 2nd Ed., p. 410. W. A. Benjamin, Inc., 1970.)

Table 22A. Nonsense Suppressors*

Suppressor gene	Phenotypic symbol	Nonsense suppressed	Amino acid inserted
supD	Su1(SuI)	UAG	serine
supE	Su2(SuII)	UAG	glutamine
supF	Su3(SuIII)	UAG	tyrosine
supC	Su4(Su_C)	{UAG, UAA	tyrosine
supG	Su5	{UAG, UAA	lysine
—	Su6	UAG	leucine
supU	Su7	UAG	glutamine
supV	Su8	{UAG, UAA	—
—	Su9	UGA	tryptophan
supB	Su_B	{UAG, UAA	glutamine

* From GORINI, L., *Ann. Rev. Genet.*, **4**, 107, 1970.

		AMBER SUPPRESSORS			OCHRE SUPPRESSORS			
Segment	Site	su_I^+	su_{II}^+	su_{III}^+	su_B^+	su_C^+	su_D^+	su_E^+
A 2	HD 120	+	+	+	+	0	+	0
A 3	N 97	+	+	+	Poor	Poor	+	0
	S 116	0	0	+	0	+	+	0
	HF103	+	+	+	Poor	0	+	0
A 4	N11	+	+	+	Poor	+	+	0
A 5	S172	+	+	+	Poor	+	+	0
	S24	+	+	+	0	0	+	0
	HB129	+	+	+	+	0	+	0
A 6	S99	+	+	+	+	0	+	0
	N19	+	Poor	+	Poor	0	+	0
	N34	+	+	+	+	+	+	0
B 1	HE122	+	+	+	Poor	0	+	0
	2074	+	+	+	+	0	+	0
	EM84	+	+	+	+	+	+	0
	HB74	+	+	+	+	+	+	+
	NT332	+	+	+	+	0	+	0
B 4	X237	+	Poor	+	0	+	+	0
	AP164	+	+	+	+	+	+	+
B 7	HB232	+	+	+	+	0	+	+
	X417	+	+	+	0	0	+	0
	HD231	+	+	+	+	+	+	+

Figure 22C. Qualitative suppression patterns of rII amber mutants. (From Brenner and Beckwith, 1965.)

THE GENETIC CODE

1st ↓ 2nd →	U	C	A	G	↓ 3rd
U	PHE	SER	TYR	CYS	U
	PHE	SER	TYR	CYS	C
	LEU	SER	Ochre	Nonsense	A
	LEU	SER	Amber	TRP	G
C	LEU	PRO	HIS	ARG	U
	LEU	PRO	HIS	ARG	C
	LEU	PRO	GLUN	ARG	A
	LEU	PRO	GLUN	ARG	G
A	ILEU	THR	ASPN	SER	U
	ILEU	THR	ASPN	SER	C
	ILEU	THR	LYS	ARG	A
	MET	THR	LYS	ARG	G
G	VAL	ALA	ASP	GLY	U
	VAL	ALA	ASP	GLY	C
	VAL	ALA	GLU	GLY	A
	VAL	ALA	GLU	GLY	G

Figure 22D. The genetic code.

as a replacement for the normal residue. Every amber mutation has a characteristic "pattern of suppression" or response to a set of nonsense suppressors. Figure 22C shows a series of nonsense mutants in the rII cistrons of phage T4 and their response to different amber and ochre suppressors (Brenner and Beckwith).

Only certain codons can mutate by a single base change to one of the three nonsense codons. These triplets, and the particular amino acids they code for, can be deduced from the diagram showing the genetic code (Figure 22D).

References

BRENNER, S. and J. R. BECKWITH. 1965. Ochre mutants, a new class of suppressible nonsense mutants. *J. Mol. Biol. 13:* 629.

GOODMAN, H. M., J. ABELSON, A. LANDY, S. BRENNER and J. D. SMITH. 1968. Amber suppression: A nucleotide change in the anticodon of a tyrosine transfer RNA. *Nature 217:* 1019.

GORINI, L. 1970. Informational suppression. *Ann. Rev. Genet. 4:* 107.

HIRSH, D. 1971. Tryptophan transfer RNA as the UGA suppressor. *J. Mol. Biol. 58:* 439.

EXPERIMENT 22

Isolation of Nonsense Mutants

In order to isolate a set of strains carrying different nonsense mutations in a particular gene, a larger collection of mutants must first be examined, since there is at present no way to select directly for ochre or amber mutants. It is important, therefore, to design screening procedures which allow the rapid testing of a large number of candidates. In a typical isolation procedure nonsense mutants constitute usually between 1% and 15% of the total mutants recovered, depending on the type of mutagenesis, conditions of selection, methods of screening, and the particular gene involved.

Consideration for Screening

Clearly, mutagens that cause base substitutions will induce nonsense mutations, while agents such as ICR 191 which induce frameshifts will not.

One practical problem is the poor efficiency of suppression of some suppressors. Most ochre suppressors are only 5–10% efficient, while the amber suppressors (Su1, Su2, Su3, and Su6) vary between 25% and 65%. Some proteins, particularly regulatory proteins, are normally synthesized in limiting amounts. In these cases, 5–10% suppression is not sufficient to restore detectable activity *in vivo*, which would prohibit the recovery of ochre mutations. However, it is sometimes possible to find conditions which result in the overproduction of certain proteins.

Table 22B. Suppression Pattern of Nonsense Mutations in the *i* Gene*

Mutation	Amber suppressors				Ochre suppressors		
	Su1	Su2	Su3	Su6	Su_B	Su_C	Su5
816	+	+	+	+	±	+	±
971	+	±	+	+	±	+	±
973	+	+	+	+	+	+	+
509	+	+	+	+	+	+	±
811	+	+	+	+	+	+	±
959	+	+	+	+	−	+	−
315	±	−	±	+	−	±	−
258	±	−	±	+	−	±	−
919	+	−	+	+	−	+	±
136	+	−	+	+	−	+	±
913	+	+	+	+	+	+	+
446	−	−	−	−	±	+	±
984	−	−	−	−	+	+	±
985	−	−	−	−	±	+	−
978	−	−	−	−	+	+	+
301	−	−	−	−	−	+	±
211	−	−	−	−	−	+	−
512	−	−	−	−	−	+	±
920	−	−	−	−	−	+	−
259	−	−	−	−	±	±	±
918	−	−	−	−	−	±	−
917	−	−	−	−	+	+	+
914	−	−	−	−	−	+	±
912	−	−	−	−	−	+	+
550	−	−	−	−	+	+	±
449	−	−	−	−	±	+	−

* Courtesy of D. Ganem.

For instance, the *lac* repressor (*i* gene product) is normally synthesized at low levels in wild-type cells, but at 10-fold higher levels in a mutant strain. Using strains containing this mutation (i^Q), a large series of amber and ochre mutations in the *i* gene have been found (see Table 22B).

Another problem often overlooked is the pattern of suppression. It is advisable to screen a collection of mutants with several different suppressors to avoid missing any strains carrying nonsense mutations which are poorly suppressed by a particular suppressor. Examine Table 22B which shows the pattern of suppression of a collection of nonsense mutations in the *i* gene. How many amber mutants would have been overlooked if just Su2 had been used for screening? How many total nonsense mutations would have been missed if just Su5 had been employed?

Strain Construction

In order to screen a set of strains for the presence of nonsense mutations a method has to be devised to introduce a suppressor gene into each mutant. The easiest way to do this is to utilize specialized transducing phage carrying suppressor genes. Stocks of ϕ80 are available which carry *supF* (Su3) and *supC* (Su_c), respectively. Colonies of the mutants are simply replicated onto a selective plate onto which a lawn of a lysate of one of these phages has been spread. Those cells containing a suppressible mutant will grow under these conditions.

If transducing phage are not available or if different suppressors are to be used, then Hfr crosses can be employed. In this case each starting strain should carry a marker linked to the suppressor gene, for instance his^- for *supD*, gal^- for *supE*, or trp^- for *supF*. However, for this method, Hfr strains carrying the suppressor genes are also required. If the mutations were isolated on an F′ factor, though, then transfer of the F′ factor to different F^-Su^+ strains could be effected.

A powerful technique for isolating Su^+ derivatives of *E. coli* strains, although of less utility for screening large numbers of mutants, is depicted in Figure 30B. This method makes use of a starting strain which already carries an amber mutation, in this case at the *argE* locus. Because of its ability to be suppressed by any of the amber or ochre suppressors, this marker enables us to introduce each of these suppressors into any strain of our choice. The introduction to Experiments 30–32 describes how to transfer this marker into different strains.

In the absence of transduction or conjugation, it is still feasible to convert strains to Su^+ derivatives. If the starting strain carries two different amber mutations, then single-step revertants recovering both activities must be due to a new suppressor (see Experiment 23). Thus, if i^- mutations are selected in a His^-Trp^- background, each i^- mutant can be reverted to His^+Trp^+ and then examined for restoration of i^+ activity (Müller-Hill, 1966).

Temperature-sensitive Suppressors

A mutation resulting in the temperature-sensitive synthesis of a suppressor tRNA (Su3) has recently been isolated (Smith et al., 1970). Here screening can be simplified by selecting mutants at high temperature and then scoring for restoration of wild-type activity at low temperature. Strains such as this can also be used to **select** amber mutants in cases where both selection and counterselection can be applied.

If we were interested in obtaining a series of amber mutations in the *lac* permease structural gene (*y*), we would mutagenize cells and grow at high temperature in the presence of TONPG + IPTG (see Experiment 19) which specifically inhibits the growth of y^+ cells. After a shift to low temperature (at which all mutagen-induced y^- cells, except those carrying temperature-sensitive and suppressed amber mutations, would still be y^-), the cells are grown in lactose medium to select for the growth of y^+ cells. This enriches for the amber mutants at the expense of the other y^- mutants. Finally, the cells are plated on plates containing TONPG + IPTG at high temperature to select for y^- colonies.

Variations of this procedure should facilitate selections for specific nonsense mutations.

References

GANEM, D., J. H. MILLER, T. PLATT and K. WEBER. 1972. *Proc. Nat. Acad. Sci.* In press. Describes isolation and characterization of a large set of amber and ochre mutations in the *i* gene.

MÜLLER-HILL, B. 1966. Suppressible regulator constitutive mutants of the lactose system in *Escherichia coli*. *J. Mol. Biol. 15:* 374.

SMITH, J. D., L. BARNETT, S. BRENNER and R. L. RUSSELL. 1970. More mutant tyrosine transfer ribonucleic acids. *J. Mol. Biol. 54:* 1. Describes the isolation and properties of the *supC* tyrosine suppressor transfer RNA gene.

Strains

Number	Sex	Genotype	Important Properties
CSH61	HfrC	*trpR thi*	Su^- Str^s Nal^s
CSH56	F^-	*ara* Δ(*lacpro*) *supD nalA thi*	Su^+ Pro^- Nal^r
CSH55	F^-	Δ(*lacpro*) *supE nalA thi*	Su^+ Pro^- Nal^r
CSH54	F^-	Δ(*lacpro*) *supF trp pyrF his strA thi*	Su^+ Trp^- Pro^- Pyr^- His^- Str^r

Method

Part A. Screening for Nonsense Mutants

All Lac^- mutants isolated in strain CSH61 (D7011) can easily be screened for response to Su3, a tyrosine-inserting amber suppressor, by transferring the *lac* region from our starting strain into an Su^+ strain. Strain CSH54 carries *supF* (Su3). This strain is also $F^-Lac^-Pro^-Str^r$. CSH61 transfers the *lacpro* region of the chromosome early in a bacterial mating, but the *supF* region very late. Therefore, all CSH61 Lac^- strains can be crossed against CSH54 in a 60-minute interrupted mating.

The *lacpro* region is transferred as a single unit to a *lacpro* deletion strain in an Hfr × F^- cross. Any Pro^+Lac^+ colonies must be due to the presence of the suppressor gene in CSH54 (see Figure 22E). Revertants should appear at too low a frequency to interfere with this test. (Which strain in the collection would you use as a control for revertants in this type of cross?) In order to screen many colonies rapidly, we recommend gridding 50 Lac^- colonies from CSH61 onto a rich master plate. This should be incubated for 6–8 hours at 37° and then replicated onto a lawn of CSH54 (6 drops of a fresh overnight culture spread directly before use) on a minimal lactose plate with streptomycin and the additional required supplements.

It is also possible to select for Pro^+ recombinants only in this type of mating. Instead of lactose streptomycin plates, glucose streptomycin plates are used. The Pro^+ colonies are then directly scored for Lac^+ or Lac^-. It is important for all types of replica matings that freshly growing colonies are gridded onto the master plate.

Part B. Testing the Pattern of Suppression

Lac^- colonies derived from CSH61, which gave Lac^+Pro^+ recombinants with the Su3 recipient, are Lac^- amber mutants. Recover from the master plate these colonies and repurify them. Recall that these are the CSH61(HfrC) copies, and are Lac^- because CSH61 is Su^-. These amber mutants can now be tested with two other amber suppressors, Su1 (CSH56), and Su2 (CSH55). Both the Su1 and the Su2 recipients contain *lacpro* deletions and are Nal^r. HfrC transfers both of these suppressor markers late. Cross the CSH61 Lac^- amber strains with each of these two recipients. Use 60-minute liquid matings as described in Unit II. The matings, **after interruption**, are then plated out on glucose minimal plates with nalidixic acid. The selection is for Pro^+. The Pro^+Nal^r colonies (take three from each

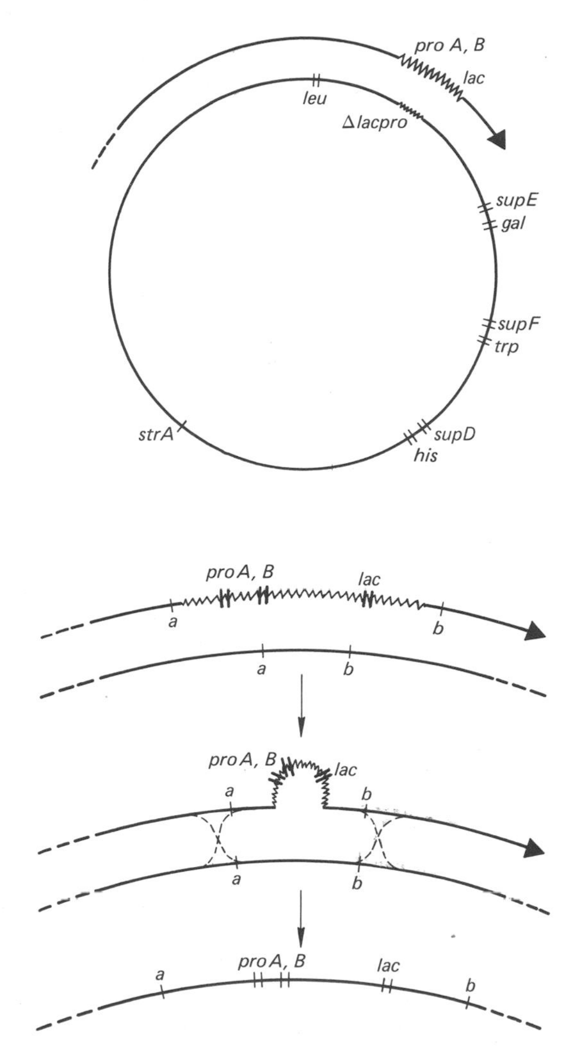

Figure 22E. Testing *lac*⁻ mutations for suppression. The top part of the diagram shows a cross of CSH61 (HfrC) against an F⁻ containing a *lacpro* deletion and any one of several suppressors. It can be seen that this donor transfers the *lacpro* region early, but the suppressor loci (*supD*, *supE*, and *supF*) late. The bottom part of the diagram depicts incorporation of part of the transferred chromosome fragment into the recipient chromosome by means of a double crossover between points of homology on either side of the *lacpro* region.

cross) are then streaked onto lactose minimal or indicator plates along with the corresponding Su3 and Su⁻ derivatives. Compare the degree of suppression by each of the three suppressors. For a more accurate determination of this, β-galactosidase assays should be done, as described in Experiment 48.

Part C. Screening for i⁻ and z⁻ Mutants on an F′*lacpro* Factor

In a similar manner the Lac⁻ and i⁻ mutants derived on F′*lacpro* factors can be screened against different suppressors (see Experiment 19). Prepare a masterplate by picking freshly growing colonies from the GalE⁻Str^s (CSH41) background or from CSH38 onto a rich plate. After 8–12 hours incubation at 37° replicate onto a lawn of the Su3 strain (CSH54) on glucose streptomycin plates, with uracil, tryptophan, and histidine, or onto a lawn of the Su1 strain on glucose nalidixic acid plates. Incubate the selective plates at 37° for 36 hours and then replicate onto the appropriate lactose minimal plates when testing Lac^- mutants, and onto Xgal glucose plates when testing i^- mutants. The i^- mutants, originally selected as blue colonies on Xgal plates, appear white on Xgal glucose when suppressed by a nonsense suppressor.

Materials

Part A **(per 50 colonies tested)**

1 LB plate; overnight culture of CSH54
1 glucose minimal plate with uracil, tryptophan, histidine and streptomycin
1 lactose minimal plate with uracil, tryptophan, histidine and streptomycin
2 replica velvets, replicating block
50 round wooden toothpicks
1 pasteur pipette

Part B **(per strain)**

overnight cultures of CSH55 and CSH56
exponential culture of strain to be tested
2 test tubes
water bath at 37°
vortex
4 dilution tubes
4 glucose minimal plates with nalidixic acid (20 μg/ml)
2 lactose minimal plates or 2 lactose tetrazolium plates
6 0.1-ml, 2 1-ml, and 3 pasteur pipettes

Part C **(per 50 i^- mutants tested)**

i^- colonies in CSH38 background
50 round wooden toothpicks
1 LB plate
2 replica velvets, replicating block
overnight culture of CSH54
1 glucose minimal plate with uracil, tryptophan, histidine and streptomycin
1 Xgal glucose plate with uracil, tryptophan, histidine and streptomycin

(per 50 Lac^- mutants tested)

Lac^- colonies in CSH41 background
50 round wooden toothpicks
1 LB plate; overnight culture of CSH54
2 replica velvets, replicating block
1 glucose minimal plate with uracil, tryptophan, histidine and streptomycin
1 lactose minimal plate with uracil, tryptophan, histidine and streptomycin
1 pasteur pipette

EXPERIMENT 23

Isolation of Nonsense Suppressor Strains

Part A

As discussed in the introduction to this section, some of the revertants of a nonsense mutant are actually strains containing nonsense suppressors. In practice, Su^+ strains are isolated by screening these revertants for the ability to suppress a different nonsense mutation. This is done by introducing a second nonsense mutation into the strain or by testing for the growth of phage (such as T4) containing nonsense mutations.

If we use a strain carrying two different nonsense mutations, then the only single-step revertant for these two lesions will be a strain containing a suppressor. In the following experiment we use a Lac^- Trp^- strain and select spontaneous revertants which are Lac^+ Trp^+. Since both the *lac* and *trp* mutations are nonsense mutations, it is very likely that Lac^+ Trp^+ colonies are Su^+. We then verify this by testing the ability of T4 amber phage to grow on these strains (see Figure 23A).

All of the F^- strains used in Experiment 21 contain, in addition to a nonsense mutation in *z*, an amber mutation in the *trp* operon. For those strains having an amber in both *trp* and *lac* (CSH4), single-step revertants will be either amber or ochre suppressor strains. For those mutants which have an amber mutation in *trp* but an ochre in *lac* (CSH11a), single-step revertants will be solely ochre

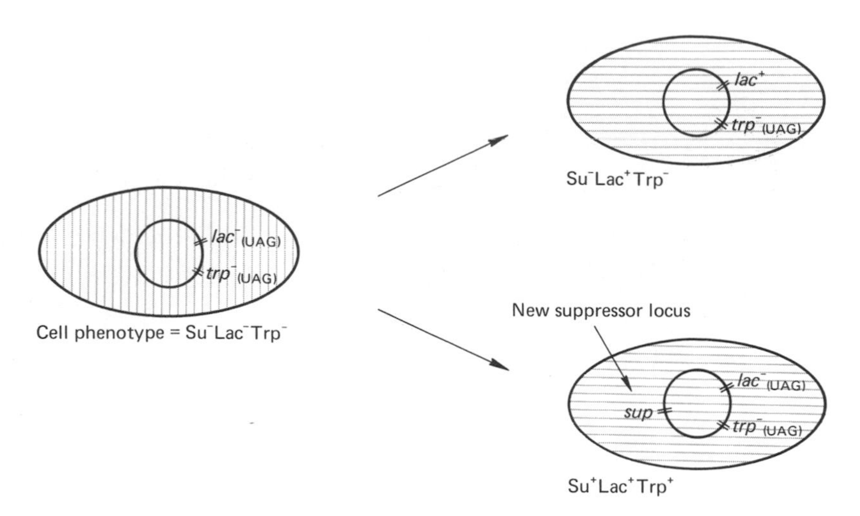

Figure 23A. Isolation of Su$^+$ strains, selection for Lac$^+$ revertants. The cell at the left contains amber mutations in both *lac* and *trp* (Lac$^-$ cells are shaded vertically and Lac$^+$ cells horizontally). Two types of Lac$^+$ revertants are shown at the right. One of them is due to the creation of a new suppressor locus, and the other due to a back reversion at the amber site.

suppressor strains. This is because ochre suppressors act on both amber and ochre mutations, whereas amber suppressors correct only amber mutations. An additional way of verifying that a suppressor is responsible for a particular revertant phenotype is a **back cross** to recover the original mutation, which should still be present. Therefore, we would be able to recover the original *lac* mutation (by crossing onto an F′*lac*, for instance) in revertants with a Lac$^+$ Su$^+$ phenotype, since the genotype is *lac*$^-$ *supE* (see Figure 23B).

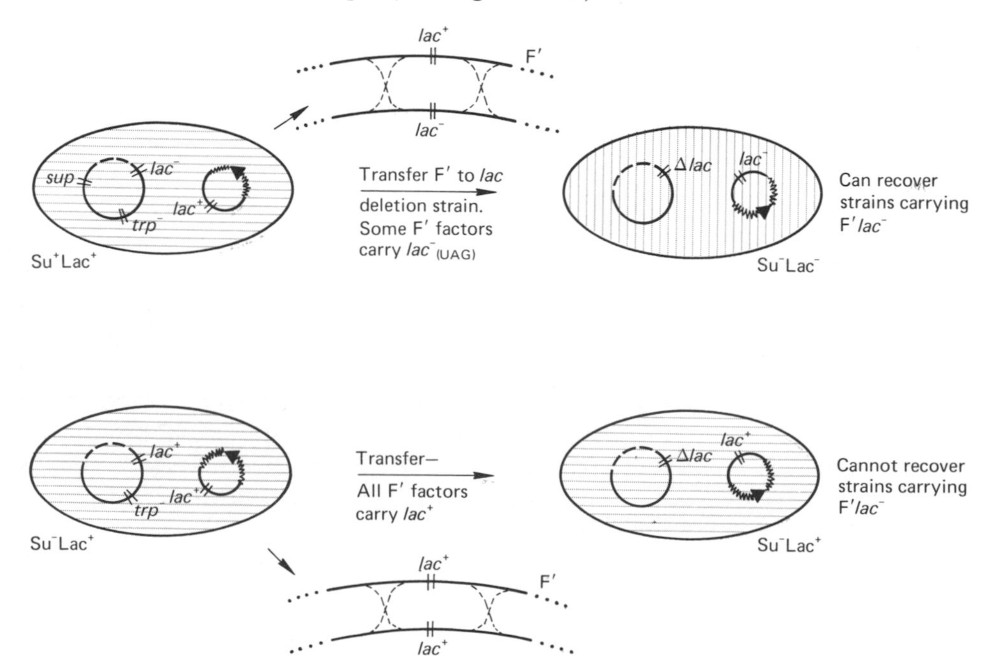

Figure 23B. Backcross. Su$^+$ and Su$^-$ revertants can be distinguished by the presence of the original *lac*$^-$ mutation. This can be recovered onto an F′*lac* which is then transferred to an F$^-$Lac$^-$ strain.

Strains

Number	Sex	Genotype	Important Properties
CSH4	F^-	*trp lacZ strA thi*	carries UAG mutation in *z* and UAG mutation in *trp*
CSH11a	F^-	*trp lacZ strA thi*	carries UAA mutation in *z* and UAG mutation in *trp*

Method

Day 1 Spread 2 drops of a fresh overnight of strain CSH11a (to isolate ochre suppressor strains) and CSH4 (to isolate both ochre and amber suppressor strains) onto different lactose minimal plates. Incubate at 30–32°. (Some of the suppressors will insert an amino acid which results in a temperature labile protein. We can recover a larger spectrum of suppressors by isolating the revertants at low temperature.) Prepare 3 plates for each strain in this manner.

Day 3 After 36 hours several small colonies per plate should be visible. If these cannot be seen, then repeat the procedure using more cells, and continue to incubate the original plates. Purify several revertant colonies onto lactose minimal plates.

Day 5 Prepare broth cultures of one revertant from each plate and when the cells have reached exponential phase ($2–3 \times 10^8$ cells/ml), test with T4 amber phage. Use the starting strains as controls. Testing with T4 amber phage can be done by cross-streaking against a lysate on rich plates (LB or 2YT). Plaque formation can also be observed by plating a 10-minute preadsorption mixture onto rich plates in H-top agar and incubating overnight at 37°. A T4 amber stock will lyse and thus make plaques on some Su^+ strains (amber or ochre) but not on Su^- strains.

Day 6 Examine the plates. Did the Lac^+ Trp^+ revertants plate the T4 amber phage? If the T4 phage were an ochre stock, what results would you expect from each set of revertants? Suppose a strain contains a UGA mutation in *lac*, in addition to a UAG mutation in *trp*. Would you expect any single-step Lac^+ Trp^+ revertants? Amongst Lac^+ Trp^- single-step revertants of the *lac* UGA strain, what percentage would you expect to plate the T4 amber phage? Using the strains described at the beginning of this manual, what experiment would you do to demonstrate that some of these revertants were actually due to a UGA suppressor?

Materials

Day 1 fresh overnight cultures of CSH11a and CSH4
6 lactose minimal plates
2 pasteur pipettes

Day 3 2 lactose minimal plates
16 round wooden toothpicks

Day 5 fresh overnight culture of CSH4
5 test tubes with 2 ml broth each
lysate of T4 amber phage
2 LB plates

Part B

As an additional exercise, we will select both spontaneous and nitrosoguanidine-induced revertants of a Lac^- amber mutant. Strain CSH39 contains a z^- amber mutation, which is carried on an F′ *lacpro* episome. Since the chromosome harbors a *lacpro* deletion, the only *lac* region in this strain is carried on the episome. A Lac^+ revertant can either be due to a mutation in the *z* gene, which converts the amber codon back to wild type or a different sense codon, or else due to an outside suppressor. Only a reversion in the *lac* region will be transferred with the episome to another strain. We then transfer the episome to a $Pro^-Su^-Str^r$ strain and examine individual Pro^+Str^r colonies to test whether they are also Lac^+. Those which are Lac^+ have received an episome which carried a reversion in the *lac* region itself. The Pro^+Str^r colonies which are Lac^- contain an episome which originated from a Lac^+ revertant that was due to an outside suppressor.

As a final confirmation, we will test each suppressor candidate against a different Lac^- amber mutation.

Strains

Number	Sex	Genotype	Important Properties
CSH39	F′*lacZ proA*$^+$,*B*$^+$	Δ(*lacpro*) *thi nalA*	Str^s Nal^r
CSH50	F^-	*ara* Δ(*lacpro*) *strA thi*	Pro^- Str^r
CSH40	F′*lacY proA*$^+$,*B*$^+$	Δ(*lacpro*) *thi*	Str^s

Method

Day 1 **Isolation of revertants of the z$^-$ amber:** Spread 3–4 drops of an overnight culture of strain CSH39 onto two lactose minimal plates. Do this for five different cultures, each arising from a different single colony. Place a crystal of nitrosoguanidine (NG) in the center of one of the plates for each culture. Incubate at 37° for 48 hours.

Day 3 **Repurification of revertant colonies:** Revertant colonies should appear at random over the surface of the plate without NG, but in a circle around the mutagen on those plates with NG. Pick and repurify ten colonies from each plate, by streaking onto lactose minimal plates. Incubate at 37° for 48 hours.

Day 5 Prepare two master plates for replication by gridding 50 colonies each onto two rich plates. Incubate for 6–8 hours at 37°.

Transfer of the F′*lacpro* episome into an Su⁻ background: Replicate the master plates onto a lawn of CSH50 (6 drops of a fresh overnight) on glucose minimal streptomycin plates. Save the master plate and incubate along with the selective plate for 24 hours at 37°.

Day 6 Replicate the glucose minimal plates, with the F′*lacpro* now in the Su^- background, onto lactose minimal plates. Incubate at 37°. an outside suppressor for the amber lac^-, then the amber will

Day 7 Examine the lactose minimal plates. If the original strain contained an outside suppressor for the amber lac^-, then the amber will not be suppressed in the new Su^- strain, and the colonies will be Lac^-; however, if the revertant was at the site of the amber mutation, then the episomal *lac* region will be lac^+ in the new Su^- background. All of the Pro^+ colonies from the original transfer (Day 5) which are Lac^- should be marked. Pick from the original master (containing the Lac^+ revertants of strain CSH39) the colonies corresponding to these positions on the plate, which gave Lac^-Pro^+ colonies in the Su^- strain. These are strains which probably have an outside amber or ochre suppressor, and we will test these further in part C.

Materials

Day 1
5 overnight cultures of strain CSH39, each arising from a different single colony
10 lactose minimal plates
nitrosoguanidine
5 pasteur pipettes

Day 3
20 lactose minimal plates
100 round wooden toothpicks

Day 5
2 LB plates
100 round wooden toothpicks
2 velvets for replication, replicating block
overnight of strain CSH50
2 glucose minimal plates with streptomycin
1 pasteur pipette

Day 6
2 lactose minimal plates
2 velvets

Part C

To verify the presence of a new suppressor in the original strain, we will cure the episome from this strain and then introduce a new F′*lacpro* episome carrying a different lac^- amber mutation. If this diploid also becomes Lac^+, then this confirms the isolation of a nonsense suppressor strain.

Method

Day 1 **Curing the episome from the Lac$^+$ revertants of strain CSH39:** Episome curing is accomplished by growth in the presence of acridine orange (see Experiment 11). Inoculate 1000–2000 cells (0.1 ml of a 10^{-5} dilution of a fresh overnight) into 5 ml LB broth pH 7.6 containing 75 μg/ml acridine orange. Grow overnight in the dark until the cells are at least several times 10^8/ml.

Day 2 Plate dilutions of the acridine orange overnights onto lactose tetrazolium plates and incubate at 37°. Plate as a control the untreated strain.

Day 3 Observe the plates. There should be predominantly Lac$^-$ colonies on the plates from the acridine orange treated cultures, and only Lac$^+$ colonies on the plates from the control. Pick and purify several Lac$^-$ colonies. These have lost the F′*lacpro* factor.

Day 4 Prepare an overnight from a purified Lac$^-$ from Day 3. Also prepare an overnight of strain CSH40.

Day 5 Subculture the donor (CSH40) into broth and allow to grow to a density of 2–4 × 10^8 cells/ml. Do likewise for the recipients. Prepare a mating mixture by mixing 0.5 ml of each, both donor and recipient, in a test tube and place on the outside ring of a 30 rpm rotor in a 37° room for 30 minutes. Plate 0.1 ml of a 10^{-2} and also of a 10^{-4} dilution on glucose minimal plates with nalidixic acid. The selection is for Pro$^+$ and nalidixic acid is used to counterselect. Streak both donor and recipient alone on a selection plate as a control.

Day 6, 7 Purify the Pro$^+$ colonies from the preceding cross. These have received an F′*lacpro* episome carrying a *lac*$^-$ amber mutation.

Day 8 Test the purified Pro$^+$ colonies for growth on lactose. If these strains actually contained a suppressor, they should also suppress the new amber mutation.

Materials

Day 1

overnight of Su$^+$ candidate
3 dilution tubes
stock solution of acridine orange (7.5 mg/ml) made up directly prior to use in sterile H_2O
4 0.1-ml and 1 1-ml pipettes
5 ml LB broth, pH 7.6; empty test tube

Day 2

2 overnight cultures of Su$^+$ candidate, one of which has been grown up in acridine orange
8 dilution tubes
6 0.1-ml, 2 1-ml, and 6 pasteur pipettes
6 lactose tetrazolium plates

Day 3
2 lactose tetrazolium plates
12 round wooden toothpicks

Day 4
strain CSH40
cured Lac^- colony from Day 3

Day 5
2 overnights from Day 4
2 tubes with 5 ml fresh LB broth
empty tube
waterbath at 37° or rotor at 30 rpm in 37° room
4 0.1-ml, 2 1-ml, and 2 pasteur pipettes
3 glucose minimal plates with nalidixic acid
2 dilution tubes

Day 7
2 glucose minimal plates with nalidixic acid
12 round wooden toothpicks

Day 8
2 lactose minimal plates
12 round wooden toothpicks

EXPERIMENT 24

Detecting Polar Mutants in the *lac* Operon

Mutations which lower the amount of enzymes synthesized by distal genes of an operon are termed polar mutants. These generally result from the generation of a nonsense codon in the middle of a gene. Suppressors for chain termination reverse the polarity effects of these mutations. Several factors are important in determining the degree of polarity of a given nonsense mutation. The distance between the site of the mutation and the next polypeptide initiation site is a critical factor, as is the efficiency of that initiation site. Generally this results in nonsense mutations early in a gene being more polar than those which occur late in the same gene. It has also been suggested that degradation of mRNA is a cause of polarity. In fact, a suppressor of polarity (SuA) which restores *y* and *a* gene activity to polar mutants with nonsense mutations in *z* (Scaife and Beckwith, 1966) has been shown to protect against the loss of mRNA that occurs after premature chain termination in the *trp* operon (Morse and Primakoff, 1970).

Internal Reinitiation Sites

It has been shown that efficient polypeptide re-initiation sites also occur in the middle of genes. One recent example is provided by studies of amber and ochre mutations in the *i* gene of the *lac* operon. Figure 24B depicts the early part of the *i* gene and its protein product, the *lac* repressor. Here early nonsense

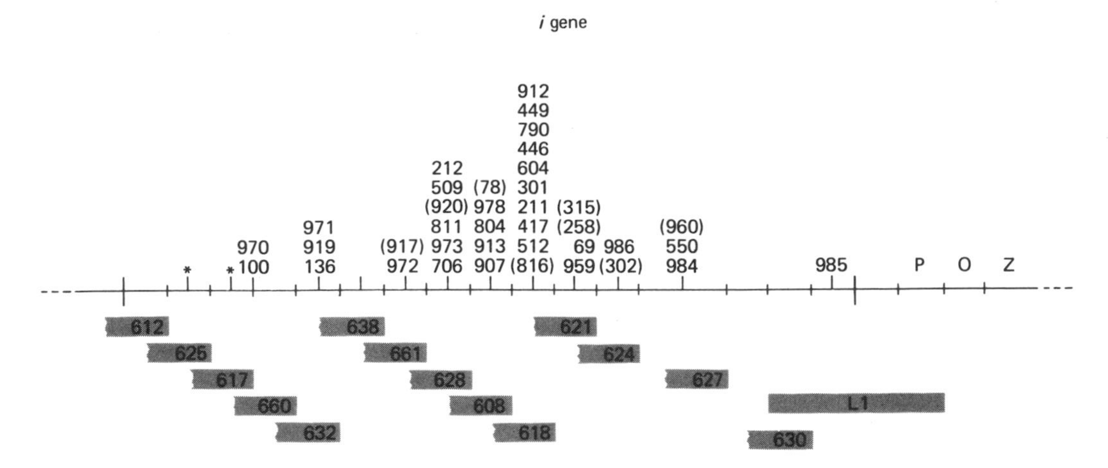

Figure 24A. Nonsense mutations in the *i* gene. Thirty-eight amber and ochre mutations in the *i* gene have been mapped against a set of 15 deletions (Platt et al., 1972). The position of the markers in parentheses has not been precisely determined. The stars represent the locations of several frameshift mutations isolated by D. Gho. The relative order of these deletions has also been independently determined by M. Pfahl. It can be seen that the amber mutation 100 maps relatively early. The *i* gene is transcribed from left to right as drawn in the diagram (Miller et al., 1968; Kumar and Szybalski, 1969).

mutations which map prior to an internal translation reinitiation site (π in Figure 24B) result in the synthesis of protein fragments missing the amino terminal region of the repressor. The particular AUG codon in this region of the message is clearly able to mimic the initiation signal at the beginning of the message.

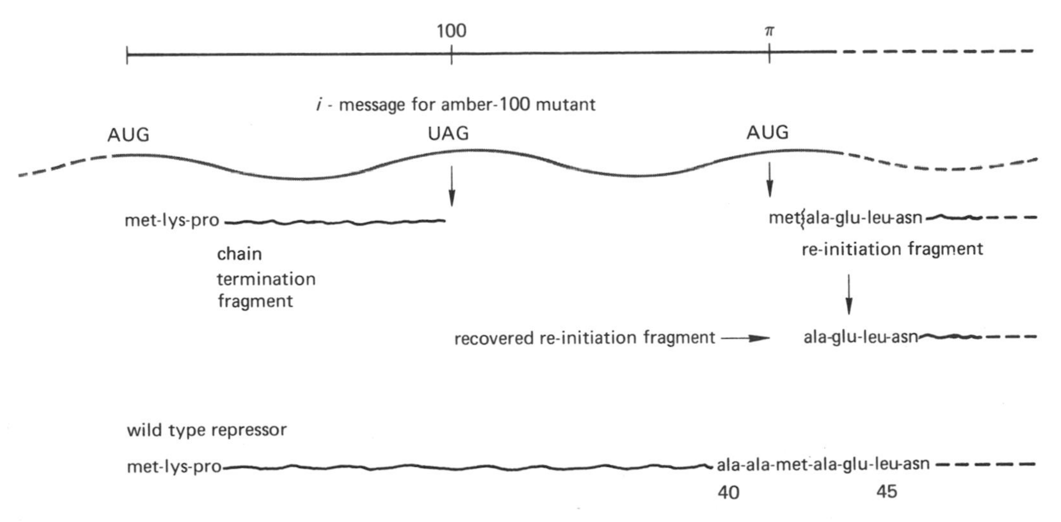

Figure 24B. Identification of a restart fragment. The messenger RNA from the amber mutant i^{100} is shown above. Translation begins at an AUG codon and proceeds as normal up to the amber block, at which point chain termination occurs. Reinitiation takes place at the next AUG which in this case codes for the methionine residue at position 42 in the wild-type protein (see bottom of figure). The recovered fragment is missing the amino-terminal end of repressor, and begins with the sequence ala-glu-leu-asn, corresponding to amino acids 43–46 of the wild-type repressor. The methionine residue is presumed to be cleaved by an amino-peptidase (Platt et al., 1972).

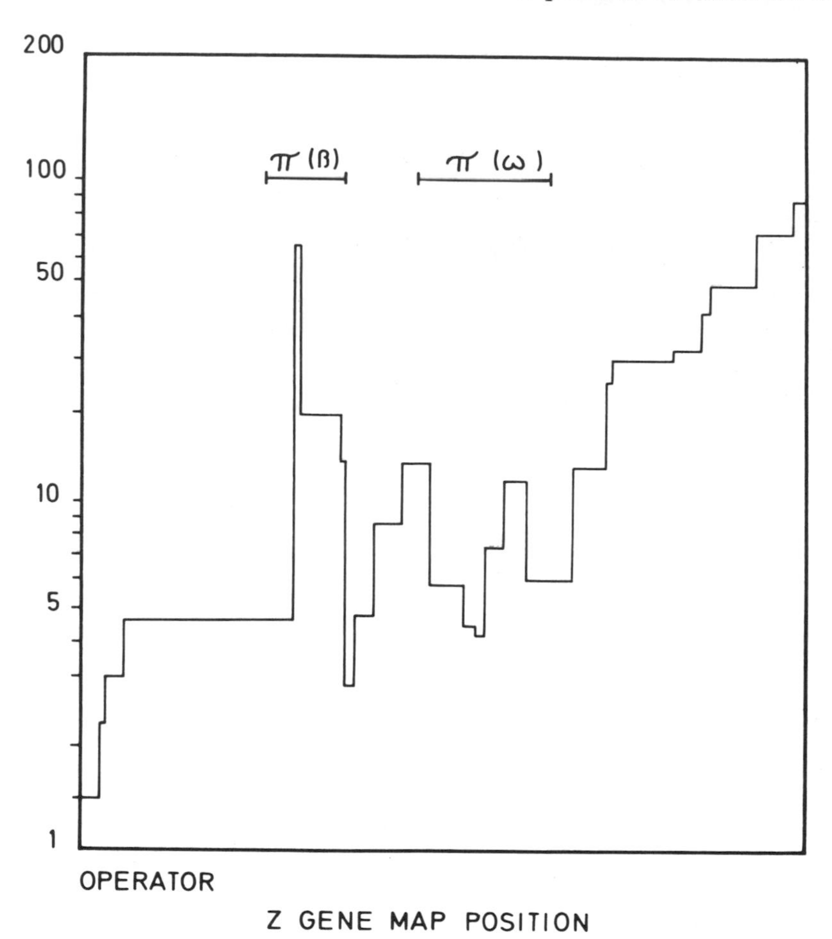

Figure 24C. This graph (from Zipser et al., 1970) plots percent transacetylase *vs* map position for different nonsense mutations in the *z* gene. The π regions indicate breaks in the polarity gradient. One interpretation of the drop after each reinitiation point is that the nonsense mutations in this region are too far away from reinitiation points to allow efficient re-starting of translation.

Although the function of these sites is not well understood, it is evident that these complicate attempts to correlate the map position of nonsense mutations with the degree of polarity. This type of study has been done extensively in the *lac* operon. Figure 24C shows the polarity gradient in the *z* gene. One can clearly distinguish three peaks in the gradient, corresponding to internal polypeptide reinitiation signals. This was done by assaying for transacetylase in strains with different z^- nonsense mutations.

Alternatively, it is possible to assay for permease. Levels of permease can also be estimated by scoring for growth on melibiose at 42°. Cells able to synthesize 6–8% of the wild-type level of permease (when fully induced) can utilize melibiose at high temperature. By characterizing z^- nonsense mutants as good, moderate, and poor growers on melibiose at 42°, the degree of polarity can be estimated.

References

Kumar, S. and W. Szybalski. 1969. Orientation of transcription in the *lac* operon and its repressor gene in *Escherichia coli*. *J. Mol. Biol. 40:* 145.

Miller, J. H., J. R. Beckwith and B. Müller-Hill. 1968. Direction of transcription of a regulatory gene in *E. coli*. *Nature 220:* 1287.

MORSE, D. and P. PRIMAKOFF. 1970. Relief of polarity in *E. coli* by suA. *Nature 226:* 28.
PLATT, T., K. WEBER, D. GANEM and J. H. MILLER. 1972. Translation restarts: AUG re-initiation of a *lac* repressor fragment. *Proc. Nat. Acad. Sci.* in press.
SCAIFE, J. G. and J. R. BECKWITH. 1966. Mutational alteration of the maximal level of *lac* operon expression. *Cold Spring Harbor Symp. Quant. Biol. 31:* 403.
ZIPSER, D. 1970. Polarity and translational punctuation. *The Lactose Operon,* p. 221. Cold Spring Harbor Laboratory.
ZIPSER, D., S. ZABELL, J. ROTHMAN, T. GRODZICKER and M. WENK. 1970. Fine structure of the gradient of polarity in the *z* gene of the *lac* operon of *Escherichia coli*. *J. Mol. Biol. 49:* 251.

Method

In this experiment, all of the nonsense mutants used for mapping in Experiment 21 will be tested for growth on melibiose at 42°. All Lac$^-$ mutants isolated in previous exercises will also be tested, to estimate the amount of residual *y* gene function. Pick a single colony from a freshly grown LB plate, or streak with a loop from a fresh overnight culture onto melibiose supplemented with tryptophan or proline when necessary. (All nonsense mutants from Experiment 21 require tryptophan, and all mutants derived in CSH51 require proline.) Streak each strain onto a section of two plates, incubating one at 30° and the other at 42°. Observe the plates after 36 and 48 hours. Does the amount of growth on melibiose at 42° have any relation to the map position for a given nonsense mutant?

Materials

Lac$^-$ mutants isolated in Experiments 13–19 and strains CSH1 to 12, freshly growing on LB plates or glucose minimal plates
melibiose minimal plates (containing proline or tryptophan when necessary)
incubator at 40–42°

Additional Exercises

1. Measure the levels of transacetylase in these nonsense mutants and determine a polarity gradient from this data. Procedures for acetylase assays are given in Experiment 52.

2. Some of the amber mutants in the F$^-$Trp$^-$Str^r background do not grow on melibiose at 42°. Plate out a lawn of cells on melibiose tryptophan plates and incubate at 42° to select revertants. Test the revertants to determine whether they are still z$^-$ (Lac$^-$). Some of the z^-y^+ strains are due to outside mutations. Others are due to internal *z* deletions which bring the nonsense mutation closer to the *y* boundary. A third class fuses the *z* gene to an outside operon. How can you distinguish between these three classes of revertants with the strains available?

Unit III Questions

The amino acid sequence of extracellular ribonuclease (barnase) of *B. amyloliquefaciens* (Hartley, R. W. and E. A. Barker, Nature New Biol. 235, 15, 1972) is given below.

10 20
Ala-Gln-Val-Ile-Asn-Thr-Phe-Asp-Gly-Val-Ala-Asp-Tyr-Leu-Gln-Thr-Tyr-His-Lys-Leu-

30 40
-Pro-Asn-Asp-Tyr-Ile-Thr-Lys-Ser-Glu-Ala-Gln-Ala-Leu-Gly-Trp-Val-Ala-Ser-Lys-Gly-

50 60
-Asn-Leu-Ala-Asp-Val-Ala-Pro-Gly-Lys-Ser-Ile-Gly-Gly-Asp-Ile-Phe-Ser-Asn-Arg-Glu-

70 80
-Gly-Lys-Leu-Pro-Gly-Lys-Ser-Gly-Arg-Thr-Trp-Arg-Glu-Ala-Asp-Ile-Asn-Tyr-Thr-Ser-

90 100
-Gly-Phe-Arg-Asn-Ser-Asp-Arg-Ile-Leu-Tyr-Ser-Ser-Asp-Trp-Leu-Ile-Tyr-Lys-Thr-Thr-

110
-Asp-His-Tyr-Gln-Thr-Phe-Thr-Lys-Ile-Arg

1. Consult Figure 22D and determine which sites in the protein can be converted to amber, ochre, and UGA in a single step.

2. At which sites would you expect amber, ochre, or UGA mutations induced by 2-aminopurine, EMS, and by the mutator gene, *mutT*?

3. Using the nonsense suppressors in Table 22A, derive a two-step method for obtaining a mutant which inserts tryptophan in place of glutamine at position 15.

4. Draw the RNA sequence for the coding for amino acids 10–20 in the above chain. Suppose that a single base (G) is inserted before the codon for valine at position 10, and a single base in the codon for proline at position 21 (the first C) is deleted as the result of frameshift mutagenesis, done in two steps. Will the resulting protein give enough information to enable you to unambiguously determine the wild-type RNA sequence for that region?

5. If two nonsense mutations map in the same deletion group, but one is found to be an ochre and one to be an amber, does that prove that these mutations are at different sites?

6. Design a procedure which enables you to directly select for i^- amber mutants.

7. In Table 22B, the amber mutation 959 is not suppressed by Su_B, which inserts glutamine. Does that prove that glutamine was not the amino acid at that point in the wild-type protein?

8. Design a selection for a mutant carrying a temperature-sensitive suppressor.

9. Suppose you wanted to devise a selection for mutants in the translation initiation codon, AUG, for the *lac* operon. Which selections in this unit would you employ and why? Is a double selection necessary (for both z^- and y^-)? What other structural gene mutations would result in the same phenotype? Which

of the mutagens discussed in the introduction to this unit would reduce the number of structural gene mutations of this type?

10. How might you decode the translation termination signal at the end of the *z* gene in the *lac* operon or at the end of the *trpA* gene?

11. Using lactose and nitrophenyl-thiogalactoside (see Experiment 19), devise a procedure which enriches for temperature-sensitive *lac* permease mutants. What special properties would a galactoside have to have to allow its use in a similar enrichment procedure for temperature-sensitive z^- (β-galactosidase) mutants?

12. What types of revertants would you expect from a temperature-sensitive z^- mutant? From a temperature-sensitive i^- mutant?

13. Suppose you streaked a single colony containing 10^7 cells, and also a loop containing 10^7 cells from a culture grown overnight onto a plate selecting for i^- and o^c mutants. Are both procedures identical in this case? Which would give the higher percentage of spontaneous o^c mutants and why? How would using i^Q affect the percentage of each class of mutant?

14. If a mutation does not revert spontaneously, what information does this really provide? Does failure to revert always mean the same thing? What are the actual reversion rates expected for amber mutants in bacteria? Can a point mutation revert with a frequency which would yield less than one revertant per 10^8 cells in a population? Less than one revertant per 10^9 cells?

UNIT IV

STRAIN CONSTRUCTION

INTRODUCTION TO UNIT IV

Strain construction is the preparation, through a series of genetic manipulations, of strains which carry certain combinations of mutations. The importance of proper strain construction cannot be overstated, for it makes possible many biochemical experiments which would not otherwise be feasible. In Unit II we were introduced to the concepts of **selection** and **counterselection** in bacterial matings. In the crosses described in that section, each donor could be counter-selected against and each recipient or recombinant selected for. This was possible because each parent strain carried just the right set of markers. Unfortunately this is not always the case, and more often than not we find that the donor and recipient strains we have available contain no markers which enable us to apply selection or counterselection. There are two basic approaches to this situation. The first is used often in F′ factor transfers (Experiments 25 and 26) and involves plating a mating mixture on either partially selective or non-selective medium and then examining individual colonies for parental markers and the presence of the F′ factor. The second approach requires the preparation of derivatives of the parent strains so that they now carry markers which can be used for selection. For instance, **drug-resistant mutants** of the donor and recipients containing nutritional requirements can be isolated easily. This is described in Experiments 30–32.

Specific selections (Experiments 30–31) exist for some **nutritional markers**, such as Thy^-, Trp^-, and His^-. Other auxotrophs, however, are isolated only after enrichment procedures such as **penicillin treatment** (Experiment 33). This unit also describes other aids to strain construction in current use. **F′ factors** which are isolated from Hfr strains (Experiment 37) are valuable tools. Specific markers can be transferred from one strain to another by using merodiploids (Experiment 38), and strains which are not Hfr's still can donate chromosomal markers when carrying an F′ factor (Experiment 29). It is possible, moreover, to convert most F^- strains to Hfr strains (Experiment 36). In addition, this section describes procedures for the use of the **generalized transducing phage**, Pl (Experiment 28).

EXPERIMENT 25

F′ Factor Transfers—Inability to Select for Recipient

A problem often faced in strain construction is that of transferring an episome from one strain to another without being able to select directly for the desired diploid. The first experiment in this series deals with this situation. Let us consider the transfer of an F′*lac*$^+$*pro*$^+$ from strain CSH28 (Trp$^-$Ura$^-$His$^-$) to strain CSH25 (wild type), pictured in Figure 25A. We can select against the donor in this case by plating the mating mixture on glucose minimal plates without uracil, histidine, and tryptophan. However, all of the prospective recipients (CSH25) will grow whether they receive the episome or not. In certain cases, it would be feasible to isolate a Lac$^-$ derivative of the F$^-$ strain, and then directly select for transfer of the Lac$^+$ character. Often this is not desirable, however. Instead, we carry out a mating and score resulting colonies for the presence of the F′ factor. Here the mating mixture should contain a high multiplicity of donor to prospective recipient (10-20:1) to increase the chances that a large percentage of the CSH25 population will receive the F′ factor. Testing for the presence of the F factor can be done by screening for sensitivity to male-specific phage as demonstrated in Experiment 9.

Alternatively, we can test for the ability to donate the *lac*$^+$ character (Figure 25B) to determine which of the CSH25 cells carry the F′*lac*$^+$*pro*$^+$. For this procedure freshly growing single colonies are picked onto a rich master plate which is

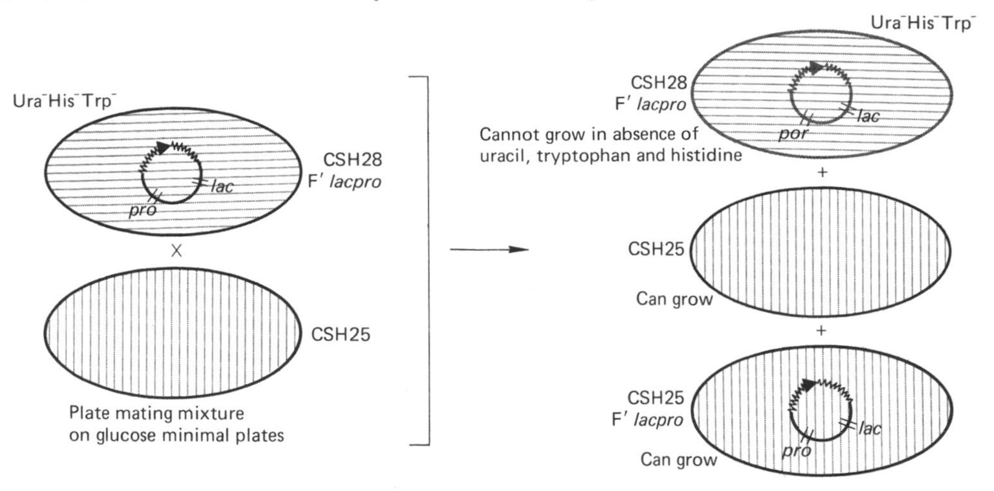

Figure 25A. Cross of CSH28 × CSH25. An F′*lacpro* factor is transferred from CSH28 background to the CSH25 background. CSH28 cells are Ura$^-$ His$^-$ Trp$^-$ and are shaded horizontally. CSH25 cells are Ura$^+$ His$^+$ Trp$^+$ and are drawn with vertical shading (see text).

then incubated to prepare for replica plating. After 8 hours of growth, the master plate is replicated onto a lawn of an F$^-$Lac$^-$Str^r strain (CSH7) on a lactose minimal streptomycin plate. The selection is for Lac$^+$, the donors being killed by the streptomycin. Only the areas of the selection plate onto which an F′*lac* donor has been replicated will show growth after 24 hours. From the master plate all of the colonies containing cells which had received the F′ factor in the original cross can be recovered.

Strains

Number	Sex	Genotype	Important Properties
CSH28	F′*lac*$^+$*proA*$^+$,*B*$^+$	Δ(*lacpro*) *supF trp pyrF his strA thi*	Ura$^-$ His$^-$ Trp$^-$
CSH25	F$^-$	*supF thi*	Str^s
CSH7	F$^-$	*lacY strA thi*	Lac$^-$ Str^r

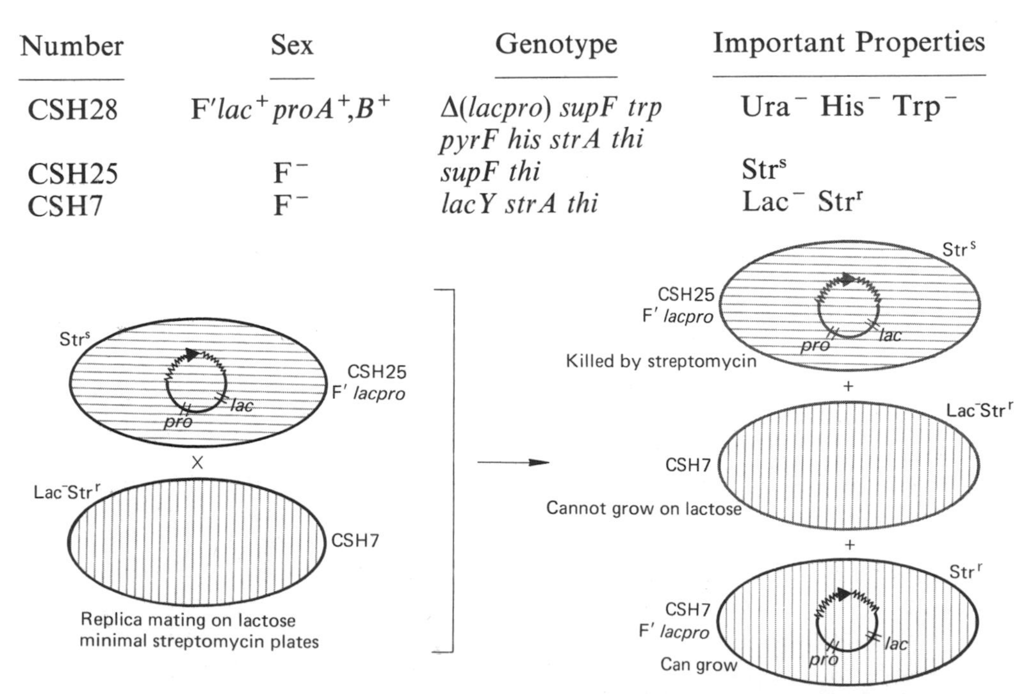

Figure 25B. Replica mating to test for CSH25 cells carrying F′*lacpro* factor. An F′*lacpro* factor is transferred from the CSH25 background to the CSH7 background. CSH25 cells are Str^s and are drawn with horizontal shading in this diagram. CSH7 cells are Str^r and are indicated by vertical shading.

Method

Transfer of F′*lacpro* from CSH28 to CSH25

Day 1

Prepare a mating mixture by mixing donor and recipient in a large test tube. Place the test tube on the outside ring of a 30 rpm rotor at 37° for 60 minutes. Use a 10:1 ratio of donor to recipient by mixing 1.0 ml of CSH28 with 0.1 ml of CSH25. Use exponentially growing cultures at 2–3 × 10^8 cells/ml for both donor and recipient. A total volume of 4–6 ml in a 100-ml or 125-ml flask can also be used. The flask should be gently shaken. Plate 0.1 ml of a 10^{-3}, 10^{-4}, and 10^{-5} dilution onto glucose minimal plates. Incubate at 37° for 30–36 hours. As a control streak the donor alone on a similar plate.

Day 2, 3

Using sterile round toothpicks, prepare a master plate by picking isolated single colonies from the glucose minimal plate onto a rich plate. It is possible to pick 50 colonies per plate. Prepare two YT or LB plates and incubate for 6–8 hours at 37°. Also prepare a culture of strain CSH7 (Lac^-Str^r).

Spread 6 drops of the culture of CSH7 onto each of two lactose minimal streptomycin plates and allow to dry. Replicate the master plate onto a sterile velvet, and then replicate onto the lactose minimal plate (onto which strain CSH7 has been spread). Do this for each master plate. Incubate at 37°. After 24–36 hours one can easily distinguish which colonies donated an F′*lac* episome in this test, and which did not.

Materials

Day 1

1 ml exponentially growing culture of strains CSH28 and CSH25
test tube
4 dilution tubes
waterbath at 37° or preferably 30 rpm rotor drum in 37° room
6 0.1-ml, 3 1-ml pipettes
4 glucose minimal plates

Day 2, 3

1 ml of a fresh overnight of strain CSH7
2 rich plates (YT or LB)
100 round wooden toothpicks
2 velvet pads and replicating block
2 lactose minimal plates with streptomycin
1 pasteur pipette

EXPERIMENT 26

F′ Factor Transfers—Inability to Select against Donor

Another problem often encountered in strain construction is that of crossing an F′ factor from one strain into another but not being able to select against the donor. First we consider the case in which it is possible to select for the females which have received the F′ factor, and then the case in which it is not.

Suppose we wish to transfer an F′*lac*$^+$*pro*$^+$ from CSH23 to CSH26 (Ara$^-$Pro$^-$Lac$^-$). We can select those CSH26 cells which receive this F′ in a cross by plating on minimal plates without proline. However, there are no markers in CSH23 (the donor) which we can use for counterselection. One way to get around this problem is to isolate a drug-resistant derivative of the recipient (CSH26), for instance nalidixic acid-resistant or streptomycin-resistant (see Experiment 32). Then we could plate cells on minimal plates containing nalidixic acid or streptomycin and select against the donor. Sometimes this is not desirable, however. It is possible, though, to carry out a mating and examine colonies for the presence of both the episome and parental markers. If in the mating mixture a large multiplicity of recipient to donor is used, the chances are high that each donor cell will mate with a different recipient. In the ideal case, the number of donor colonies should equal the number of CSH26 colonies which have received an F′ factor. The diploids are then scored for a parental marker, in this case Ara. The Ara$^-$ colonies are F′*lac*$^+$*pro*$^+$/CSH26 (see Figure 26A).

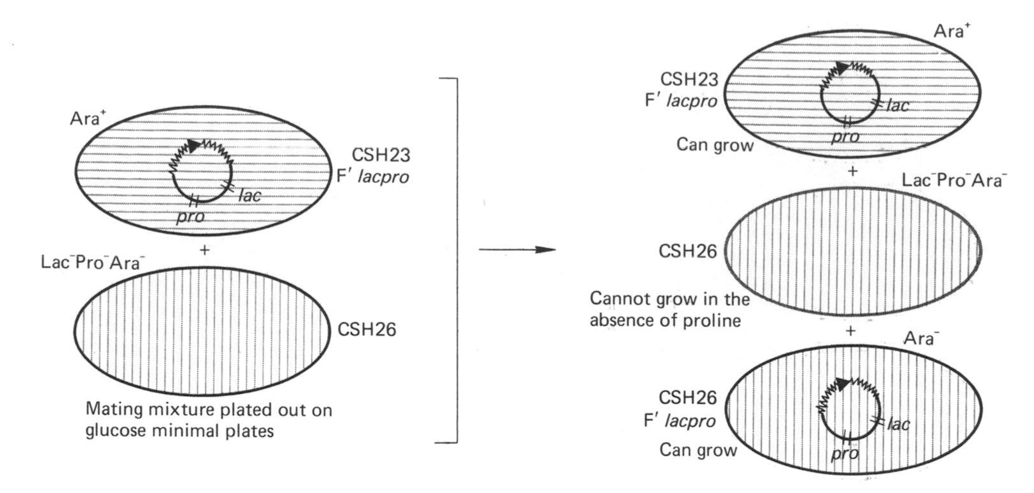

Figure 26A. Cross of CSH23 × CSH26. An F′*lacpro* factor is transferred from the CSH23 background to the CSH26 background. CSH23 cells are Ara$^+$ and are indicated by horizontal shading. CSH26 cells are Ara$^-$ and are shown with vertical shading (see text).

In the cross of CSH23 with CSH27 it is not possible to select for the recipients or against the donor (Figure 26B). Therefore, a mating mixture (approximately 1 : 1) should be plated out for single colonies, which are then scored for both parental markers and the ability to donate the episome (or sensitivity to male phage). Sometimes the colony morphology of the two strains differs enough to enable one to distinguish between them. If this is not the case, it is better to plate the mating mixture on rich plates and pick and grid colonies to replicate for Trp$^-$ (in this example) and also the ability to donate an F′*lac* episome. Sometimes this can be done with the same replica grid.

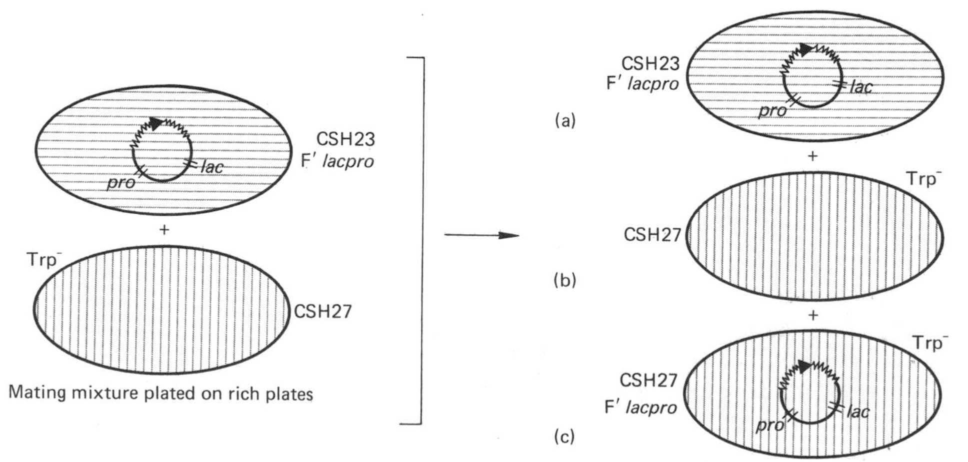

Figure 26B. Cross of CSH23 × CSH27. An F′*lacpro* factor is transferred from the CSH23 background to the CSH27 background. CSH23 cells are Trp$^+$ and are indicated by horizontal shading. CSH27 cells are Trp$^-$ and are shown with vertical shading (see text).

Strains

Number	Sex	Genotype	Important Properties
CSH23	F′*lac*$^+$ *proA*$^+$,*B*$^+$	Δ(*lacpro*) *supE spc thi*	Ara$^+$
CSH26	F$^-$	*ara* Δ(*lacpro*) *thi*	Pro$^-$ Ara$^-$

Method

Transfer of F′*lacpro* from CSH23 to CSH26

Day 1

Mix 0.1 ml of a fresh log culture of CSH23 with 1.0 ml of a log culture of CSH26 in a large test tube and place at 37° on a 30 rpm rotor for 60 minutes. The ratio of donor to recipient (23:26) should be 1:10.

After 60 minutes remove the test tube from the waterbath. Plate out 0.1 ml of both a 10^{-4} and a 10^{-5} dilution onto two glucose minimal plates. Also plate out 0.1 ml of a 10^{-2} dilution of CSH26 as a control. Incubate at 37° for 36 hours.

Day 3

Record the number of colonies appearing on each plate and pick about 100 onto two rich plates as a master plate for replication. Incubate at 37°. After about 12–15 hours the plates will be ready for replication to test for the Ara$^-$ phenotype. It is permissible for the plate to incubate longer.

Day 4

Replicate the Pro$^+$ colonies from the master plate onto an arabinose tetrazolium plate and incubate this at 37°.

Day 5

Record the percentage of Pro$^+$ colonies which were Ara$^-$.

Materials

Day 1

1 ml log culture of CSH23 at about 2 × 10^8 cells/ml
1 ml log culture of CSH26 at 2–5 × 10^8 cells/ml
1 test tube
rotor in 37° room or shaking waterbath at 37°
4 dilution tubes
7 0.1-ml and 2 1-ml pipettes
3 glucose minimal plates
vortex

Day 3

100 round wooden toothpicks
2 LB plates

Day 4

2 velvet pads with replicating block
2 arabinose tetrazolium plates

EXPERIMENT 27

Construction of a Temperature-sensitive Episome

It is often desirable to use episomes which can be easily cured from the host strain. One procedure is to employ acridine orange (see Experiment 11). Another technique makes use of temperature-sensitive F′ factors. In this case we can isolate segregant clones merely by streaking several times at high temperature (40–43°). These F′ factors carry a mutation in an F factor gene which renders the episome unable to replicate at high temperature. This temperature-sensitive replication marker can be crossed from one F′ factor to another, just as in a cross between two bacterial chromosomes. To allow recombination between F′ factors, we must bring about a situation in which the two episomes coexist in the cell, at least temporarily. Then we score for the desired recombinant episome.

Normally a male strain will not accept another episome in a bacterial cross because of the surface exclusion which prevents pairing between two males. To overcome this block we make **phenocopies** of the male strain which is to act as recipient (Experiment 10). An effective method is the aeration of saturated cultures at low temperature for a long time to increase the proportion of cells in the population which have lost their pili and cell surface antigens and not reformed them.

Even after an F′ factor is transferred to a male strain, the two F factors cannot

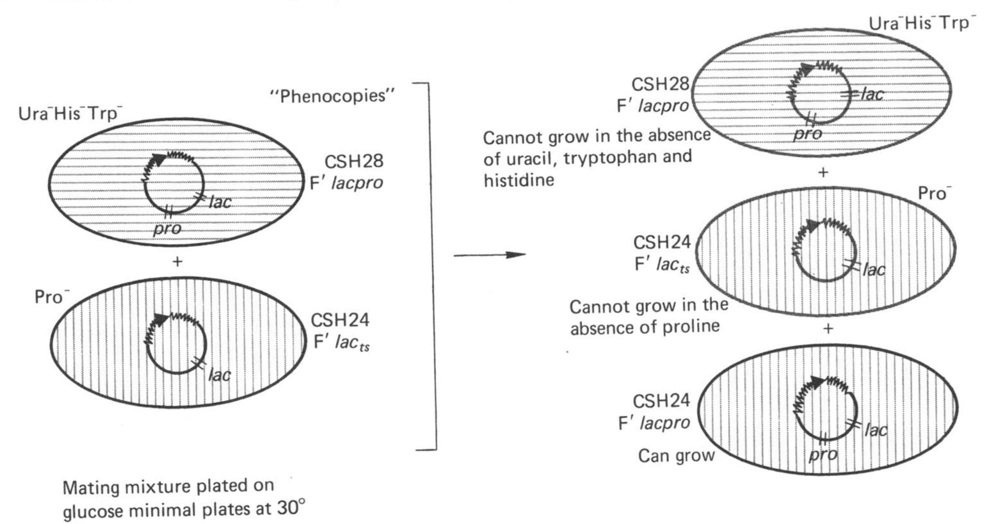

Figure 27A. Cross of CSH28 × CSH24. The CSH28 background (Ura$^-$His$^-$Trp$^-$) is represented by horizontal shading, and the CSH24 background (Ura$^+$His$^+$Trp$^+$) is depicted by vertical shading. Both strains, in the absence of an F′*lacpro* factor, are Pro$^-$. After a cross of these two strains, only the CSH24 cells which have received an F′*lacpro* factor will form colonies on glucose minimal medium. Some of these will have had a recombination event between the F′*lacpro* factor and the F'_{ts}*lac* factor. These will be temperature sensitive for both the Lac$^+$ and the Pro$^+$ character (see text).

coexist in a wild-type *E. coli*, since one F factor restricts the replication of the other. Some of the cells with the two F′ factors will throw off one of the F′ factors, and some will lose the other. After a cross involving phenocopies, we can isolate males which have lost the original F factor and retained the newly transferred one.

During the period of time in which both episomes coexist in the cell, recombination events between episomal genes can take place. The experiment in this section involves a cross of strain CSH28 (F′*lac*$^+$*pro*$^+$/*trp his pyrF*) with CSH24 (F'_{ts} *lac*$^+$/Δ*lacpro*). We wish to convert the F′*lac*$^+$*pro*$^+$ to a temperature-sensitive (ts) F'_{ts} *lac*$^+$*pro*$^+$. Strain CSH24 is to be the recipient in this cross, so phenocopies are made of this strain. The F′*lac*$^+$*pro*$^+$ is transferred from CSH28 to CSH24 at 30° selecting for His$^+$ Ura$^+$ Trp$^+$ Pro$^+$ (Figure 27A). Only the CSH24 cells which have lost the F'_{ts} *lac*$^+$ and gained the F′*lac*$^+$*pro*$^+$ will form colonies on glucose minimal plates.

Directly prior to the mating there will be in the population of CSH24 cells both phenocopies and some cells which have lost the F'_{ts}*lac*$^+$. F′*lac*$^+$*pro*$^+$ factors which are transferred to this latter class will not have an opportunity to recombine with the temperature-sensitive episome. Therefore, it is important to create as many phenocopies as possible. The colonies which grow up after this cross at 30° are then replicated onto glucose minimal plates at 30° and 43° to test for temperature sensitivity of the proline requirement. Several hundred should be tested since the observed frequency of conversion to sensitivity to high temperature in crosses of this type is on the order of 1%.

References

Gottesman, S. and J. R. Beckwith. 1969. Directed transposition of the arabinose operon: A technique for the isolation of specialized trnasducing bacteriophages for any *Escherichia coli* gene. *J. Mol. Biol. 44:* 117.

Jacob, F., Brenner, S. and F. Cuzin. 1963. On the regulation of DNA replication in bacteria. *Cold Spring Harbor Symp. Quant. Biol. 28:* 329.

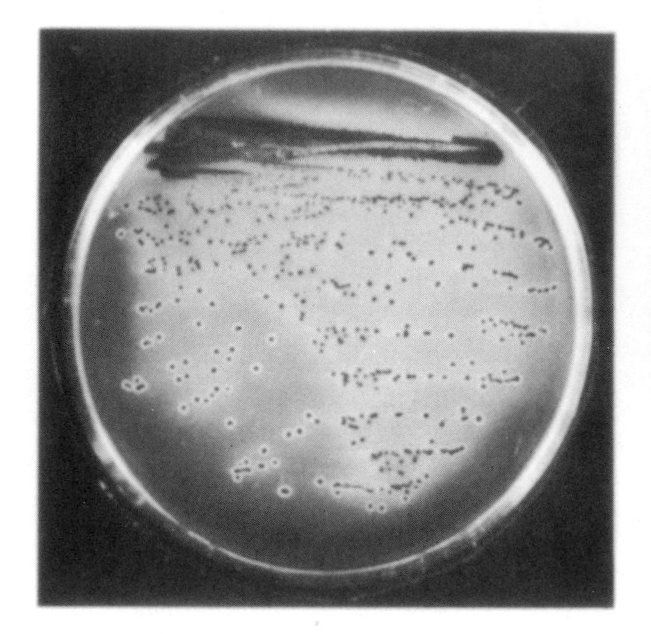

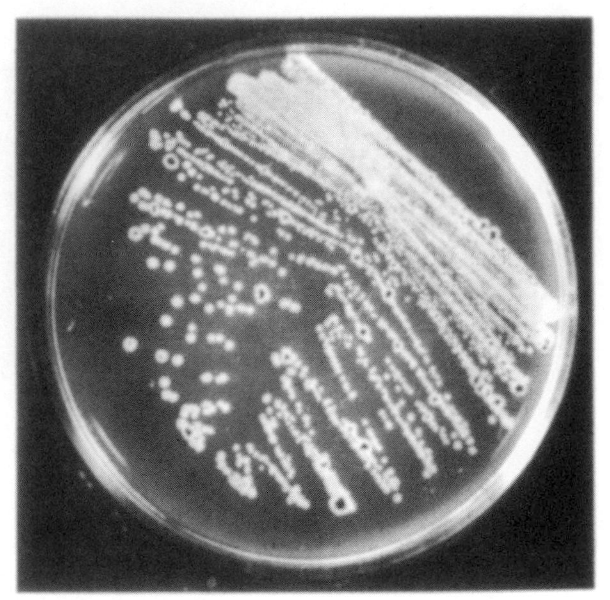

Figure 27B. Incubation at 30° (left) and 42° (right) of lactose EMB agar spread with bacteria carrying temperature-sensitive F′*lac* factor. (From Jacob et al., 1963.)

Strains

Number	Sex	Genotype	Important Properties
CSH24	F'_{ts} *lac*$^+$	Δ(*lacpro*) *supE thi*	Pro^-
CSH28	F′*lac*$^+$*proA*$^+$,*B*$^+$	Δ(*lacpro*) *supF trp pyrF his strA thi*	Ura^- His^- Trp^-

Method

Day 1

Transfer of F′*lacpro* from CSH28 to CSH24

The first part of this experiment is to make phenocopies of strain CSH24, so that it may act as a recipient even though it harbors an F'_{ts}*lac* episome. The best way to do this is to aerate at room temperature (although 30° is sufficient) a saturated culture of CSH24 in tryptone or B broth for 24 hours.

Day 2

Make a mating mixture of the CSH24 phenocopy culture and a fresh log culture of CSH28, the F′*lacpro* donor. Use a log culture of the donor (CSH28) at about 2×10^8 cells/ml. Shake the recipient culture (CSH24) vigorously on the vortex for 30 seconds just before using. Mix about 5 drops of each culture together in a large test tube and place in a waterbath at 30° for 2 hours. This will give a ratio of donor to recipient of about 1:20.

Stop the mating after several hours and plate dilutions on medium selecting for Pro^+ colonies. Plate out 0.1 ml of a 10^{-1}, a 10^{-2}, and a 10^{-3} dilution onto glucose minimal plates. Also plate out 0.1 ml of a 10^{-2} dilution of each parent. Only CSH24 cells which have received the *lacpro* episome from CSH28 can grow in the absence of proline, histidine, uracil, and tryptophan. Incubate at 30° for 36 hours.

Day 4 **Screening for F'_{ts}*lacpro***

The temperature sensitivity of the *ts* episome is not absolute. The best procedure is to pick colonies from the cross with toothpicks onto 2 glucose minimal plates, incubating one at 30° and the other at 43°. The 43° plate should be pre-warmed at that temperature for several hours. Pick first onto the 43° plate, and then, using the same toothpick, stab the 30° plate in the corresponding position on the grid. Pick as many as possible, since the percentage of CSH24 cells which have received and maintained a recombinant F'_{ts}*lacpro* episome is low. Test at least 300 colonies, and 500, if possible. Alternatively, lactose tetrazolium plates can be used instead of minimal plates.

Day 5 Examine the plates after 24 and 36 hours. Any colonies (or spots) which look larger at 30° than at 43° should be tested further. Pick these from the 30° plate again onto 2 glucose minimal plates. This time streak for single colonies. Streak only 4 to each plate to avoid cross feeding. There should be a marked difference between the temperature-sensitive F′*lacpro* episome strain and the wild type.

Day 6 Examine the re-test plates and record the number of colonies which harbored a temperature-sensitive *lacpro* episome from the original cross.

Materials

Day 1 overnight of CSH24

Day 2
culture of CSH24 from Day 1
exponentially growing culture of CSH28
test tube
waterbath at 30°
2 pasteur pipettes
5 dilution tubes
8 0.1-ml, 2 1-ml pipettes
5 glucose minimal plates
incubator at 30°

Day 4
12 glucose minimal plates
300 round wooden toothpicks
incubators at 30° and 43°

Day 5
4 glucose minimal plates
8 round wooden toothpicks
incubators at 30° and 43°

Questions

What percentage of the colonies which received the F′*lacpro* in the cross maintained an F′*lacpro* episome which was temperature-sensitive? How might this percentage be affected by the percentage of phenocopies in the recipient population? What other factors might cause this frequency to vary? What would you expect to happen if the donor for this cross were Rec^-? If the recipient were Rec^-? Is this a general method for making a temperature-sensitive derivative of an episome? What special conditions, if any, must exist? As an alternative, one might mutagenize a culture of CSH28 and then examine the survivors for mutants which have a temperature-sensitive episome. Why is this method not preferred?

EXPERIMENT 28

Generalized Transduction; Use of P1 in Strain Construction

Generalized transduction is mediated by phage P1 in *E. coli*. Lysates of this phage can transduce any marker on the *E. coli* chromosome at a frequency of 10^{-4}–10^{-5} per infected cell. Labeling studies have shown that P1 lysates contain occasional phage particles in which bacterial DNA has completely replaced the normal complement of phage DNA. These particles are responsible for generalized transduction. In addition, rare particles of P1 can be isolated which behave as specialized transducing phage (Luria et al.), but their use has been limited up to now. We should distinguish **generalized transduction** from **specialized transduction** at this point. The former involves the presence in a phage population of a random collection of transducing particles carrying different regions of the host chromosome. While these phage are useful for mapping and strain construction, the resulting transducing particle in this case carries practically no phage genes and is thus not of use for preparing enriched DNA (as described in Unit V), since transduction occurs by recombination and the transduced bacteria no longer give rise to phage. Bacteria transduced by specialized transducing phage, on the other hand, do give rise to phage carrying the transduced marker, since transduction occurs by lysogenization of an entire phage, rather than by recombination with a single marker.

Co-transduction

Markers which are very closely linked on the *E. coli* chromosome (within 1.8 minutes) can be co-transduced because they can be carried on in the same P1 phage particle. In fact, the frequency of co-transduction often serves as the best genetic estimate of the distance between two markers. For instance, *purB* and *galU* are 2% co-transducible. This means that if we transduce a $PurB^-GalU^-$ strain with a P1 lysate from a wild-type strain and select for $PurB^+$, then 2% of the $PurB^+$ transductants will also be $GalU^+$ (Figure 28A). Correlations of co-transduction frequencies for two loci with time units* for the same markers show that markers which are separated by less than 0.5 min are co-transduced 35–95%, markers which are separated by 1 min are co-transduced 4–26%, and markers which are separated by 1.5–1.8 min are co-transduced 1% or less (Taylor and Trotter). Many examples of the use of P1 in mapping studies can be found in Taylor and Trotter.

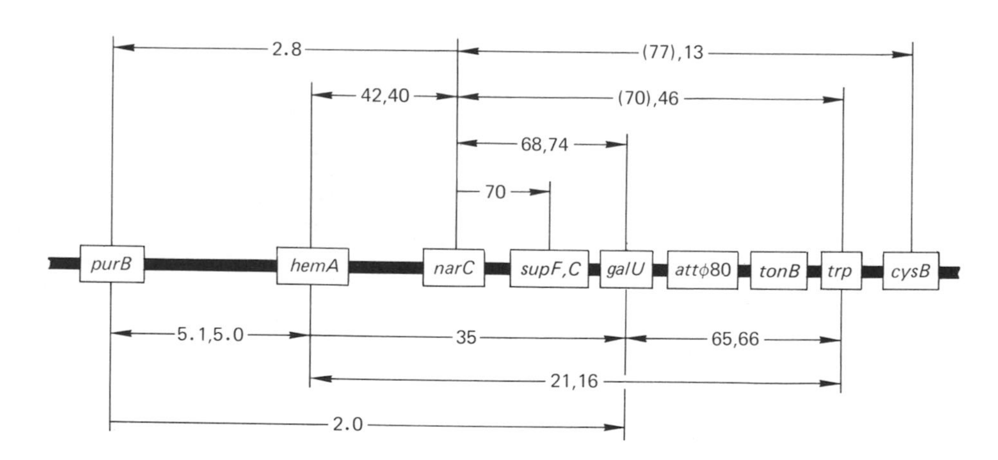

Figure 28A. Genetic map of the *purB* to *cysB* region determined by P1 co-transduction. The numbers given are the averages in percent for co-transduction frequencies obtained in several experiments. Where transduction crosses were performed in both directions, the head of each arrow points to the selective marker and the corresponding linkage nearest to it. The values in parentheses are considered unreliable due to interference from the non-selective marker. (Figure redrawn from Guest, 1969.)

Strain Building

P1 is an invaluable tool in strain construction. Often it is the best method for the construction of complicated strains. In preparing strains by P1 transduction, it is desirable to prevent non-defective P1 phage particles from lysogenizing the recipient strain. This is because P1 lysogens restrict foreign DNA introduced during conjugation or during phage infection and interfere with genetic crosses. By using low multiplicities of infection and by adding citrate to prevent re-adsorption (by complexing Ca^{++} required for adsorption), we can virtually eliminate this problem. Also, we can use a virulent mutant of P1 ($P1_{vir}$) which is unable to form lysogens.

The original P1 isolate plated poorly on *E. coli* K12. Now a mutant, P1*kc*, which plates well on *E. coli* K12 and with clearer plaques, is used by most workers (Lennox).

* As determined by interrupted mating experiments.

Clear Mutants of P1 Carrying the *cml* Region

Recently a mutant of P1 which carries the genes for chloramphenicol resistance (*cml*) and which makes clear plaques at high temperature (42°) but turbid plaques at low temperature (32°) has been isolated (Rosner). This derivative, P1*cml,clr100*, has several technical advantages over wild-type P1 phage. Lysogens are easily prepared by infecting cells and then selecting for chloramphenicol resistance. High titer lysates are made from lysogens by heat induction, and this allows storage of lysogens rather than lysates. Transduction can be carried out at high temperature to avoid creating lysogens, or at low temperature (allowing high moi's) with subsequent curing of transductants by streaking at high temperature (see addendum to this experiment).

References

CAMPBELL, A. M. 1969. *Episomes*, p. 25. Harper and Row.

GUEST, J. R. 1969. Biochemical and genetic studies with nitrate reductase *C*-gene mutants of *E. coli*. *Mol. Gen. Genet. 105:* 285.

IKEDA, H. and J.-I. TOMIZAWA. 1965. Transducing fragments in generalized transduction by phage P1. *J. Mol. Biol. 14:* 85.

LENNOX, E. S. 1955. Transduction of linked genetic characters of the host by bacteriophage P1. *Virology 1:* 190.

LURIA, S. E., J. N. ADAMS and R. C. TING. 1960. Transduction of lactose-utilizing ability among strains of *E. coli* and *S. dysenteriae* and the properties of the transducing phage particles. *Virology 12:* 348.

ROSNER, J. 1972. Personal communication.

TAYLOR, A. L. and C. D. TROTTER. 1967. Revised linkage map of *Escherichia coli*. *Bacteriol. Rev. 31:* 332.

Strains

Number	Sex	Genotype	Important Properties
CSH62	HfrH	*thi*	Lac^+
CSH7	F^-	*lacY strA thi*	Lac^-
lysate of $P1_{vir}$			

Method

Part A Preparation of Lysates

Day 1 Subculture one drop of an overnight of CSH62 (Lac^+) into 5 ml of LB broth containing 5×10^{-3} M $CaCl_2$. Continue to aerate until the cells are growing exponentially and have reached a density of 2×10^8 cells/ml. Preadsorb $P1_{vir}$ by adding 10^7 phage to 1 ml of this culture and incubating for 20 minutes in a 37° waterbath. Then add 2.5 ml R-top agar (kept at 45°) and immediately plate onto a freshly made R plate. Incubate at 37°, face up, for 8 hours. At the end of this time scrape the soft agar layer into a small centrifuge tube. Wash the surface of the plate with 1 ml broth and add the wash to the centrifuge tube. Add 5 drops of chloroform and vortex vigorously for 30 seconds. Allow to stand at room temperature for 10 minutes and then centrifuge down the cell debris and save the supernatant, which contains the P1 lysate.

The lysate can be titered in a manner similar to that used for titering ϕ80v phage, except that R plates and R-top agar are used (preadsorb in 5×10^{-3} M $CaCl_2$). Rather than titer a lysate, it is often easier to use different dilutions of a P1 lysate to transduce a marker.

Part B **Transduction with P1 Lysates**

Day 1

Resuspend 5 ml of a fresh overnight culture of the Lac^- strain to be transduced (CSH7, 22, or 34) in 5 ml of MC buffer (0.1 M $MgSO_4$, 0.005 M $CaCl_2$). Aerate at 37° for 15 minutes. Add 0.1 ml of the suspended cells to each of 5 small test tubes. To the first tube add 0.1 ml of the P1 lysate (in this case a lysate made on CSH62). To the second add 0.1 ml of a 10^{-1} dilution of the lysate, to the third 0.1 ml of a 10^{-2} dilution, and to the fourth 0.1 ml of a 10^{-3} dilution. Add nothing to the fifth tube, which serves as a control.

When using $P1_{vir}$ it is necessary to prevent excessive killing of the recipient strain. This is done by using a very low multiplicity of infection (1:100) or by adding sodium citrate to the mixture, which prevents readsorption of the P1 phage by removing Ca^{++} ions. With this latter procedure, multiplicities of infection of 1:1 can be used.

Prepare a sixth tube with 0.1 ml of the P1 lysate, but no cells, as another control. Preadsorb the P1 by incubating at 37° in a water bath for 20 minutes. Add 0.2 ml 1 M sodium citrate to each tube. Plate the entire contents of each of the six tubes on lactose minimal plates with 2.5 ml F-top agar. Incubate at 37° for 48 hours.

Day 3

Examine the plates and determine whether P1 transduction to Lac^+ occurred.

Materials

Part A **Preparation of Lysates**

Day 1

overnight of CSH62
5 ml LB broth with 5×10^{-3} M $CaCl_2$ (50 ml 0.1 M $CaCl_2$ per liter added after autoclaving)
water bath at 37°
2.5 ml R-top agar kept at 45°, 1 ml LB broth
R-top agar is per liter: 10 g tryptone, 8 g NaCl, 1 g yeast extract, 8 g agar (2 ml 1 M $CaCl_2$ + 5 ml of 20% glucose added after autoclaving)
Medium for R plates is identical to that for R-top agar except that 12 grams of agar per liter are used instead of 8 grams.
chloroform
desk-top centrifuge, 2 small centrifuge tubes
2 0.1-ml, 2 1-ml, and 1 5-ml pipettes
empty test tube
lysate of $P1_{vir}$

Part B **Transduction with P1 Lysates**

Day 1

5 ml of overnight of CSH7, 22, or 34
5 ml MC buffer (0.1 M $MgSO_4$, 0.005 M $CaCl_2$)
P1 lysate of CSH62 or of any Lac^+ strain
3 dilution tubes
6 small test tubes
water bath at 37°
1.2 ml 1 M sodium citrate
6 lactose minimal plates
15 ml F-top agar kept at 45°
5 0.1-ml, 5 1-ml, and 2 10-ml pipettes
centrifuge, centrifuge tube

Additional Method—P1*cml,clr100*

Strains lysogenic for P1*cml,clr100* grow well at 32° but not at 34°. To prepare a lysate, grow cells with aeration in Luria broth (see Appendix) containing 10^{-2} M $MgSO_4$ and no glucose to a density of 2×10^8 cells/ml. Then shift to 40° for 35 minutes with good aeration, and shift to 37° for an additional 60 minutes. Add chloroform, shake, and spin down the cell debris.

To prepare a lysogen, infect by preadsorbing cells in medium containing 5×10^{-3} M $CaCl_2$ and streak on Luria plates containing 12.5 μg/ml chloramphenicol at 32°. All colonies which grow on these plates should be lysogens, and this can be confirmed by testing for temperature-sensitive growth.

The *cml* region of this phage is derived from an R factor. Single-step mutants of *E. coli* resistant to this level of chloramphenicol do not occur.

EXPERIMENT 29

F′ Factor-mediated Chromosome Transfer

We have previously described (see introduction to Unit II) how the F′ factor can mediate chromosome transfer, that unlike F^+-mediated chromosome transfer, this occurs from the same fixed point for every cell in the population, and that the point of origin of transfer is dependent on the location of regions of homology between the F′ factor and the bacterial chromosome (see Figure 29A). Part B of this experiment attempts to show that F′-mediated chromosome mobilization is dependent on the host recombination enzymes.

The following experiment will utilize diploid strains prepared in previous exercises. Other diploid strains (which are used as optional controls) are constructed either by the instructor beforehand or by the students as an exercise. Part A requires CSH25 carrying an F′*lac*$^+$*pro*$^+$ factor. This is prepared in Experiment 25. For an optional exercise, CSH25 carrying an F^+ factor should be prepared. This can be constructed by crossing CSH42 with CSH25 in the exact manner described in Experiment 25, although at 33°. Part B requires CSH51 and CSH52 strains carrying the F′*lac*$^+$*pro*$^+$ factor. These are prepared in Experiment 12 by transferring the F′ factor from CSH23. Part B also requires the isolation of a Nal^r derivative of CSH59, and also of CSH57 if this latter strain is to be used (optional). This simple procedure is described in Experiment 32. The preparation

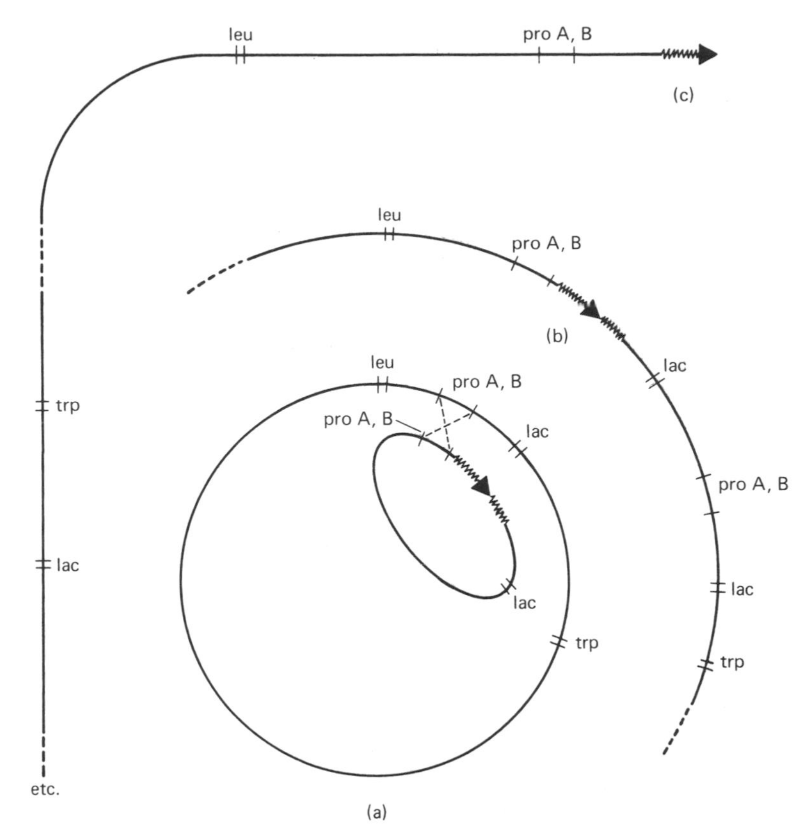

Figure 29A. F′ factor-mediated chromosome transfer. Part **(a)** shows an F′*lacpro* factor integrating into the chromosome by a reciprocal crossover in the region of homology provided by the bacterial genes incorporated into the F factor DNA. This produces the Hfr depicted in **(b)**. The origin of transfer is at a point within the F factor, and the resulting Hfr transfers markers in the order shown in **(c)**.

of a $\phi 80h$ lysogen from CSH59 is also a prerequisite for Part B. This is easily accomplished, as outlined in Experiment 39.

Strains

Number	Sex	Genotype	Important Properties
CSH70	HfrP4X	*metB argE thi*	Met$^-$ Arg$^-$ Str^s
CSH57	F$^-$	*ara leu lacY purE gal trp his argG malA strA xyl mtl ilv metA* or *B thi*	Ara$^-$ Leu$^-$ Lac$^-$ Pur$^-$ Gal$^-$ Trp$^-$ His$^-$ Arg$^-$ Mal$^-$ Ilv$^-$ Met$^-$ Str^r
CSH25/F′		*supF thi*	Str^s

Method

Part A **Test for Transfer of Chromosomal Markers Mediated by F′ Factors**

Day 1 Prepare a mating mixture of the multiply marked recipient, CSH57, with each of the three donors: F′*lac*$^+$*pro*$^+$/CSH25,

F^+/CSH25, and HfrP4X (CSH70). Mix 2 ml of CSH57 with 0.2 ml of each donor strain in a test tube and place on a 30–33 rpm rotor at 37°. Both donor and recipient should be growing exponentially (see Unit II for methods of preparing mating mixtures). After 60 minutes plate 0.1 ml of both a 10^{-1} and a 10^{-2} dilution onto each of the selective plates A–I (Table IIC, Unit II). Also plate 0.1 ml of a 10^{-1} dilution of the recipient onto each type of selective plate. This is a control for the number of revertants for each marker in the CSH57 population. In these crosses streptomycin is used as the counterselective agent. Incubate at 37°.

Doing the experiment in this manner should demonstrate the relative efficiencies of transfer for each marker by each of the three types of male strains (Hfr, F′, and F^+). Due to the gradient of transmission (Experiment 8) both the point of origin and the direction of transfer can be approximated, and more precise determination achieved by subsequent interrupted mating experiments. Alternatively, it is possible to interrupt the initial matings (by vigorous agitation) after 30 minutes, which will give less ambiguous results as to the approximate point of origin.

Day 2, 3

Examine the selection plates. Determine which markers were inherited from the F′ donor by CSH57. Because of the gradient of transmission you should be able to determine the direction of transfer and the approximate location of the origin. Is there an apparent gradient of transfer with the F^+ donor? Would you expect to see one? Can you detect transfer at all by the F^+ donor? Compare the frequency of transfer for different markers by each of the three donors. The F′$lac^+ pro^+$ was derived from an Hfr strain which is similar to P4X (Figure IIE). What direction of transfer of markers would you therefore expect from chromosome mobilization mediated by the F′$lac^+ pro^+$ factor?

Optional Exercises

1. Use interrupted matings to determine more precisely the point of origin of transfer of chromosomal markers from the F′ donor.

2. Repeat the experiment with a different episome. Use CSH33 which carries a $trp^+ cys^+$ colicinogenic factor and determine the direction of transfer and also the approximate point of origin of transfer.

Part B **Influence of *recA***

Day 1

The recipient for this cross is a Nal^r derivative of CSH59 which also carries $\phi 80h$ as a prophage (to avoid complications resulting from zygotic induction). The F′ merodiploids, to be used as donors, are CSH51 and CSH52 carrying an F′$lac^+ pro^+$ factor. Strain CSH52 is a Rec^- derivative of CSH51. Mating mixtures should be prepared for each donor with the CSH59 derivative, exactly as in Part A. Interrupt the matings after 30 minutes and plate 0.1 ml of a 10^{-1} and a 10^{-2} dilution of each mating onto glucose minimal plates with tryptophan and nalidixic acid (to select for PyrC$^+$) and onto glucose minimal plates with uracil and nalidixic acid (to select for Trp$^+$). Also plate 0.1 ml of a 10^{-1} dilution of both the donors and the recipient on each type of plate as a control for revertants. Incubate at 37°.

Day 2, 3

Examine the plates. Can you determine the direction of transfer of chromosomal markers from the strain in which the *lac* region is transposed (CSH51)? Should both the direction of transfer and the point of origin be the same as that in Part A? Does the Rec$^-$ derivative of CSH51 show a decreased amount of transfer? What can you conclude from this result? Refer to Figures IIE and 42A and determine whether zygotic induction would interfere with the inheritance of the PyrC marker, the Trp marker, or both. The ϕ80d*lac*$^+$ prophage carried by CSH51 is defective for most of the late phage functions. Would you expect zygotic induction to occur at all? Test this by repeating the experiment with a non-lysogen (CSH59, Nal^r).

Materials (per group)

Part A

Day 1

exponentially growing cultures of F′*lac*$^+$*pro*$^+$/CSH25, F$^+$/CSH25, and CSH70
3 test tubes
7 dilution tubes
7 0.1-ml, 10 1-ml, and 1 5-ml pipettes
7 each of plates A, B, C, D, E, F, G, H, I (Table IIC, Unit II)
waterbath at 37° or rotor drum in 37° room

Part B

Day 1

exponentially growing cultures of a Nal^r, ϕ80*h* lysogen derived from CSH59, of F′*lac*$^+$*pro*$^+$/CSH51, and of F′*lac*$^+$*pro*$^+$/CSH52
2 test tubes
vortex mixer
7 dilution tubes
7 0.1-ml, 9 1-ml, and 1 5-ml pipettes
7 glucose minimal plates containing tryptophan and nalidixic acid
7 glucose minimal plates containing uracil and nalidixic acid
waterbath at 37° or rotor drum in 37° room

INTRODUCTION TO EXPERIMENTS 30–32

Manipulation of Genetic Markers

Constructing the proper strain is all too often the rate limiting step in the completion of important experiments. It is the purpose of this section to show how the preparation of seemingly complicated strains can be greatly simplified, and even made routine.

The key to strain construction is to have a set of markers mapping at different locations which can be introduced into strains easily by simple operations. This is important because some strains require many steps before the final combination of markers is present. Often each step calls for just the right set of selective and counterselective markers. Consider the following two examples.

Example 1: Construct Su1$^+$ and Su3$^+$ derivatives of an F$^-$ strain carrying an amber mutation in the *i* gene (*lacI*)

(A) We have an HfrH which is SuI$^+$ (Figure 30A). In order to utilize this strain we would like to: (1) have a drug-resistance marker in the F$^-$ (Str^r, Spc^r, Nal^r, Val^r, Rif^r, etc.) to enable counterselection against the Hfr; and (2) have a marker such as His$^-$ in the F$^-$ strain to enable the selection of His$^+$ recombinants, since the *his* locus and the *supD* (SuI) locus are closely linked.

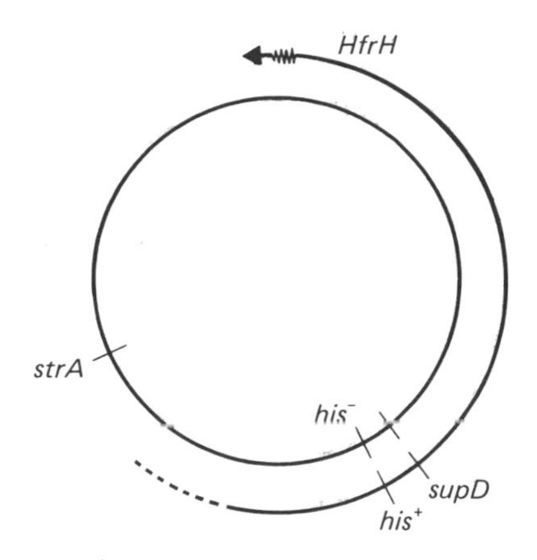

Figure 30A. Linkage of *supD* to *his.*

(B) We also have an HfrP4X which carries an amber mutation in *argE* (Figure 30B). We would utilize this strain by crossing the Arg^- marker into our F^- strain and then using P1 phage (Experiment 28) grown on an Su^+ strain to transduce the recipient to Arg^+. Some of the Arg^+ transductants will be cells which received the suppressor and will be Su^+. This can be verified by checking for suppression of the *i* mutation or by using T4 amber phage (Experiment 23).

However, in order to cross the Arg^- marker into the F^- strain, we must find a linked marker to use for our selection. Although *argE* is linked to *metB* and *purD*, it is easier to use the nearby *rif* locus. A Rif^r derivative of the Hfr P4X can easily be isolated and the selection in a cross of P4X against the F^- strain can be for rifampicin resistance. Rif^r recombinants are then scored for Arg^-.

Example 2: Construct a Rec^- (*recA*) derivative of an unmarked F^- strain

We have an Hfr which transfers *thyA* and *recA* very early and *nalA* somewhat later. We should therefore:

(A) Isolate a drug-resistant derivative of the F^- to enable counterselection against the donor.

(B) Isolate a Thy^- (*thyA*$^-$) derivative of the F^- strain. The *thyA* region is closely linked to *recA* and Thy^+ recombinants can then be examined for the Rec^- character.

Alternatively, a Nal^r (*nalA*) derivative of the Rec^- Hfr can be used. After allowing several generations for segregation of the recessive *nal*^{r} allele, cells are plated on medium selecting for Nal^r colonies (an additional counterselective marker must also be used). Since *recA* is transferred before *nalA* by this Hfr, a percentage of the Nal^r colonies will also be Rec^-.

In the preceeding crosses, seemingly difficult strain constructions are simplified by the ability to prepare, in this case, (A) drug-resistant mutants, (B) His^- mutants, and (C) Thy^- mutants. There are straightforward procedures available for isolating strains carrying each of these markers and many others. In Experiments 30–32 we have cataloged methods for putting these markers into strains. These are not complicated exercises, but simple techniques often requiring a single operation. Using phage resistance, drug resistance, and nutritional markers **we can mark a strain easily at many different places on the genetic map** (see Figure 30C). The following table lists all the markers referred to in this section.

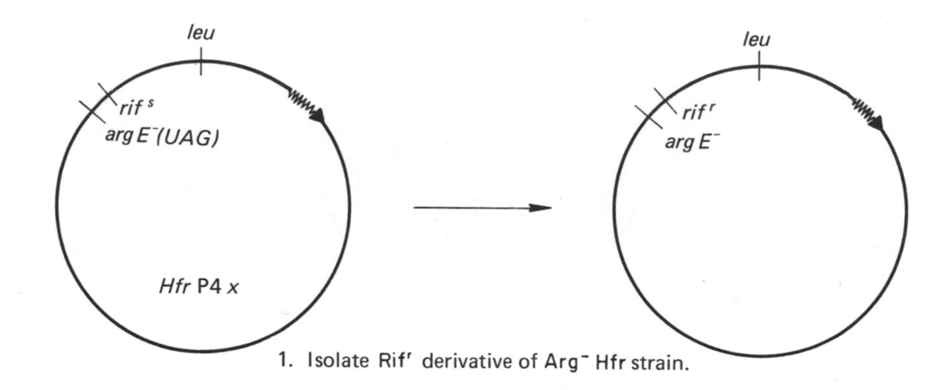

1. Isolate Rif^r derivative of Arg$^-$ Hfr strain.

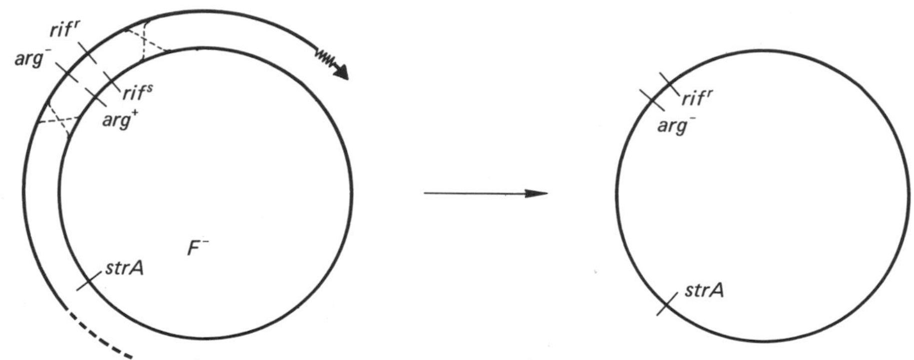

2. Cross Arg$^-$ Rif^r Str^s Hfr with Arg$^+$ Rif^s Str^r F$^-$ and select for Rif^r Str^r recombinants. Score for Arg$^-$.
3. Prepare Pl lysate on Su$^+$ strain.
4. Transduce F$^-$Arg$^-$Su$^-$ strain to Arg$^+$ and score for Su$^+$

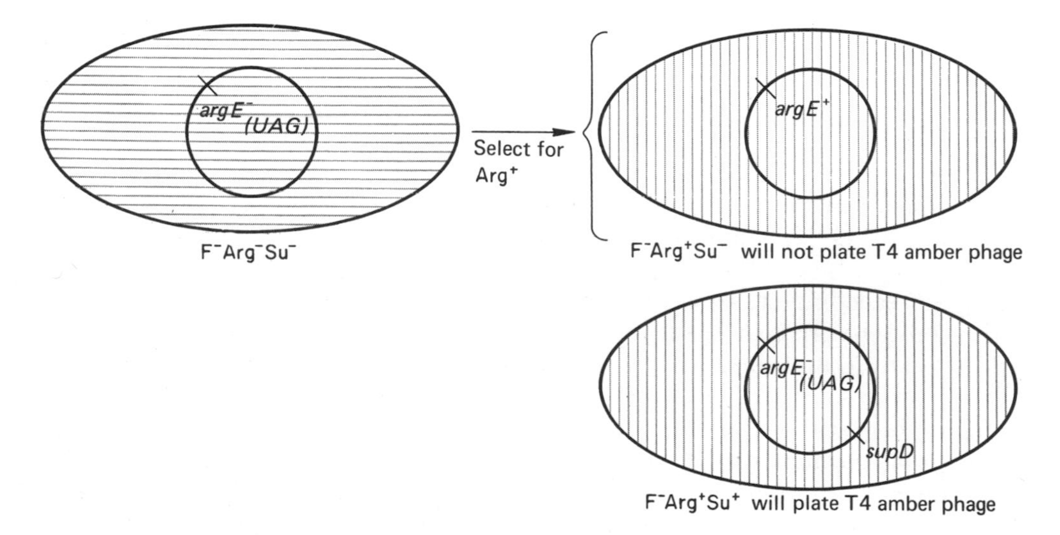

Figure 30B. Method for construction of Su$^+$ derivative of *E. coli* strains.

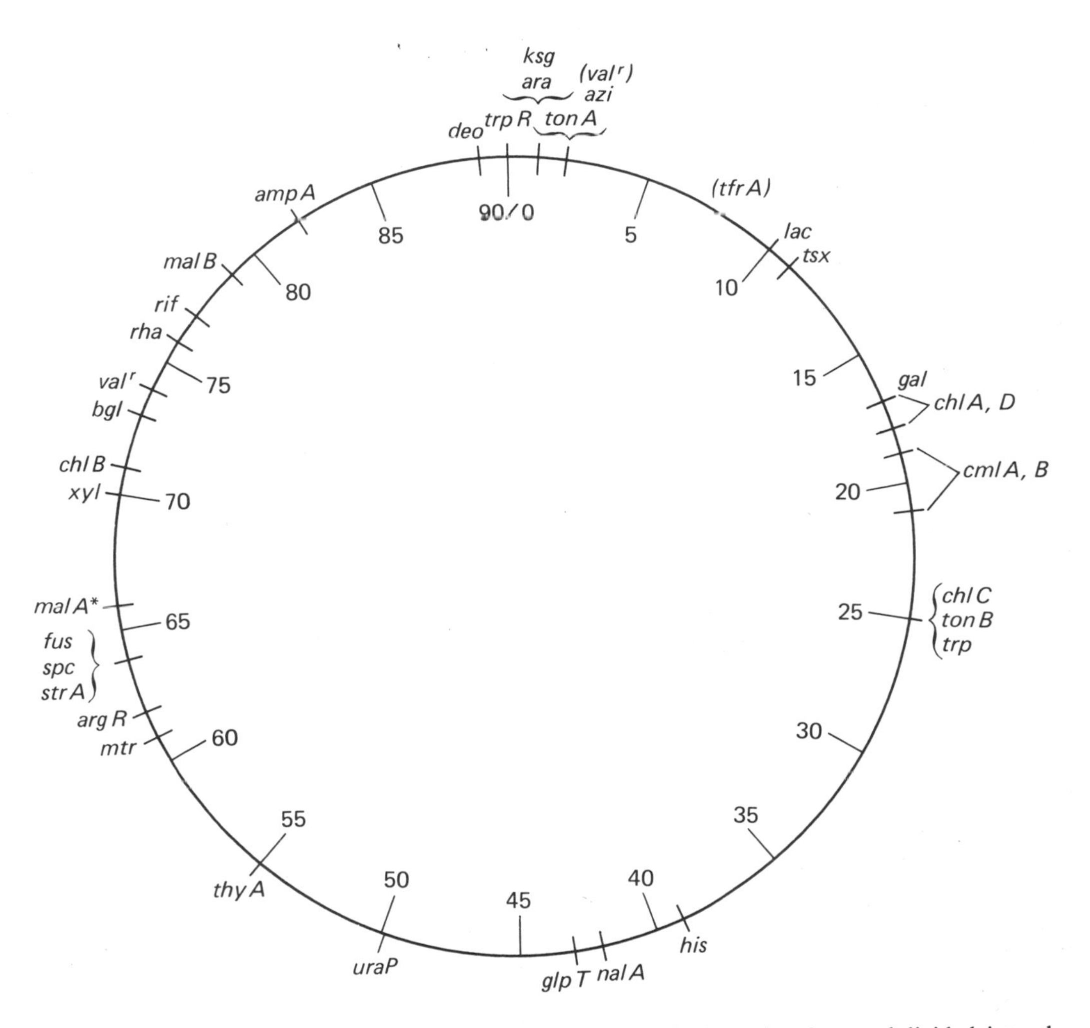

Figure 30C. Markers used for strain construction. *The *malA* locus has been subdivided into the *malP*, *malQ*, and *malT* loci.

It should be pointed out that each marker can be used either for selection or for counterselection. Thus, nalidixic acid can be used to counterselect against a different sensitive Hfr, or else (if the Hfr is Nal^r and the $F^- Nal^s$) selection for Nal^r recombinants is possible, provided the Hfr is counterselected against in some other way. Some markers, particularly drug resistance markers such as Nal^r and Rif^r, are recessive. Thus, when selecting Nal^r recombinants of a Nal^s strain, several generations of growth are required before plating on selective medium to allow for the segregation of the recessive allele.

Table 30A. Markers Used for Strain Construction

Marker	Locus	Map position (min)
	Sugar markers	
L-arabinose	*ara*	1
lactose	*lac*	10
galactose	*gal*	17
maltose	*malA,B*	66, 79
D-xylose	*xyl*	70
rhamnose	*rha*	76
	Drug resistance	
5-methyltryptophan	*trpR,trp,mtr*	0, 25, 61
kasugamycin	*ksg*	1
azide	*azi*	2
chlorate resistance	*chlD,A,C,B*	17, 18, 25, 71–73
chloramphenicol	*cmlA,B*	19, 21
nalidixic acid	*nalA,B*	42, 51
phosphonomycin	*glpT*	43
6-azauracil	*uraP*	50
canavanine	*argR*	62
streptomycin	*strA*	64
spectinomycin	*spc*	64
fusidic acid	*fus*	64
rifampicin	*rif*	77
ampicillin	*ampA*	82
	Nutritional requirement	
tryptophan	*trp*	25
histidine	*his*	39
high thymine	*thyA*	55
low thymine	*deo*	89
	Phage resistance	
T1, T5, ϕ80	*tonA*	2
T3, T4, T7	*tfrA*	5–10
T6	*tsx*	11
T1, ϕ80, colV,B	*tonB*	25
λ	*malA,B*	66, 79
BF23	—	77
	Misc. selections	
arabinose + fucose	*araC*	1
valine resistance	*val^r*	1–2, 74
phenyl-β-D-galactoside	*lacI*	10
nitrophenyl-β-D-thiogalactoside	*lacY*	10
β-glucosides	*bgl*	73

EXPERIMENT 30

Use of P2 Phage to Isolate His$^-$ Strains of *E. coli*

Phage P2 can integrate at several different chromosomal sites in *E. coli* K12, one of which (location H) is closely linked to the histidine operon. These lysogens spontaneously segregate His$^-$ cells. This provides a convenient method for constructing His$^-$ derivatives of *E. coli* K12 strains, since the His$^-$ segregants are easily detected as small translucent colonies on medium with limiting histidine. Approximately 0.3% of the colonies are His$^-$ (Kelly and Sunshine, 1967).

An analysis of the His$^-$ mutants derived from P2 lysogens indicates that P2 frequently excises from the chromosome in an abnormal fashion (Figure 30D). All the resulting cured cells, termed **eductants**, are deleted for a region extending from the P2 attachment site through *sbcB*, *his*, *gnd*, and *rfb* (Figure 30E). Some of the deletions extend into *mglP* (Sunshine and Kelly, 1971).

References

KELLY, B. L. and M. G. SUNSHINE. 1967. Association of temperate phage *p2* with the production of histidine negative segregants by *Escherichia coli*. *Biochem. Biophys. Res. Commun. 28:* 237.

SUNSHINE, M. G. and B. KELLY. 1971. Extent of host deletions associated with bacteriophage P2-mediated eduction. *J. Bacteriol. 108:* 695.

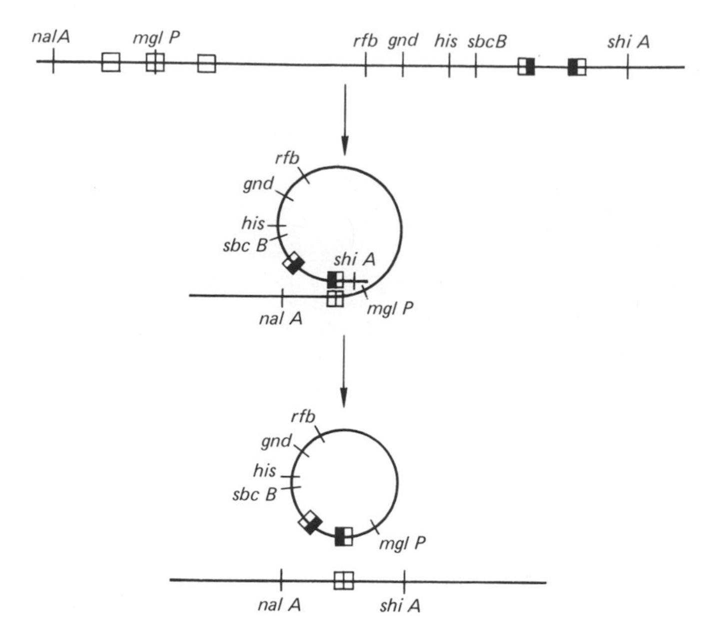

Figure 30D. Formation of eductants by P2 phage. (Redrawn from Sunshine and Kelly, 1971.)

Method

Day 1 **Preparation of P2 lysogens:** Mix 0.1 ml of the appropriate dilution of a P2 lysate with 0.1 ml of a fresh overnight of the strain to be tested in a small tube. Add 2.5 ml H-top agar and plate out onto an LB plate. Incubate overnight at 37°.

Day 2 Single plaques should be visible, with faint bacterial growth in the center of the plaque. Pick from the center of the plaque into 3 ml LB broth and grow up overnight at 37°.

Day 3 Plate 0.1 ml of a 10^{-4}, 10^{-5}, and 10^{-6} dilution onto a glucose minimal plate containing 0.5 μg/ml histidine. Incubate at 37° for 36–48 hours.

Day 4, 5 Observe the plates. His$^-$ colonies appear as small translucent colonies, barely visible to the naked eye, but easily distinguishable under a viewing scope. Pick with a fine wire several of these colonies onto LB plates. Streak for single colonies.

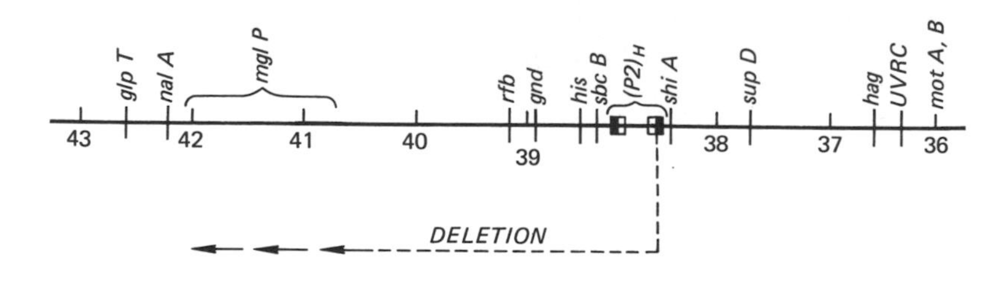

Figure 30E. Extent of deletions in P2 eductants. (Redrawn from Sunshine and Kelly, 1971.)

Day 5, 6 Test these colonies for His^- by streaking on glucose minimal plates with and without histidine (50 μg/ml).

Materials

Day 1
overnight of any *E. coli* K12 strain
1 LB plate
2.5 ml H-top agar at 45°
4 0.1-ml, 1 1-ml, and 1 5-ml pipettes
1 small test tube
P2 lysate prepared as described below
3 dilution tubes

Day 2
1 test tube with 3 ml LB broth

Day 3
3 glucose minimal plates with 0.5 μg/ml histidine
6 0.1-ml pipettes and 1 1-ml pipette
4 dilution tubes

Day 4, 5
platinum wire
1 LB plate

Day 5, 6
1 glucose minimal plate
1 glucose minimal plate with 50 μg/ml histidine
6 round wooden toothpicks

Preparation of a P2 Lysate

Strains lysogenic for P2 phage yield low titer lysates when grown up overnight in broth, due to spontaneous lysis occuring in a fraction of the population. Titers on the order of 10^6 phage/ml can be obtained by this procedure. Higher titer lysates can be obtained by infecting a growing culture of a sensitive strain with a high multiplicity (5:1) and then growing for several hours with good aeration (see Experiment 2).

EXPERIMENT 31

Selection of Thy$^-$ Strains with Trimethoprim

Trimethoprim (and also aminopterin) depresses the growth of Thy$^+$ cells, but not that of Thy$^-$ cells, in the presence of high amounts of thymine or thymidine (greater than 50 μg/ml). After many generations of growth in thymine and trimethoprim, cultures of *E. coli* contain predominantly Thy$^-$ cells. The Thy$^-$ mutants obtained by this procedure all map at the *thyA* locus, at approximately 55 min. These mutants are termed "high thymine" requiring strains, since 50 μg/ml of thymine is needed to satisfy the growth requirement. Second-step mutations can occur which result in a lower level thymine requirement. These map in the *deo* operon, 89 min (Lomax and Greenberg).

Trimethoprim can also be used effectively in plates to select Thy$^-$ strains.* Colonies growing on plates with 10 μg/ml trimethoprim and 50 μg/ml thymine, which are repurified on the same selective plate, are predominantly Thy$^-$ (*thyA*).

Although 50 μg/ml thymine can satisfy the growth requirement of *thyA* strains, we use 200 μg/ml in the liquid culture selection.

* This is now the most widely used method.

References

ALIKHANIAN, S. I., T. S. ILJINA, E. S. KALIAEVA, S. V. KAMENEVA and V. V. SUKHODOLEC. 1965. Mutants of *Escherichia coli* K12 lacking thymine. *Nature 206:* 848.

LOMAX, M. S. and G. R. GREENBERG. 1968. Characteristics of the *deo* operon: Role in thymine utilization and sensitivity to deoxyribonucleosides. *J. Bacteriol. 96:* 501.

OKADA, T., K. YANAGISAWA and F. J. RYAN. 1961. A method for securing thymineless mutants from strains of *E. coli. Z. Vererbungslehre 92:* 403.

Method

Day 1 Subculture an overnight of CSH51 (any of the provided strains are adequate) into 5 ml of 1 × A minimal medium containing proline, thymine, and trimethoprim. Inoculate 3 drops of a saturated culture into this medium, and allow to grow at 37° overnight to saturation.

Day 2 Subculture 1 drop of the culture from Day 1 into 5 ml of the identical medium. Allow to grow overnight to saturation.

Day 3 Plate dilutions of the culture onto LB plates supplemented with thymine, in order to obtain single colonies. Plate 0.1 ml of a 10^{-4} and 10^{-5} dilution. Save the cultures for possible future use.

Day 4 Grid 50 colonies onto a master plate (LB + thymine) and grow up at 37° overnight.

Day 5 Replicate onto glucose minimal plates* with thymine (200 μg/ml) and without thymine. Incubate at 37°.

Day 6 Examine the minimal plates. How effective was the procedure? If no Thy$^-$ colonies were obtained, subculture the cells a third time into the trimethoprim + thymine medium and repeat the procedure.

Materials

Day 1 overnight of CSH51 (grown in glucose minimal medium + proline)
5 ml 1 × A medium supplemented with: glucose (0.4%), B1 (2 μg/ml), $MgSo_4$ (10^{-3} M), proline (50 μg/ml), thymine or thymidine (200 μg/ml), and trimethoprim (200 μg/ml, obtained from Hoffman-La Roche Inc., Nutley, New Jersey 07110)

Day 2 5 ml of 1 × A medium identical to that used in Day 1

Day 3 2 LB plates containing 200 μg/ml thymine or thymidine
4 0.1-ml and 1 1-ml pipettes
3 dilution tubes

* If CSH51 is used, the plates should contain proline.

Day 4 1 LB plate containing 200 μg/ml thymine or thymidine
50 round wooden toothpicks

Day 5 1 replica velvet pad, replicating block
1 glucose minimal plate with B1 and proline (50 μg/ml)
1 glucose minimal plate with B1, proline, and thymine or thymidine (200 μg/ml)

Alternate Method

Spread a drop of an overnight culture onto a glucose minimal plate containing 50 μg/ml thymine and 10 μg/ml trimethoprim. Incubate for 36 hours at 37°. Pick and purify small colonies onto the same type of plate. Test the colonies which grew on these plates for the Thy^- character.

Note: Some strains are sensitive to high concentrations of trimethoprim, even in the presence of thymine. For the liquid culture selection, lower amounts of this compound may have to be employed. The sensitivity point for a particular strain can be determined by inoculating a series of test tubes containing different concentrations of trimethoprim (50–200 μg/ml) and an excess of thymine.

EXPERIMENT 32

Isolation of Valine-resistant and Antibiotic-resistant Mutants

A. Valine Resistance

E. coli K12 is sensitive to exogenous valine. When a lawn of cells is spread on a minimal plate containing 40 μg/ml L-valine (or 80 μg/ml D,L-valine), a clear background with no growth is observed after 48 hours. Only a few resistant colonies appear (less than 1×10^{-6}). The mutants selected in this manner map predominantly at two loci. About 50% of these are in the *ilva* region (74 min) and 50% near the *leu* locus (1 min). In the following experiment Val^r mutants of an HfrH are selected and as an optional exercise are tested for linkage with *ara* and *pro*. Because the presence of isoleucine prevents inhibition by valine, we use minimal glucose plates in this selection instead of LB plates.

Reference

GLOVER, S. W. 1962. Valine-resistant mutants of *Escherichia coli* K12. *Genet. Res. 3:* 448.

Strains

Number	Sex	Genotype	Important Properties
CSH62	HfrH	*thi*	Str^s
CSH50	F^-	*ara* Δ(*lacpro*) *strA thi*	Str^r

Method

Day 1 Spin down 5 ml of a fresh overnight of CSH62, wash, and then resuspend in 5 ml 1 × A buffer. Spread 0.2 ml of this onto each of 3 plates containing 40 μg/ml L-valine. Incubate at 37° for 36–48 hours.

Day 3 Repurify several resistant colonies.

Day 5 **(optional)** Cross CSH62 Val^rStr^r against an $F^-Ara^-Pro^-Str^r$ (CSH50) strain in a 30-minute interrupted mating. Select for $Ara^+Pro^+Str^+$ colonies of CSH50 by plating the mating mixture on arabinose minimal plates with streptomycin. Test purified colonies from this cross for val^r and determine whether the val^r marker being tested was in the *leu* region or not.

We suggest isolating several independent Val^r mutants of CSH62 and testing them all in this manner. Compare your results with those of Glover (*Gen. Res.*, **3**, 448, 1962).

Materials

Day 1
- fresh overnight of CSH62 (or any strain to be tested)
- desk-top centrifuge
- 10 ml buffer
- 2 small centrifuge tubes
- 3–5 plates with 80 μg/ml D,L-valine (these should be, in the case of CSH62, glucose minimal plates)
- 1 1-ml, 2 5- or 10-ml pipettes

Day 3
- 3 glucose minimal plates with valine, as above
- 15–20 round wooden toothpicks

B. Nalidixic Acid Resistance

Nalidixic acid is an inhibitor of DNA replication in *E. coli*. Two loci are involved in conferring resistance. Mutants resistant to high levels (up to 100 μg/ml) map at *nalA*, at 42.5 min. *NalB*, which maps at 51 min, confers low level resistance (up to 4 μg/ml). The following experiment demonstrates the technique used for isolating *nalA* mutants. We use CSH51 in this experiment, but the method is the samc for any *E. coli* strain. It should be noted that sensitivity is dominant to resistance. Therefore, in Hfr crosses involving the selection of Nal^r recombinants, several generations of outgrowth are necessary to allow the segregation of Nal^r recombinants.

References

HANE, M. W. 1971. Some effects of nalidixic acid on conjugation in *Escherichia coli* K-12. *J. Bacteriol. 105:* 46.

HANE, M. W. and T. H. WOOD. 1969. *Escherichia coli* K12 mutants resistant to nalidixic acid: Genetic mapping and dominance studies. *J. Bacteriol. 99:* 238.

O'NEIL, D. M., L. S. BARON and P. S. SYPHERD. 1969. Chromosomal location of ribosomal protein cistrons determined by intergeneric bacterial mating. *J. Bacteriol. 99:* 242.

Method

Day 1 Spread several drops of a fresh overnight of CSH51 onto each of three rich plates (YT or LB) containing nalidixic acid (20 μg/ml). Incubate at 37°.

Day 2 Purify resistant colonies onto the same type of selective plate. If too few mutants appear, repeat the selection using more cells.

Materials

Day 1

overnight of CSH51
pasteur pipette
3 YT or LB plates with 20 μg/ml nalidixic acid
Stock solutions of nalidixic acid are 100 mg/ml in 1 N NaOH. Add 0.2 ml of this after autoclaving (per liter).

Day 2

1 LB plate containing 20 μg/ml nalidixic acid
6 round wooden toothpicks

Questions

1. Mutations conferring resistance to high levels of nalidixic acid map at *nalA*. What conditions would you employ to isolate and verify *nalB* mutants? (These are resistant to 4 μg/ml but not to 10 μg/ml).

2. Sensitivity to nalidixic acid is dominant to resistance. Suppose you want to cross Nal^r into a strain, selecting directly for Nal^r. Would you be able to plate the mating mixture directly on plates containing nalidixic acid?

C. Rifampicin Resistance

Rifampicin is a strong inhibitor of *E. coli* RNA polymerase. Mutants of *E. coli* which are resistant to this antibiotic have been isolated and shown to have an altered RNA polymerase. The mutations map at the *rif* locus, at 77 min. Rifampicin-resistant mutants are easy to obtain, and in the following experiment we select for resistance in an HfrH strain. We then use this mutant in subsequent experiments. The vast majority of *rif*^{r} mutations are recessive to the wild type, *rif*^{s}.

References

AUSTIN, S. and J. SCAIFE. 1970. A new method for selecting RNA polymerase mutants. *J. Mol. Biol. 49:* 263.

TOCCHINI-VALENTINI, G. P., P. MARINO and A. J. COLVILL. 1968. Mutant of *E. coli* containing an altered DNA-dependent RNA polymerase. *Nature 220:* 275.

WEHRLI, W., F. KNÜSEL, K. SCHMID and M. STAEHELIN. 1968. Interaction of rifamycin with bacterial RNA polymerase. *Proc. Nat. Acad. Sci. 61:* 667.

Method

Day 1 Spread 6 drops of a fresh overnight culture of CSH62 onto each of 3 LB plates containing rifampicin (100 μg/ml). Incubate at 37°.

Day 2 Examine the rifampicin plates. Ideally there should be a clean background with approximately 1 to 40 colonies per plate. Pick several of these and repurify on rifampicin plates.

Materials

Day 1
fresh overnight of CSH62
3 LB plates with 100 μg/ml rifampicin
(Dissolve 100 mg rifampicin in 2 ml 100% methanol: add to LB medium after autoclaving. Mix well before pouring plates.)
1 pasteur pipette

Day 2
1 LB plate with 100 μg/ml rifampicin
6 round wooden toothpicks

D. Ampicillin Resistance

A single mutation in *E. coli* can provide resistance to 10 μg/ml D, L-ampicillin. This locus, *ampA*, maps at approximately 82 min. In the following experiment, ampicillin-resistant mutants are selected in CSH51. However, any of the provided strains are adequate. Ampicillin is D (−)-α-aminobenzylpenicillin.

Reference

ERIKSSON-GRENNBERG, K. G. 1968. Resistance of *Escherichia coli* to penicillins. II. An improved mapping of the ampA gene. *Genet. Res. 12:* 147.

Method

Day 1 Plate 6 drops of a fresh overnight of CSH51 onto each of 3 LB plates with 10 μg/ml ampicillin. Incubate at 37°.

Day 2 Repurify ampicillin-resistant colonies.

Materials

Day 1
overnight of CSH51
3 LB plates with 10 μg/ml ampicillin
Add ampicillin, dissolved in the minimum amount of DMSO necessary, to medium directly before pouring.

Day 2
1 LB plate with 10 μg/ml ampicillin
6 round wooden toothpicks

Additional Markers Used for Selection

1. **Kasugamycin resistance:** Kasugamycin is an aminoglycoside antibiotic which inhibits protein synthesis. Mutants resistant to this drug have been shown to have an altered 30S ribosomal subunit due to a change in the methylation of the 16S RNA. Mutations conferring resistance map at the *ksg* locus, at approximately **1 min.** The selection is simple; rich plates containing 1000 μg/ml kasugamycin are used. This drug can be obtained from Bristol Labs, East Syracuse, New York.

References

HELSER, T. L., J. E. DAVIES and J. E. DAHLBERG. 1972. Mechanism of kasugamycin resistance in *E. coli*. *Nature New Biol. 235:* 6.
HELSER, T. L., J. E. DAVIES and J. E. DAHLBERG. 1971. Change in methylation of 16S ribosomal RNA associated with mutation to kasugamycin resistance in *Escherichia coli*. *Nature New Biol. 233:* 12.
SPARLING, P. F. 1970. Kasugamycin resistance: 30S ribosomal mutation with an unusual location on the *Escherichia coli* chromosome. *Science 167:* 56.

2. **Fucose resistance:** The growth of *E. coli* is inhibited on minimal arabinose medium by D-fucose. Mutants which grow on plates containing 0.1% L-arabinose and 0.2% D-fucose have mutations in *araC* (C^c), which map at **1 min**.

Reference

ENGLESBERG, E., J. IRR, J. POWER and N. LEE. 1965. Positive control of enzyme synthesis by gene C in the L-arabinose system. *J. Bacteriol. 90:* 946.

3. **Azide resistance:** Mutations conferring resistance to sodium azide map at approximately **2 min**. LB plates with 2×10^{-3} M azide are used, although the background is often heavy for this selection if a lawn of cells is directly plated. Stock solutions are prepared directly before use and filter-sterilized. Add 10 ml from a 0.2 M stock solution of sodium azide in water to one liter of medium, directly before pouring.

References

LEDERBERG, J. 1950. The selection of genetic recombinations with bacterial growth inhibitors. *J. Bacteriol. 59:* 211.
YURA, T. and C. WADA. 1968. Phenethyl alcohol resistance in *Escherichia coli*. I. Resistance of strain C600 and its relation to azide resistance. *Genetics 59:* 177.

4. **Constitutive *lac* mutants:** Phenyl-β-D-galactoside is a substrate for β-galactosidase, but it is not an inducer of the *lac* operon. Smith and Sadler have determined optimal conditions for the use of this compound in selecting mutations in the *i* gene. These involve the use of minimal medium containing 0.7–1.0 grams of phenylgalactoside per liter. Pure agar is recommended, and alternative carbon sources from the plates are scavenged by spreading Lac^- bacteria on the plates several hours in advance of their use.

Reference

SMITH, T. F. and J. R. SADLER. 1971. The nature of lactose operator constitutive mutations. *J. Mol. Biol. 59:* 273.

5. **Lac^- mutants:** Selection against the Lac permease function can be achieved by using minimal agar containing $1–2 \times 10^{-3}$ M *o*-nitrophenyl-β-D-thiogalactoside, available from Cyclo Chemical Co. Since this compound is not an inducer, IPTG must be present in the plates (5×10^{-4} M). It has been reported that succinate is the best carbon source to use for this type of selection. The majority of mutants resistant to this compound will be either y^- or z^- polar mutants (**10 min**).

References

GILBERT, W. and B. MÜLLER-HILL. 1966. Isolation of the *lac* repressor. *Proc. Nat. Acad. Sci. 56:* 1891.
SMITH, T. F. and J. R. SADLER. 1971. The nature of lactose operator constitutive mutations. *J. Mol. Biol. 59:* 273.

6. **Chloramphenicol resistance:** Mutations at several loci confer resistance to chloramphenicol. Some of these mutations result in mucoid colonies. Resistance mutations which do not result in mucoidy map principally at two loci, *cmlA* and *cmlB* (**19 min** and **21 min**, respectively). Mutants are isolated by plating approximately 10^7 cells on nutrient agar plates containing 5 μg/ml chloramphenicol. By their characteristic pattern of resistance to other antibiotics, such as tetracycline and puromycin, the mutants can be assigned to either *cmlA* or *cmlB*.

References

REEVE, E. C. R. 1966. Characteristics of some single-step mutants to chloramphenicol resistance in *Escherichia coli* K12 and their interactions with R-factor genes. *Genet. Res. 7:* 281.
REEVE, E. C. R. and P. DOHERTY. 1968. Linkage relationships of two genes causing partial resistance to chloramphenicol in *Escherichia coli*. *J. Bacteriol 96:* 1450.

7. **Chlorate resistance:** Mutations in the *chlA*, *chlB*, or *chlC*, or *chlD* locus (also referred to as the *nar* locus) confer resistance to chlorate during anaerobic growth. These mutants lack the enzyme nitrate reductase and cannot use nitrate as a terminal electron acceptor during anaerobic growth. These mutants are selected on plates containing nutrient broth to which 0.2% glucose and 0.2% $KClO_3$ (autoclaved separately) are added. The plates must be incubated under anaerobic conditions, which can be achieved by using a desiccator at 37° under an atmosphere of 90% N_2 and 10% CO_2. After 24 hours, further incubation overnight in the presence of air is carried out (Adhya et al., 1968).

In addition, *chl* mutants will not form colonies on a special lactate-nitrate medium under anaerobic conditions. Per liter this medium contains: 6.08 g

$NaH_2PO_4{\cdot}2H_2O$; 10.6 g K_2HPO_4; 2 g $(NH_4)_2SO_4$; 4 g KNO_3; 7 ml of a 50% solution of potassium lactate; 0.4 g vitamin-free casamino acids (Difco); and 5 ml of a stock salts solution containing, per liter, 10 g $MgSO_4{\cdot}7H_2O$; 1 g $MnCl_2{\cdot}4H_2O$; 0.05 g $FeSO_4{\cdot}7H_2O$; 0.1 g $CaCl_2$ and a "trace of HCl to clarify" (Venables and Guest, 1968).

The four *chl* loci map at **18 min**, **71 min**, **25 min**, and **17 min**, respectively. The *chlA* locus has now been subdivided into two loci, *chlA* and *chlE* (Guest, 1969).

References

ADHYA, S., P. CLEARY and A. CAMPBELL. 1968. A deletion analysis of prophage lambda and adjacent genetic regions. *Proc. Nat. Acad. Sci. 61:* 956.

GUEST, J. R. 1969. Biochemical and genetic studies with nitrate reductase *C*-gene mutants of *E. coli*. *Mol. Gen. Genet. 105:* 285.

VENABLES, W. A. and J. R. GUEST. 1968. Transduction of nitrate reductase loci of *Escherichia coli* by phages P1 and λ. *Mol. Gen. Genet. 103:* 127.

8. **5-Methyltryptophan resistance:** Mutations conferring resistance to 5-methyltryptophan map at the *trpR* locus (**90/0 min**; Cohen and Jacob), in the *trp* operon (**25 min**; Moyed; and Somerville and Yanofsky) and at the *mtr* locus (**61 min**; Hiraga et al.). Plates containing glucose minimal medium and 100 μg/ml of 5-methyltryptophan (and no tryptophan) are used. The 5-methyltryptophan should be prepared directly before use, and preferably filter-sterilized. There is often a background of slowly growing cells in this selection.

References

COHEN, G. and F. JACOB. 1959. Sur la répression de la synthèse des enzymes intervenant dans la formation du tryptophane chez *Escherichia coli*. *Compt. Rend. Acad. Sci. 248:* 3490.

HIRAGA, S., K. ITO, T. MATSUYAMA, H. OZAKI and T. YURA. 1968. 5-Methyltryptophan-resistant mutations linked with the arginine G marker in *Escherichia coli*. *J. Bacteriol. 96:* 1880.

MOYED, H. S. 1960. False feedback: Inhibition of tryptophan biosynthesis by 5-methyltryptophan. *J. Biol. Chem. 235:* 1098.

SOMERVILLE, R. L. and C. YANOFSKY. 1965. Studies on the regulation of tryptophan biosynthesis in *Escherichia coli*. *J. Mol. Biol. 11:* 747.

9. **Glycerol-phosphate transport mutants:** Mutants lacking the glycerol-phosphate transport system (*glpT*, **43 min**) are isolated by plating cells on medium containing phosphonomycin (Merck).

Reference

Quoted in LIN, E. C. C. 1970. The genetics of bacterial transport systems. *Ann. Rev. Genet. 4:* 234.

10. **Canavanine resistance:** *E. coli* mutants resistant to 100 μg/ml L-canavanine sulfate are isolated using arginine-free minimal medium (Jacoby and Gorini). The medium is enriched with a mixture of amino acids and other nutrients (Novick and Maas). Strains isolated in this manner have mutations in *argR* (**62 min**).

References

JACOBY, G. A. and L. GORINI. 1967. Genetics of control of the arginine pathway in *Escherichia coli* B and K. *J. Mol. Biol. 24:* 41.

NOVICK, R. P. and W. K. MAAS. 1961. Control by endogenously synthesized arginine of the formation of ornithine transcarbamylase in *Escherichia coli*. *J. Bacteriol. 81:* 236.

11. **Streptomycin and spectinomycin resistance:** Mutants resistant to 100 μg/ml of either antibiotic are rare and at least 10^{10} cells must be plated in order to find spontaneous mutants. A preferred method for converting a strain to either Str^r or Spc^r is to transduce it with P1 phage grown on a resistant strain (several generations of growth must occur before the transduced cells are plated on selective medium). These markers are widely used for counterselection, since the background is clean and revertants to Str^r or Spc^r are rare. Mutations to resistance map at **64 min**.

References

ANDERSON, P., J. DAVIES and B. D. DAVIS. 1967. Effect of spectinomycin on polypeptide synthesis in extracts of *Escherichia coli*. *J. Mol. Biol. 29:* 203.

LEDERBERG, J. 1950. The selection of genetic recombinations with bacterial growth inhibitors. *J. Bacteriol. 59:* 211.

12. **Fusidic acid resistance:** Mutants resistant to levels between 100 μg/ml and 1000 μg/ml of this drug have altered G factors. The mutations map at **64 min.** Fusidic acid is added to rich medium at 800 μg/ml, to which 2×10^{-3} M EDTA has been added.

Reference

KUWANO, M., D. SCHLESSINGER, G. RINALDI, L. FELICETTI and G. P. TOCCHINI-VALENTINI. 1971. G factor mutants of *Escherichia coli:* Map location and properties. *Biochem. Biophys. Res. Commun. 42:* 441.

13. **β-glucoside fermentors:** *E. coli* cannot ferment β-glucosides such as salicin or arbutin. However, single-step mutants can grow on these sugars. Selection is carried out on minimal medium supplemented with 0.075% yeast extract, 0.5% of either salicin or arbutin, and 0.02% bromothymol blue. Mutants are detected as fermenting papillae on the colonies. There is also a change in the indicator; for instance, arbutin fermentors turn black on this medium. The mutations allowing growth on β-glucosides map in the *bgl* region, at **73 min**. This marker is co-transducible with the *ilv* region. It is particularly useful in transferring Ilv^- markers from one strain to another. An Ilv^- strain is converted to a derivative which can grow on salicin, and P1 lysates are made on this strain. Transduction of an Ilv^+ strain with the lysate, selecting for recombinants which can grow on salicin, yields strains which are Ilv^-.

Reference

SCHAEFLER, S. 1967. Inducible system for the utilization of β-glucosides in *Escherichia coli*. I. Active transport and utilization of β-glucosides. *J. Bacteriol. 93:* 254.

14. **Low thymine-requiring strains:** As described in Experiment 31, mutants with lesions in the *deo* operon (**89 min**) can be isolated starting from *thyA*$^-$ strains, by selecting for strains which can grow in the presence of 4 μg/ml of thymine.

15. **Methylglyoxal resistance:** Exogenous methylglyoxal is toxic to *E. coli*. Cells are plated on glucose minimal medium containing 1 mM methylglyoxal (Sigma Chemical Co.). Resistant mutants appear at a frequency of approximately 10^{-7}. All of thc resistant mutants studied so far show an increased level of a detoxification enzyme, which is probably glyoxalase.

Reference

FREEDBERG, W. B., W. S. KISTLER and E. C. C. LIN. 1971. Lethal synthesis of methylglyoxal by *E. coli* during unregulated glycerol metabolism. *J. Bacteriol. 108:* 137.

16. **6-Azauracil resistance:** Mutants of *E. coli* which are resistant to 6-azauracil have a defect in the uracil permease system and map at the *uraP* locus at **50 min.** Wild-type strains are tolerant to levels lower than 4 μg/ml. Selection for resistant mutants should be done in minimal medium in the absence of uracil. Plates should contain 20–40 μg/ml of 6-azauracil. (R. Lavalle, personal communication).

17. **5-Fluorouracil resistance:** Mutants at the *udp* locus **(74–75 min)** can be found as two-step 5-fluorouracil-resistant mutants. Single-step mutants are resistant to 2.5 μg/ml 5-fluorouracil. These are also found to be resistant to 6-azauracil (100 μg/ml) and are probably identical to the mutants mapping at *uraP* discussed above. These strains can be re-sensitized to 5-fluorouracil by the addition of 200 μg/ml of adenosine. Second-step mutants resistant to 2.5 μg/ml 5-fluorouracil and 200 μg/ml adenosine map primarily at the *udp* locus. Mutants which are *udp*$^-$ cannot use uridine as the sole carbon source.

For a full discussion of the pathways of pyrimidine metabolism and the use of analogs for isolating mutants lacking enzymes involved in these pathways, see Pritchard and Ahmad (1971) and the review by O'Donovan and Neuhard (1970).

References

O'DONOVAN, G. A. and J. NEUHARD. 1970. Pyrimidine metabolism in microorganisms. *Bacteriol. Rev. 34:* 278.

PRITCHARD, R. and S. I. AHMAD. 1971. Fluorouracil and the isolation of mutants lacking uridine phosphorylase in *E. coli*. Location of the gene. *Mol. Gen. Genet. 111:* 84.

18. **Isolation of Trp$^-$ mutants:** Experiment 42 details the use of T1 resistance in obtaining deletions of the *E. coli* chromosome which extend into the *trp* operon. The procedure can be used for any strain of *E. coli* K12. Treat cultures with lysates of $\phi 80v$ and colicinV, B, and test the survivors for Trp$^-$ by replicating onto minimal medium with and without tryptophan.

19. **Sugar markers:** In Unit II methods are presented for the mutagenesis of strains and the isolation of Lac$^-$ mutants on lactose indicator plates. In an identical manner Ara$^-$, Mal$^-$, Gal$^-$, Rha$^-$, and Xyl$^-$ mutants can be selected. Tetrazolium plates with arabinose, galactose, maltose, rhamnose, or xylose (50 ml of a 20% solution per liter) should be used. EMB plates work equally well and are preferable for isolating Rha$^-$ and Xyl$^-$ mutants.

20. **Phage resistance:** Phage-resistant mutants can usually be selected by spotting a drop of a concentrated lysate of the phage onto a lawn of cells in a soft agar layer. Resistant colonies grow up out of the spot. It is often found that pre-adsorption of a concentrated suspension of cells followed by plating on rich medium gives a cleaner background. For full details of selection for phage resistance, see Experiment 42A.

EXPERIMENT 33

Penicillin and Ampicillin Treatment for the Isolation of Auxotrophic Mutants

Penicillin is a bacteriocidal agent which kills only growing cells; non-growing cells are not affected by this drug. This makes possible a selection technique which enriches for auxotrophic mutants in the population (Davis; Lederberg and Zinder). Cells requiring histidine, for instance, cannot grow in minimal medium without histidine. Treatment with penicillin after transfer of a culture to minimal medium without histidine will result in the death of all His^{+} cells but not His^{-} cells, since only the former class are growing. Normally, cells growing in the presence of penicillin lyse, because the action of this antibiotic interferes with cell wall synthesis. However, if we use a hypertonic protective medium (with added sucrose and Mg^{++}), then the cells are converted to protoplasts and do not lyse (Gorini and Kaufman). Sometimes this is necessary because a large population of lysed cells can cross-feed nutrients to auxotrophs, which will then begin to grow and subsequently be killed by the penicillin (Rossi and Berg).*

Since all auxotrophs survive the treatment, the cells are grown up in minimal medium with histidine (to select against other auxotrophs) and then plated out, and individual colonies are tested for His^{-}. The technique enriches for non-growing cells (in this case His^{-} cells), but it is not completely selective, and many

* The majority of investigators do not employ the sucrose procedure.

His^+ cells "leak" through. Following a single cycle of penicillin treatment, a typical enrichment for the desired mutant is 10^3- to 10^4-fold. This means that if the desired mutant occurs spontaneously at a frequency of 10^{-7}, we must score several thousand of the penicillin survivors in order to find one colony with the desired mutation.

It is, however, possible to increase substantially this frequency by either of two methods:

(1) Perform the penicillin treatment on cells which have previously been mutagenized.

(2) In the event that mutagenesis is undesirable, it is advantageous to carry out repeated penicillin treatments.

However, due to the simultaneous enrichment for penicillin-resistant mutants, enrichment for a particular penicillin-sensitive auxotroph decreases with each additional cycle. Three penicillin treatments are usually the limit. This problem can be overcome by combining in sequence penicillin treatment with other similar enrichment procedures, such as nalidixic acid or cycloserine treatment (see addendum to this experiment).

Ampicillin

An analog of penicillin, ampicillin (Experiment 32D), has several advantages over penicillin for use in enrichment procedures. Since gram-negative bacteria such as *E. coli* are more sensitive to this drug, lower concentrations are required. Also, resistant mutants are less of a problem with ampicillin than with penicillin. Because cleaner results are consistently obtained with ampicillin, it is the agent of choice for auxotroph enrichment procedures (Molholt).

Other Mutants

Penicillin or ampicillin treatment can also be used to isolate non-auxotrophic mutants. For example, *E. coli* is typically impermeable to 100 μg/ml of the antibiotic streptolydigin (and the cells are therefore resistant to its action). Using penicillin we can isolate mutants which are highly permeable to this drug. A broth culture of the cells is first treated with 100 μg/ml of streptolydigin and then shortly thereafter with penicillin. In this case a rich medium without sucrose is desirable and the culture can be agitated during the penicillin step (see methods), since cross-feeding is not a complication. The cells which are impermeable to the drug grow freely in the presence of the antibiotic, only to be subsequently killed during the penicillin step. The permeable mutants whose growth is blocked by the drug survive the penicillin, and these mutants can be identified by their failure to grow after replica plating onto rich medium containing 100 μg/ml streptolydigin (Schleif). It is important to note that the penicillin method can be used here only because the action of streptolydigin is **reversible.** If the antibiotic acted irreversibly to halt the growth of the permeable mutants, then these mutants could never be recovered in a viable form after the penicillin treatment.

In the following experiment His^- mutants are isolated from a His^+ strain (CSH7, although any strain can be used) after mutagenesis with NG followed by two cycles of ampicillin treatment. Protocols for penicillin and cycloserine treatment are also provided.

References

CURTISS, R., III, L. J. CHARAMELLA, C. M. BERG and P. E. HARRIS. 1965. Kinetic and genetic analyses of D-cycloserine inhibition and resistance in *Escherichia coli*. *J. Bacteriol. 90:* 1238.

DAVIS, B. D. 1948. Isolation of biochemically deficient mutants of bacteria by penicillin. *Amer. Chem. Soc. J. 70:* 4267.

GORINI, L. and H. KAUFMAN. 1960. Selecting bacterial mutants by the penicillin method. *Science 131:* 604.

HOFFMAN, E. P., R. C. WILHELM, W. KONIGSBERG and J. R. KATZ. 1970. A structural gene for seryl-tRNA synthetase in *Escherichia coli* K12. *J. Mol. Biol. 47:* 619.

LEDERBERG, J. and N. ZINDER. 1948. Concentration of biochemical mutants of bacteria with penicillin. *Amer. Chem. Soc. J. 70:* 4267.

MOLHOLT, B. 1967. Isolation of amino acid auxotrophs of *E. coli* using ampicillin enrichment. *Microb. Genet. Bull. 27:* 8.

ROSSI, J. J. and C. M. BERG. 1971. Differential recovery of auxotrophs after penicillin enrichment in *Escherichia coli*. *J. Bacteriol. 106:* 297.

SCHLEIF, R. 1969. Isolation and characterization of a streptolydigin resistant RNA polymerase. *Nature 223:* 1068.

Strains

Number	Sex	Genotype	Important Properties
CSH7	F^-	*lacY strA thi*	Lac^- Su^- Str^r

Method

Day 1

Mutagenize a culture of strain CSH7 with NG, exactly as prescribed in Experiment 13. After washing the cells, grow overnight in minimal medium with 100 μg/ml of histidine. If 5 ml of an exponentially growing culture are used, then resuspend in 5 ml buffer and inoculate 1 ml into 20–25 ml of medium in a 125- or 250-ml flask and grow overnight.

Day 2

Spin down 5 ml of the minimal overnight, wash twice in 5 ml buffer, and resuspend in 5 ml buffer. Inoculate 5 or 10 ml of prewarmed minimal medium (1 × A supplemented with glucose, B1, and 10^{-3} M $MgSO_4$) with enough bacteria to give a final concentration of 1–2 × 10^7 cells/ml and shake in a flask at 37°, to starve for histidine. It is convenient if the flasks have side arms which can fit into a klettmeter. If not, measure the OD_{550} and continue to aerate until the optical density increases 4- to 5-fold.

At this point, add ampicillin at a final concentration of 20 μg/ml (or penicillin at a concentration of 10,000 units/ml) and continue to shake for about 60–90 minutes until lysis is complete. (If the cells are not shaken, this will take somewhat longer, 2–2½ hours.) This can be seen by observing loss of turbidity of the culture, and the optical density can be followed in a klettmeter.

As soon as the culture has lysed, centrifuge, wash, and resuspend in the same minimal medium plus histidine. Since the number of surviving cells will be small, spinning at a minimum of 5000 rpm for 20 minutes is necessary to obtain a pellet. As an alternative to centrifugation, add a crystal of penicillinase and continue to shake for 20 minutes. Grow overnight at 37° in histidine minimal medium.

In procedures calling for the use of hypertonic protective medium (20% sucrose and 10^{-2} M $MgSO_4$), lysis is not observed. The duration of the treatment is for one generation time, determined by optical density readings during the starvation period. Also, cells are not shaken, but allowed to sit in a large erlenmeyer flask (10 ml in 250-ml flask) to achieve aeration.

Day 3 Repeat the procedure exactly as in Day 2. After washing out the ampicillin, grow the cells in minimal medium plus histidine until a density of 10^8 cells/ml is reached. (It is perhaps simplest to allow the cells to grow overnight.) At this point plate dilutions on glucose minimal plates with histidine and incubate at 37°.

Day 5 When the colonies are at least 2 mm in diameter, replicate them in succession onto glucose minimal plates with histidine, glucose minimal plates without histidine, and again glucose minimal plates without histidine.

Day 6, 7 Retest several repurified His^- colonies for growth with and without histidine.

Materials

Day 1

exponentially growing culture of strain CSH7
nitrosoguanidine
15 ml citrate buffer, pH 5.5
10 ml phosphate buffer, pH 7.0
2 small centrifuge tubes, centrifuge
6 5-ml or 10-ml pipettes
25 ml 1 × A minimal medium with proline, L-histidine (100 μg/ml), glucose, B1, and $MgSO_4$ in a 125- or 250-ml flask

Day 2

100-ml flask with 10 ml 1 × A medium supplemented with glucose, B1, and 10^{-3} M $MgSO_4$
ampicillin
sodium penicillin G, (1965 units/mg, Nutritional Biochemicals Corp., Cleveland, Ohio)
spectrophotometer
2 centrifuge tubes
centrifuge (at least 5000 rpm)
20 ml phosphate buffer
test tube with 5 ml 1 × A minimal medium supplemented with glucose, histidine, B1, and $MgSO_4$
5 5-ml pipettes

Day 3

same materials as in Day 2
3 dilution tubes
4 0.1-ml pipettes, 1 1-ml pipette
6 glucose minimal plates with histidine

Day 4, 5 **(per 2 plates replicated)**

3 glucose minimal plates with histidine
6 glucose minimal plates
2 replica velvets, replicating block

Other Enrichment Techniques

D-Cycloserine: Like penicillin and ampicillin, D-cycloserine lyses growing *E. coli*. Enrichment procedures using D-cycloserine have proved as effective as penicillin treatment (Curtiss et al.). Mutagenized cells, growing overnight in appropriately supplemented minimal medium, are washed and resuspended in medium lacking the required nutrient, at a density of 5×10^6 cells/ml. After two doublings in this medium, D-cycloserine is added to a final concentration of 2×10^{-3} M and the culture aerated for 2–3 hours. Cells are then washed and resuspended in supplemented minimal medium, and either plated out directly or grown up and recycled. D-Cycloserine is obtained from Eli Lilly and Co., Indianapolis, Indiana. It is prepared directly before use by making a 0.2 M solution in a 0.1 M phosphate buffer, pH 8.0.

Nalidixic acid: Treatment for 3 hours at 37° with 100 μg/ml of nalidixic acid has also proved to be an effective enrichment procedure (Hoffman et al.).

Use of assay medium: Rich medium lacking individual amino acids is now available from Difco, as special "assay medium." Although not originally prepared for use with *E. coli*, a 1% solution of this medium in M9 buffer is used with *Salmonella* and should suffice for *E. coli* as well. (This medium should not be autoclaved, but filter-sterilized instead.) The advantage of, for instance, histidine assay medium is evident in penicillin selections. Killing is more complete, and enrichments of 10^6 in a single cycle are possible.

EXPERIMENT 34

Isolation and Mapping of Temperature-sensitive Lethal Mutants

A necessary part of the study of important functions in bacteria is the isolation of mutants unable to carry out essential steps in growth, cell division, or macromolecular synthesis, and also mutants lacking factors required for the control of these functions. In many cases these mutants are lethal, since for instance, a cell lacking RNA polymerase cannot grow. It is, therefore, important to be able to find conditions in which these mutants survive to allow their isolation and preservation. Mutants which grow under one set of conditions but not another are termed **conditional lethals**.

Temperature-sensitive (*ts*) Mutations

Originally studied extensively by Horowitz and Leupold (1951) in bacteria and by Edgar and Lielausis (1964) in T4 phage, temperature-sensitive mutants have facilitated the study of DNA synthesis (Kohiyama et al., 1966; Gross, 1971; Gefter et al., 1971), ribosomes (Phillips et al., 1969; Tai et al., 1965; Guthrie et al., 1969), messenger RNA breakdown (Lennette et al., 1971) and activating enzymes (Neidhardt, 1966). Temperature-sensitive mutants grow at low (permissive) temperatures but not at high (non-permissive) temperatures.

Cold-sensitive mutants have been isolated also, in phage T4 (Scotti, 1968), phage T1 (Christensen and Saul, 1966), phage ϕX174 (Dowell, 1967), and *E. coli* (Reid, 1971).

Nonsense Mutations—Streptomycin Suppression

Nonsense mutations in essential genes are also conditional lethal mutations. In the appropriate Su^+ strain these mutations are suppressed and the missing function is partially restored. An advantage of this type of conditional lethal over a temperature-sensitive mutation is the total loss of the respective protein in the case of most nonsense mutations. The elimination of residual activity facilitates biochemical studies and reduces the background in genetic reversion tests.

A disadvantage of lethal nonsense mutations in bacteria is the difficulty in isolating these mutants, since alternating between the Su^+ and the Su^- state normally involves genetic constructions which are inconvenient for screening a large number of strains. However, the recent isolation of a temperature-sensitive amber suppressor (Smith et al., 1970) makes the testing of a large number of strains for nonsense mutants lacking important functions more feasible. In this case the initial screening can be carried out by incubating colonies at both high and low temperature (see Experiment 22).

Murgola and Adelberg (1970) have made use of streptomycin suppression in the detection of lethal mutations. In the presence of this antibiotic, bacterial ribosomes make occasional errors in translation. Often these result in the suppression of a mutation if an acceptable amino acid is now substituted for the altered one.

Screening Procedures

To look for strains which are temperature sensitive, different colonies are picked onto master plates which are grown overnight at low temperature. These are subsequently replicated onto two plates, one of which is incubated at high temperature and the other at low temperature. Although 25–32° is often used as the permissive temperature and 40–43° as the non-permissive temperature, there are no limits to the temperatures which can be used. Colonies showing significantly less growth at the non-permissive temperature are then retested. Alternatively, dilutions of the culture can be plated for single colonies and incubated at low temperature for part of the incubation period until microcolonies appear, and then further incubated at high temperature. Small colonies are candidates for temperature-sensitive lethals. This technique avoids the necessity of replica plating and allows the screening of a larger number of colonies.

It is important to note the difference between mutants which stop growing under non-permissive conditions and those which are killed under these conditions. The latter class can be recovered only if a copy grown solely under permissive conditions is saved. In this case the replica plating method mentioned above is applicable, but the small colony technique is not. Also prohibited for this latter class of mutants are enrichment procedures which require incubation of cultures under non-permissive conditions.

Enrichment Procedures—Tritium Suicide

Penicillin treatment (see Experiment 33) has been used effectively to enrich for cells which stop growing at high temperature, as has tritium suicide (Lubin, 1959; Person, 1963). With this latter technique cultures are allowed to incorporate tritiated precursors and then stored in the cold. Over the course of several weeks the tritium decay preferentially kills those cells which had incorporated the labeled precursor. When labeling is carried out under restrictive conditions, such as high temperature (Tocchini-Valentini et al., 1969), there is an enrichment for conditional lethal mutants. This technique has also been used to select cold-sensitive mutants (Reid, 1971).

Mutagenesis

Which mutagen to use and how much mutagenesis to apply are problems confronting every investigator embarking on a search for new mutants. There is no set answer to these questions, for each situation requires special consideration. Compounds such as EMS and nitrosoguanidine (NG) are certainly the most powerful mutagenic agents. However, because of the number of secondary mutations created by treatment with these mutagens, their use is not recommended for studies in which mutant proteins are to be sequenced. This is particularly true for NG which has been shown to cause multiple mutations within a small region (see Experiment 14).

For isolating mutants that require difficult screening procedures, however, NG is still the agent of choice. One suggestion is to use heavy mutagenesis initially, then, after the mutant has been isolated and characterized, to go back and isolate additional mutants using lower levels of mutagenesis and different mutagens. One can then take advantage of information on the phenotype of the original mutant in refining the screening procedure. Complete protocols for mutagenesis are given in Experiments 13–16.

Mapping

It is advisable to move any new mutation out of the original strain as soon as possible, particularly when the strain has undergone heavy mutagenesis. Ideally this should be done with P1, since only a small region of the bacterial DNA is co-transduced with a particular marker (Experiment 38). In order to use P1, and to facilitate many later studies (such as the preparation of merodiploids or specialized transducing phage lines, Unit V), it is desirable to map the new mutation as precisely as possible (within 1–2 min of a known marker). This task is simplified by the proper choice of the starting strain. It is tempting to use an Hfr. However, this is risky for two reasons. If the new marker is transferred very late by the Hfr, it makes genetic manipulations difficult. Also, heavy mutagenesis often results in a loss of donor ability of Hfr strains. They do not become recipients, though, since they retain the surface exclusion properties of male strains (Experiment 10).

An F^- strain which is Str^r and is multiply marked is a better choice. If the marker to be mapped is a non-leaky temperature-sensitive lethal, then it can be mapped by using a set of streptomycin-sensitive Hfr strains, such as those depicted in Figure 34A. A short mating is carried out with each donor, and the selection is for growth at high temperature. More precise mapping is subsequently done with interrupted matings, using the other recipient markers as internal controls. This is necessary because the mating must be carried out at temperatures below 37°* and the rate of transfer is therefore lower. If the phenotype of the marker is difficult to select for, then the recipient markers are used as selected markers, and the new mutant phenotype as the unselected marker (see Experiment 7).

If it is undesirable to have multiple nutritional markers in the starting strains,

* In many instances the mating can be carried out at 35°.

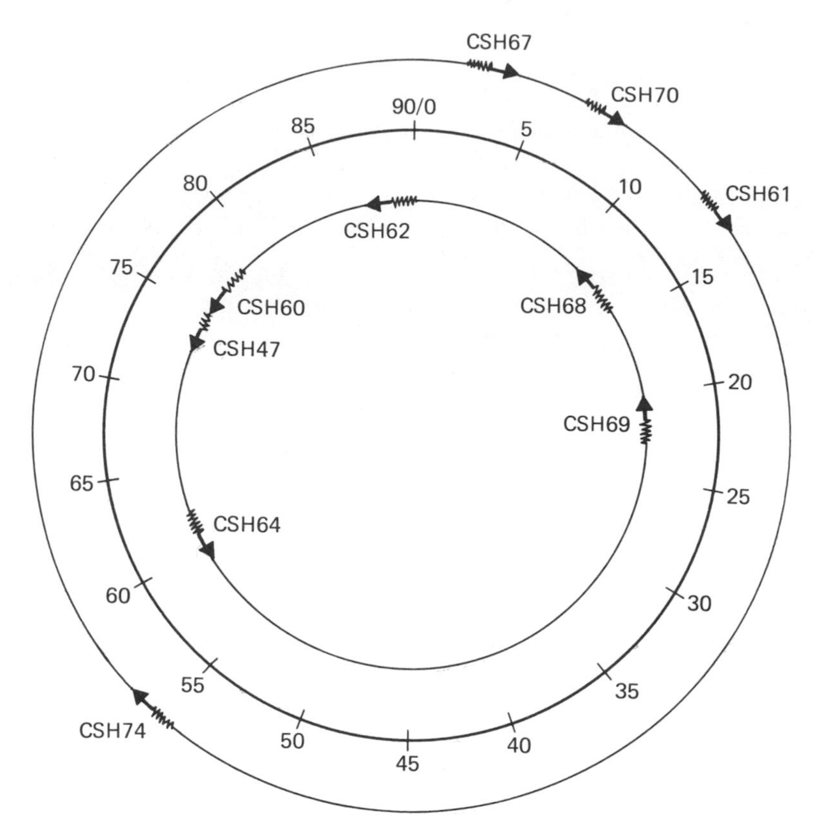

Figure 34A. Hfr strains used for mapping temperature-sensitive lethal mutants.

an alternative is to use a wild-type starting strain, and after introducing an F′ factor, to use the merodiploid as a donor in a cross with a marked recipient, since markers can be transferred from one F^- strain to another using F′-mediated chromosome mobilization (Experiment 29). This strategy was used to map the *polA* locus, which determines the structure of the DNA polymerase I (De Lucia and Cairns, 1969). An F′*lac*(i^s) was crossed into this strain, selecting for Lac^- colonies. The merodiploid was then mated with a multiply marked recipient (Gross and Gross, 1969).

Localized Mutagenesis

A technique recently developed by Hong and Ames (1971), termed localized mutagenesis, makes possible the isolation of temperature-sensitive mutations in any specific region of the chromosome. The principle of this method is to mutagenize generalized transducing phage DNA rather than host bacterial DNA. Bacteria with auxotrophic markers are then transduced to prototrophy at low temperature. When the transductant colonies are still small, the selection plates are transferred to high temperature. Colonies which remain small after continued incubation are purified and tested for temperature sensitivity.

With this procedure one selects beforehand the region of the DNA in which mutations are desired, since the resulting mutations are closely linked to the original marker. This eliminates the necessity of mapping new temperature-sensitive lethals.

Hong and Ames studied colonies of *S. typhimurium* arising from transduction with phage P22 mutagenized *in vitro* with hydroxylamine. This method should also be applicable to the P1-*E. coli* generalized transduction system (see Experiment 28). These authors examined *ts* mutants found after transduction of three different regions of the chromosome. The number of small colonies varied between 2.4% and 7% of the total transductants. Of these approximately 1% were true temperature-sensitive mutants. In all cases the mutations obtained by this procedure were found to be co-transducible with the marker used in the isolation.

Sample Procedures

Specific procedures have been used to enrich for particular temperature-sensitive mutants. The following examples describe selections which have been used by different investigators. Understanding the rationale of each approach should facilitate the beginning student in devising his own strategies for isolating particular mutants.

Nitrosoguanidine mutagenesis. As mentioned previously, nitrosoguanidine leads to many secondary mutations even under conditions in which the survival is high and the percentage of induced auxotrophs relatively low. The following example of a typical mutagenesis experiment shows what to expect.

Karamata and Gross used nitrosoguanidine to induce temperature-sensitive mutants in *B. subtilis*. A culture at a density of 2×10^7 cells/ml was treated with NG. The survival was approximately 50%. One ml was then transferred to 100 ml of broth and grown overnight to a density of 5×10^8 cells/ml, at 30°. Dilutions of the cultures were then plated on rich plates and the resulting colonies were replicated onto different media to test for temperature-sensitive growth, and also for the percentage of auxotrophs. Among the colonies tested, 2% were auxotrophs and 0.2% temperature-sensitive lethals. However, 10% of the temperature-sensitive mutants were also auxotrophs at the permissive temperature.

Questions: Does the increase in percentage of auxotrophs among the lethal mutants indicate that most of the temperature-sensitivity mutants carry secondary mutations? The authors, using the 0.2% value, reasoned that the initial culture after mutagenesis contained approximately 2×10^4 temperature-sensitive mutants, each of which gave rise to a clone of 5×10^3 cells (1×10^7 cells/100 ml increased to 5×10^{10} cells/100 ml). From the Poisson distribution they calculated that in this situation an isolate of 200 temperature-sensitive mutants from one culture should yield only one or two from the same clone. Since they picked 50 mutants from each original culture, they reasoned that the chances that all were of independent origin were high. Do you agree? Is the estimate of 2×10^4 mutants in the original culture a minimum or a maximum estimate? How does this affect the final determination?

Transfer-deficient mutants of the F factor. Achtman and coworkers looked for a series of F mutants which were deficient in the ability to transfer the F factor. In addition, nonsense mutants of this type were also sought. Here cultures were treated with EMS or NG and then immediately dispensed into 50–100 different tubes and grown up overnight. The starting strain carried an $F'lac^+$ factor and harbored a *lac* deletion on the chromosome. Dilutions of the cultures were plated on lactose indicator plates and the colonies were tested for the ability to transfer the F′*lac*$^+$ factor. Consider the following four strains:

(A) The starting strain	Lac$^-$ Su1$^+$ Str^s Spc^s
(B)	Lac$^-$ Su1$^+$ Str^r Spc^s
(C)	Lac$^-$ Su$^-$ Str^r Spc^s
(D)	Lac$^-$ Su$^-$ Str^r Spc^r

The F′*lac*$^+$ factor was transferred by replica matings from strain A into strains B and C. Mutants which failed to transfer at all were detected at this step. Strains B and C which received an F′*lac*$^+$ in the first mating were then mated with strain D. Amber mutations which were suppressed by Su1$^+$ in strains A and B were not suppressed in the strain C background. These were detected as mutants which did not transfer the F′ factor to strain D from strain C, but did from strain B. (Note how the use of combinations of two counterselective markers allows transfer between several strains.) Using this procedure 28 transfer-deficient colonies were found among 5000 EMS-treated colonies and 155 out of 10,000 NG-treated colonies. Six of this latter class were found to be ambers.

Questions: Why were the cultures diluted into different tubes immediately after being treated with the mutagen? Why were colonies initially plated on lactose indicator plates? Suppose you cured the starting strain (A) of the F′ factor and wished to transfer the F′*lac*$^+$ factor from strain D into the F$^-$ strain A background. How would you do this? Does failure to transfer from strain A to either strain B or C prove the existence of an F′ factor mutation? Was it necessary to test transfer from both strain B and strain C into strain D in searching for ambers? Wouldn't the failure to transfer from strain C to strain D be sufficient proof? How do you move a transfer-deficient F′*lac*$^+$ factor out of the strain A background?

Protein synthesis mutants. Tocchini-Valentini and coworkers used tritium suicide to isolate temperature-sensitive mutants of *E. coli* which were specifically blocked in protein synthesis. Mutagenized cells were incubated in minimal medium at 42° for 10 minutes and then labeled with ^{3}H-tyrosine and ^{3}H-phenylalanine for 30 minutes. The cells were then washed, resuspended in minimal medium, and allowed to sit at 4°. After two weeks the survival was 10^{-6}, and 15% of the survivors were temperature-sensitive. Mutants with unaltered RNA synthesis at high temperature were studied further. Those unable to support the polyU-directed polyphenylalanine synthesis *in vitro* were analyzed in detail. From these strains, a mutant containing a temperature-sensitive G factor was isolated.

Questions: As the survival of the tritium-labeled cells in the cold fell, did the percentage of temperature-sensitive mutants among the survivors increase, decrease, or remain the same? What other classes of mutants would be enriched for? Why were two different labeled amino acids used?

Phospholipid mutants. An example of how effective the tritium suicide method can be is provided by the work of Cronan, Tapas, and Vagelos (1970). These authors designed a protocol for isolating temperature-sensitive mutants defective in phospholipid biosynthesis. They used a strain which was defective in alkaline phosphatase and also the catabolic L-glycerol-3-phosphate dehydrogenase. In addition, the strain was constitutive for the transport of L-glycerol-3-phosphate. These properties lead to a large percentage (>98%) of added ^{3}H-L-glycerol-3-phosphate being incorporated into the phospholipid of growing cells.

After mutagenesis, cultures were grown overnight in broth at 25° and then subcultured into fresh medium. After reaching log phase, the cells were incubated at 37° for 30 minutes and then labeled with ^{3}H-L-glycerol-3-phosphate for 70 minutes. The cultures were stored at 4° and periodically plated for survivors at 25°. Cell survival dropped exponentially, the fraction of survivors falling to less than 10^{-7} in 22 days. Of these, 75% were temperature sensitive in their growth behavior, and practically all of the *ts* mutants were defective in their ability to incorporate labeled phosphate into phospholipid at 37°.

DNA synthesis mutants. Bonhoeffer and Schaller took advantage of the observation that cells with DNA containing 5-bromouracil were more sensitive to ultraviolet light (313 mμ) than cells without this analog. They were able to isolate mutants which were temperature-sensitive for DNA synthesis. Mutagenized cultures were labeled with 5-bromouracil and after several doublings the bacteria were resuspended in buffer and irradiated with ultraviolet light (filtered to diminish radiation outside the 313 mμ range). Two successive treatments resulted in a 10^5-fold enrichment for temperature-sensitive mutants for DNA synthesis.

Design of Experiment

It can be seen from the preceeding examples how important proper planning is in searching for new mutants, regardless of whether they are conditional lethals or not. Which selection or screening procedure to use is often the only problem that investigators focus on. However, no selection or screening procedure is perfect, and methods should be devised beforehand for the analysis of the many candidates which pass through the screening process.

In the following experiment we mutagenize an F^- Lac^- Str^r strain and search for conditional lethals. Any strain which contains many nutritional markers can be used however, for instance CSH57. Temperature-sensitive mutants are then mapped with a set of Hfr strains which have different points of origin (Figure 34A).

References

ACHTMAN, M., N. WILLETS and A. J. CLARK. 1971. Beginning a genetic analysis of conjugational transfer determined by the F factor in *E. coli* by isolation and characterization of transfer-deficient mutants. *J. Bacteriol. 106:* 529.

BONHOEFFER, F. and H. SCHALLER. 1965. A method for selective enrichment of mutants based on the high UV sensitivity of DNA containing 5-bromouracil. *Biochem. Biophys. Res. Commun. 20:* 93.

CHRISTENSEN, J. R. and S. H. SAUL. 1966. A cold-sensitive mutant of bacteriophage T1. *Virology 29:* 497.

CRONAN, J. E., JR., R. K. TAPAS and R. P. VAGELOS. 1970. Selection and characterization of an *E. coli* mutant defective in membrane lipid biosynthesis. *Proc. Nat. Acad. Sci. 65:* 737.

DE LUCIA, P. and J. CAIRNS. 1969. Isolation of an *E. coli* strain with a mutation affecting DNA polymerase. *Nature 224:* 1164.

DOWELL, C. E. 1967. Cold-sensitive mutants of bacteriophage φX174. *Proc. Nat. Acad. Sci. 58:* 958.

EDGAR, R. S. and I. LIELAUSIS. 1964. Temperature-sensitive mutants of bacteriophage T4: Their isolation and genetic characterization. *Genetics 49:* 649.

GEFTER, M. L., H. YUKINORI, T. KORNBERG, J. A. WECHSLER and C. BARNOUX. 1971. Analysis of DNA polymerase II and III in mutants of *E. coli* thermosensitive for DNA synthesis. *Proc. Nat. Acad. Sci. 68:* 3150.

GROSS, J. D. 1971. *Current Topics in Microbiology and Immunology.* Springer-Verlag, Berlin.

GROSS, J. and M. GROSS. 1969. Genetic analysis of an *E. coli* strain with a mutation affecting DNA polymerase. *Nature 224:* 1166.

GUTHRIE, C., H. NASHIMOTO and N. NOMURA. 1969. Structure and function of *E. coli* ribosomes. VIII. Cold-sensitive mutants defective in ribosome assembly. *Proc. Nat. Acad. Sci. 63:* 384.

HONG, J. and B. N. AMES. Localized mutagenesis of any specific small region of the bacterial chromosome. *Proc. Nat. Acad. Sci. 68:* 3158.

KARAMATA, D. and J. D. GROSS. 1970. Isolation and genetic analysis of temperature-sensitive mutants of *B. subtilis* defective in DNA synthesis. *Mol. Gen. Genet. 108:* 277.

KOHIYAMA, M., D. COUSIN, A. RYTER and F. JACOB. 1966. Mutants thermosensibles *d'Escherichia coli* K12. Isolement et caractérisation rapid. *Ann. Inst. Pasteur 110:* 465.

LENNETTE, E. T., L. GORELIC and D. APIRION. 1971. An *E. coli* mutant with increased messenger ribonuclease activity. *Proc. Nat. Acad. Sci. 68:* 3140.

LUBIN, M. 1959. Selection of auxotrophic bacterial mutants by tritium-labeled thymidine. *Science 129:* 838.

MURGOLA, E. J. and E. A. ADELBERG. 1970. Streptomycin-suppressible lethal mutations in *Escherichia coli. J. Bacteriol. 103:* 20.

NEIDHARDT, F. C. 1966. Roles of amino-acid activating enzymes in cellular physiology. *Bacteriol. Rev. 30:* 701.

PERSON, S. 1963. Comparative killing efficiencies for decays of tritiated compounds incorporated into *E. coli. Biophys. J. 3:* 183.

PHILLIPS, S. L., D. SCHLESSINGER and D. APIRION. 1969. Mutants in *E. coli* ribosomes: A new selection. *Proc. Nat. Acad. Sci. 62:* 772.

REID, P. 1971. Isolation of cold sensitive-rifampicin resistant RNA polymerase mutants of *E. coli. Biochem. Biophys. Res. Commun. 44:* 737.

SCOTTI, P. D. 1968. A new class of temperature conditional lethal mutants of bacteriophage T4D. *Mutation Res. 6:* 1.

SMITH, J. D., K. ANDERSON, A. CASHMORE, M. L. HOOPER and R. L. RUSSELL. 1970. Studies on the structure and synthesis of *E. coli* tyrosine transfer RNA. *Cold Spring Harbor Symp. Quant. Biol. 35:* 21.

TAI, P. C., D. P. KESSLER and J. INGRAHAM. 1969. Cold-sensitive mutations in *S. typhimurium* which affect ribosome synthesis. *J. Bacteriol. 97:* 1298.

TOCCHINI-VALENTINI, G. P., L. FELICETTI and G. M. RINALDI. 1969. Mutants of *E. coli* blocked in protein synthesis: Mutants with an altered G factor. *Cold Spring Harbor Symp. Quant. Biol. 34:* 463.

Strains

Number	Sex	Genotype	Important Properties
CSH7	F^-	*lacY thi strA*	Lac^- Str^r

Method

Day 1

Mutagenize CSH7 (or any desired strain) with nitrosoguanidine as described in Experiment 14. Inoculate as many different test tubes as possible directly after mutagenesis, or dilute 1 ml into 100 ml broth in a 500-ml flask. Grow overnight in rich (LB) broth at 30°.

Day 2

Plate dilutions of the overnight culture on rich plates (0.1 ml of a 10^{-4}, 10^{-5}, and 10^{-6} dilution). Aim for 200–300 colonies per plate. Incubate at 30° for approximately 10 hours. When very small colonies are visible, transfer the plates to 42° and incubate overnight. If mutants are sought which would be killed at 42°, allow the colonies on the initial plate to grow to normal size and then replicate onto 2 fresh plates, incubating one at 30° and the other at 42°. Always replicate onto the 42° plate first in this case. If penicillin enrichment is to be carried out, subculture the overnight and proceed as in Experiment 33.

Dilutions of the overnight from Day 1 should also be plated on selective plates to determine the extent of mutagenesis by examining independent markers. CSH7 is Lac^-, therefore the number of Lac^+

revertants in the culture can be found by plating on lactose minimal plates and also on glucose minimal plates. Alternatively, galactose tetrazolium or arabinose tetrazolium plates can be used to detect Gal^- or Ara^- mutants, respectively. The Val^r marker (Experiment 32) is also convenient in this regard.

In addition, the percentage of auxotrophs induced by the mutagenesis can be determined by replicating several hundred colonies onto glucose minimal plates and then onto rich plates.

If complicated screening procedures are to be carried out in the search for a temperature-sensitive mutation, it is better to monitor the level of mutagenesis first with one of these markers before proceeding. If the mutagenesis was too low or too high, then the mutagenesis can be repeated without wasting effort by screening these cells. If you intend to test for auxotrophs, then incubate several plates from the overnight culture solely at 30°.

Day 3

Examine the plates. Pick each small colony from the 42° plate and streak for single colonies on an LB plate. Incubate at 30°.

Test for auxotrophs by replicating the colonies from one of the plates incubated at 30° onto a glucose minimal plate. Incubate at 30°.

Day 4

Retest a single colony from each strain by picking with a toothpick and stabbing into two plates and then streaking for single colonies. The first plate should be pre-incubated at 42° and then placed at the same temperature. Incubate the second plate at 30°.

Day 5

Examine the plates from Day 4. All mutants which grow significantly more slowly at high temperature should be saved for further characterization. Record the percentage of colonies from the mutagenized population which were temperature-sensitive, and compare this with the frequency of cells which were mutant for the other markers tested (see Day 2).

If you compare the frequency of Lac^+ revertants, Ara^- mutants, auxotrophs, and temperature-sensitive lethals from the same mutagenized population, what would you expect the relative frequencies of each to be? Why? What procedures would you employ to increase the proportion of each one of these particular classes? What changes in the protocol would you add which would specifically decrease the proportion of each of these classes of mutants?

Mapping

The set of Hfr strains provided with this manual (strain list; Figure IIE) allow rapid mapping of all temperature-sensitive markers. Grid the mutants on an LB plate and incubate at 30° overnight.

Day 6

Prepare fresh cultures of each Hfr to be tested. When the density reaches 2–3 × 10^8 cells/ml, chill, and spin down and resuspend in $\frac{1}{3}$ to $\frac{1}{2}$ the original volume of buffer. Spread 6 drops of each Hfr on a fresh LB plate. A more even spreading of the donor can be achieved by adding 1 ml to the plate and pouring this off after 5 minutes.

Replicate the gridded mutants onto this last plate. Incubate at 35° (use 30–32° if desired) for 60 minutes and then spray with streptomycin (2% in an atomizer) and incubate at 42°.

If replication is done carefully, each master plate can be replicated onto 3 different velvets. If two sets of master plates are made for each group of 50 temperature-sensitive mutants, then rapid screening against 6 Hfr's can be accomplished. The mutants which show the least residual growth at 42° will give the clearest results for this test.

Day 7

Record the results. For mutants containing a temperature-sensitive mutation donated early by a given Hfr strain, you should see confluent growth in the replicated patch. Otherwise, only a few colonies will appear. Clearer results can be obtained by 30 minute interrupted mating experiments in ambiguous cases.

Once the approximate location of a marker is known, it can be transferred to a multiply marked recipient such as CSH57 and then mapped more precisely by interrupted matings. (This would be facilitated by the construction of a Nal^r derivative of CSH57.)

Materials

Day 1

exponentially growing cultures of CSH7 or CSH57
nitrosoguanidine
15 ml each of phosphate buffer, pH 7.0 and citrate buffer, pH 5.5
500-ml flask containing 100 ml LB broth
shaker at 30°
centrifuge, centrifuge tube
5 5-ml pipettes

Day 2

(per culture tested)
overnight culture of mutagenized CSH7
9 0.1-ml, 3 1-ml, and 6 pasteur pipettes
10–30 LB plates, 3 glucose minimal plates
3 lactose minimal plates or 3 glucose minimal plates containing 50 μg/ml L-valine
3 galactose or arabinose tetrazolium plates
6 dilution tubes (for a 10^{-1}, 10^{-2}, 10^{-3}, 10^{-4}, 10^{-5}, and 10^{-6} dilution). Use a 10^{-1}, $10^{-}2$, and a 10^{-3} dilution for valine or lactose minimal plates and a 10^{-3}, 10^{-4}, and 10^{-5} dilution for tetrazolium plates and glucose minimal plates.
incubators at 30° and 42°

Day 3

5–10 LB plates
40–80 round wooden toothpicks
glucose minimal plate
velvet, replicating block
incubator at 30°

Day 4

10–20 LB plates, half of which are pre-incubated at 42°
40–80 round wooden toothpicks
incubators at 30° and 42°

Day 5 **(per 50 mutants mapped)**

50 round wooden toothpicks
1 LB plate
incubator at 30°

Day 6 exponential cultures of strains CSH60–64; CSH67–70 (use 4–6 of these strains)
4–6 small centrifuge tubes
15 ml dilution buffer
1 10-ml pipette, 6 pasteur pipettes
6 LB plates pre-incubated at 35°
atomizer containing 2% solution of streptomycin
6 velvets, replicating block
incubators at 35° and 42°

Questions

1. It has been observed that under certain conditions strains which are "stringent" (for RNA synthesis after amino acid starvation) resume growth after a 60-minute lag after starvation for a required amino acid, whereas "relaxed" strains resume growth after a 120-minute lag. How could you use this information to design an enrichment procedure for temperature-sensitive relaxed mutants? Almost all of the relaxed mutants isolated so far have mutations which map at the *rel* locus, at 54 min. If the allele responsible for the relaxed phenotype were recessive to wild-type, how would you increase the chances of finding a mutation outside the *rel* locus which results in a temperature-sensitive relaxed phenotype?

2. Suppose you wished to isolate a mutant which at high temperature failed to terminate RNA synthesis. Would the lesion responsible cause cessation of growth or death at high temperature? Upon return to low temperature, would you predict a greater lag before resuming growth than in other temperature-sensitive mutants? Could you devise an enrichment procedure which would take advantage of this property? Would this lag time be increased or decreased by the presence of the *RC* (relaxed) mutation? How would the presence of the *supA* allele (affecting degradation of mRNA) affect the selection?

INTRODUCTION TO EXPERIMENTS 35 AND 36

Construction of Hfr Strains

Screening Methods

At a low frequency the F factor can integrate into the host chromosome, resulting in an Hfr strain (Figure 35A). A variety of techniques have been used to isolate Hfr strains arising in an F^+ population. A variation of the fluctuation test (Luria and Delbruck) involves screening individual F^+ colonies for their donor properties by cross-streaking against mutant recipients. Approximately 1% of the cultures show a marked increase in the number of recombinants. These "jackpots" are then streaked for single colonies and these are gridded and replica plated onto a lawn of recipients to isolate the Hfr strain (Berg and Curtiss).

A simpler technique, suggested by Dr. B. Low, is to irradiate with UV an F^+ population to 1–3% survival (to increase the frequency of integration) and plate the survivors onto rich plates, aiming for 300–1000 colonies per plate. After 20 hours at 37° the colonies are suitable for replica-mating. The plates are then replicated directly onto a lawn of a recipient. Ideally, streptomycin is used to counterselect. Markers transferred in the first 30 minutes should be used in the recipient. Screening with different recipients enables one to isolate Hfr's with points of origin in different places. The original rich plate is re-incubated, together

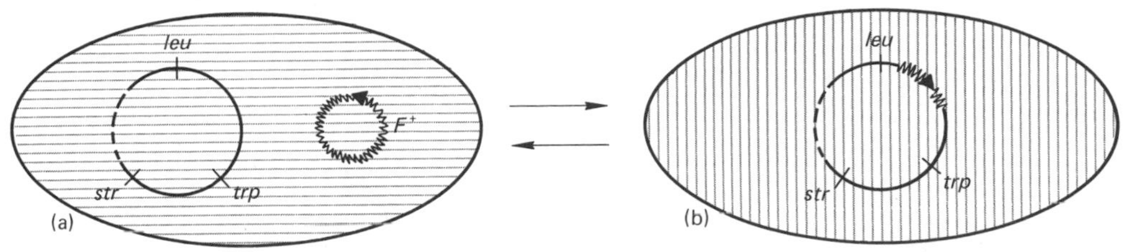

The majority of cells in an F^+ population are as depicted in (a), with the *F* factor in a non-integrated state. These cells donate chromosomal markers at a low frequency. Rare cells in an F^+ population are as depicted in (b), with the *F* factor integrated into the host chromosome. In this example the *F* factor has integrated near the *leu* region. These cells donate the *leu* region of the chromosome at a high frequency.

Figure 35A. Hfr formation.

with the selective plate, at 37°. Colonies on the rich plate, corresponding to the positions on the selective replication plate which showed growth of recombinants, are then repurified and retested (see Figure 35B).

Selection for Integration of F Factor

It is also possible to apply selective pressure for the integration of the F factor. Experiment 35 deals with a selection based on a host requirement for the F replication system which can operate on the host chromosome only if the F factor is integrated.

Experiment 40 uses an F′ factor containing bacterial genes (*lac*) which are deleted on the host chromosome. Under conditions promoting curing of the F′ factor, the only cells still maintaining these bacterial genes are those which have had an integration of the F factor.

Finally, we can convert F^- strains to Hfr's by crossing against other characterized Hfr strains (Experiment 36).

References

Berg, C. M. and R. Curtiss, III. 1967. Transposition derivatives of an Hfr strain of *Escherichia coli* K12. *Genetics 56:* 503.

Luria, S. E. and M. Delbrück. 1943. Mutations of bacteria from virus sensitivity to virus resistance. *Genetics 28:* 491.

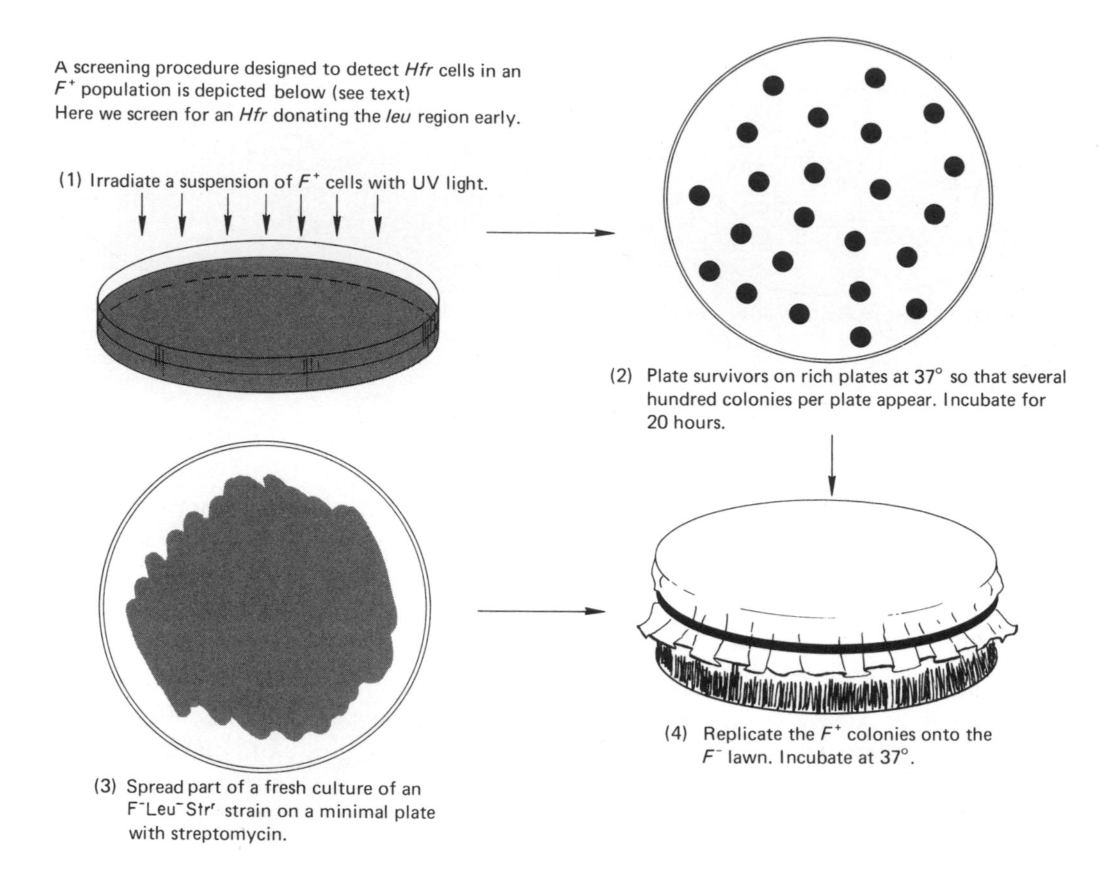

(5) After 36 hours compare the master plate with the selection plate.

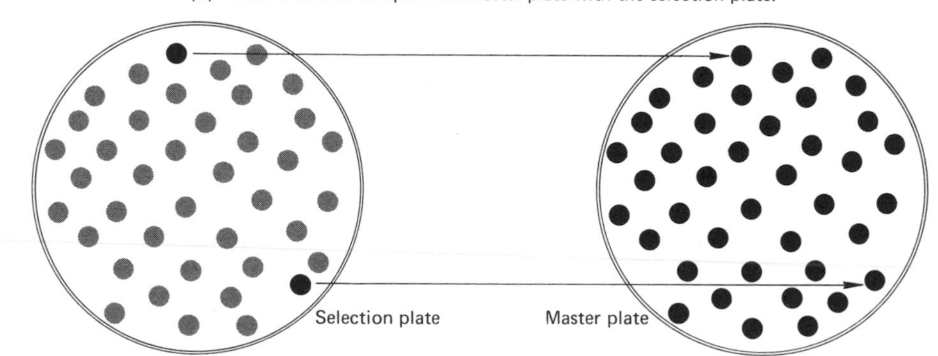

Areas of the minimal plate which show heavy growth due to many Leu^+Str^r recombinants are the result of an *Hfr* which donated the *leu* region early having been replicated on that spot. The *Hfr* can then be repurified by picking from the corresponding region of the master plate. Regions of the selection plate onto which F^+ cells have been replicated show little or no growth.

Figure 35B. Method for isolating Hfr strains.

EXPERIMENT 35

Selection for Integration of F Factor

Strain CSH42 contains a mutation which renders the cell unable to initiate DNA synthesis at high temperature (40–42°). Although the DNA replication system of the F factor cannot complement this defect when the F factor is in the autonomous (non-integrated) state, it can compensate for it when the F factor is integrated into the host chromosome. Therefore, among the revertants of CSH42 which can grow at high temperature will be a class of strains which have the F factor integrated at various points along the chromosome. These are now Hfr strains. Figure 35C shows some of the Hfr's which have been isolated by this technique. Although there are many points at which the F factor can integrate into the host chromosome, these appear to be limited since one finds many independently isolated Hfr's to be identical with respect to the point of origin.

In this experiment we will find revertants at 40° of CSH42 and test by the cross-streaking method (Experiment 8) whether they are Hfr's and if so determine the point of origin and direction of transfer.

Reference

NISHIMURA, Y., L. CARO, C. M. BERG and Y. HIROTA. 1971. Chromosome replication in *Escherichia coli*. IV. Control of chromosome replication and cell division by an integrated episome. *J. Mol. Biol.* *55:* 441.

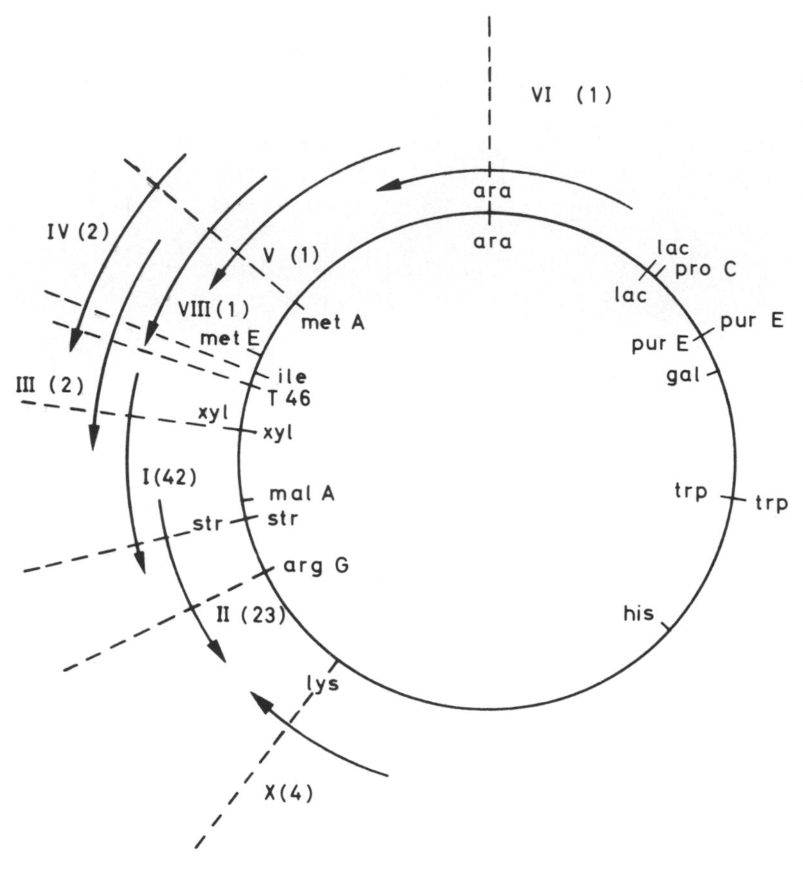

Figure 35C. Hfr strains isolated by integrative suppression. (Reprinted with permission from Nishimura et al., 1971.)

Strains

Number	Sex	Genotype	Important Properties
CSH42	F^+	*thr leu lac thyA mal ilv thi*(T46)	Str^s, does not form colonies at 40–42°
CSH57	F^-	*ara leu lacY purE gal trp his argG malA strA xyl mtl ilv metA* or *B thi*	Ara$^-$ Leu$^-$ Lac$^-$ Pur$^-$ Gal$^-$ Trp$^-$ His$^-$ Arg$^-$ Mal$^-$ Ilv$^-$ Met$^-$ Str^r

Method

Day 1 **Selection of Revertants of CSH42**

Pick 5 colonies of strain CSH42 from a plate which was grown at 30° and inoculate into 5 test tubes containing LB broth supplemented with 50 μg/ml thymine. Incubate the cultures at 30° overnight.

Day 2 Plate dilutions of each culture on LB plates containing 50 μg/ml thymine. Plate 0.1 ml directly from each culture, and also 0.1 ml from a 10^{-1} and a 10^{-2} dilution. Incubate the plates at 40°. It has been found that colonies appearing at 40° are mostly Hfr strains of the type we seek in this experiment, whereas those appearing at higher temperatures (such as 42°) are not.

Day 3 Pick 8 temperature-independent colonies from each plate (8 per original culture) and purify by streaking for single colonies on the same type of plate. Incubate at 40°.

Day 4 **Mapping Revertants**

Inoculate 5 ml of LB broth containing 50 μg/ml thymine with a single colony of each strain to be tested, and shake at 40° until a density of approximately 1×10^8 cells/ml is reached. At this point allow the cultures to sit at 37° for 60–90 minutes without aeration. Test for the ability to donate chromosomal markers by mating with strain CSH57 using the cross-streaking technique described in Experiment 8. The donor strain, CSH42, is Thr^-, Leu^-,Thy^-,Ilv^-. We can use the Ara^-,$PurE^-$,Trp^-,His^-,Arg^-, and Met^- markers of the recipient, CSH57.

In order to use the Ara^- marker, both threonine and leucine should be present in the plates, since these markers will be inherited by most of the Ara^+ recombinants. For the Ara marker use plates of type D (Table IIC, Unit II) but instead of lactose use arabinose, and in addition add threonine to these plates. Incubate the plates at 37°.

Day 4, 5 Examine the plates. The temperature-sensitive marker is near the *ilv* region. Hfr strains with points of origin in this region should be retested and the plates should be incubated at 30° after several hours at 37°.

Materials

Day 1

LB plate containing 50 μg/ml thymine with colonies of CSH42
5 test tubes with 5 ml LB broth containing 50 μg/ml thymine
shaker at 30°

Day 2

10 dilution tubes
5 1-ml and 20 0.1-ml or pasteur pipettes
15 LB plates containing 50 μg/ml thymine
incubator at 40°

Day 3

40 toothpicks
5 LB plates containing 50 μg/ml thymine
incubator at 40°

Day 4 **(per 40 strains tested)**

40 test tubes with 5 ml LB broth containing 50 μg/ml thymine
exponential culture of CSH57

41 small (100 μl) capillary pipettes
5–8 plates each of the types shown below
incubator at 37°
shaking waterbath at 40°

Plate type	Selected marker	Map position (min)
D*	Ara^+	1
E	$PurE^+$	13
G	Trp^+	25
H	His^+	39
A	$ArgG^+$	61
B	Met^+	77–78

* Arabinose substituted for lactose; threonine added.

EXPERIMENT 36

Conversion of an F^- Strain to an Hfr Strain

The origin of an Hfr is actually the last marker to be transferred in a bacterial cross. This marker is transferred after 90–100 minutes at 37°, under optimal conditions. Although we cannot directly select for recombinants that have received the origin, we can score recombinants that have received a very late marker and ask whether they have in addition become males (see Figure 36A).

In the following section we use this technique to convert F^- strains to HfrH and HfrC strains, making use of the locations of the *rif* and *trp* regions of the chromosome. The *rif* locus is transferred after approximately 80 minutes by HfrH. In part A we perform a prolonged mating of an F^-Rif^s strain with an HfrH Rif^r strain and examine Rif^r recombinants for the ability to donate chromosomal markers.

Since the *trp* region is transferred by an HfrC very late, F^-Trp^- strains can be converted to HfrC strains by selecting for Trp^+ recombinants (see Figure 36A) after a long mating and then screening for parental markers and donor properties (part B).

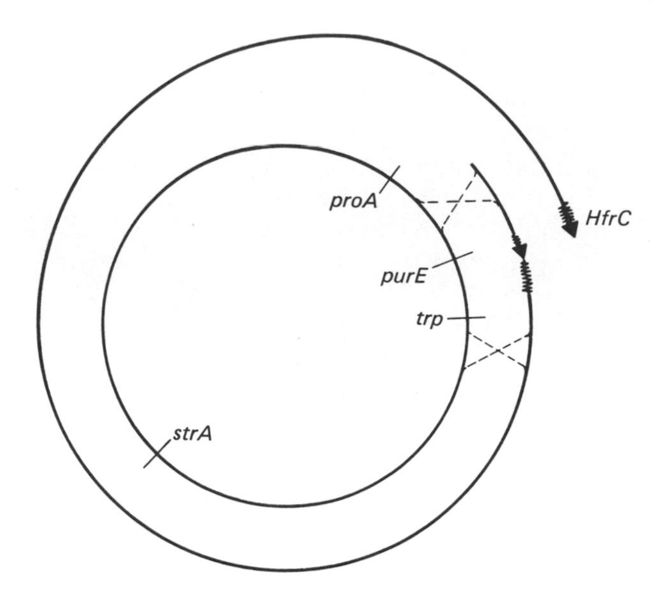

Figure 36A. Conversion of an F^- to an HfrC strain. By depicting more than one unit length of Hfr chromosome being transferred to a recipient cell, we can envision how selection for a late, or for both an early (*purE*) and a late (*trp*) marker lead to inheritance of the Hfr property.

A. Conversion of an F^- Strain to HfrH

In this experiment an HfrH which is Str^sRif^r is crossed with a $Pro^-Str^rRif^s$ recipient. The idea is to select for a late marker, Rif^r (and also an early marker, Pro^+) and then to score the recombinants for Hfr's. After a 3-hour uninterrupted mating, streptomycin is added to prevent further growth of the donor and the mating mixture is grown for several doublings to allow segregation of the recessive rifampicin resistance allele (*rif*). Dilutions are then plated out on glucose minimal plates with rifampicin and streptomycin to select $Pro^+Str^rRif^r$ recombinants. These are then scored for parental markers and also tested for the ability to donate Pro^+ in an Hfr cross. Nalidixic acid is used to counterselect in the second cross.

Use the Rif^r derivative of CSH62 (HfrH) prepared in Experiment 32C. The recipient for this cross is CSH54 which requires (in addition to proline) histidine, uracil, and tryptophan to be able to grow on glucose minimal plates.

Strains

Number	Sex	Genotype	Important Properties
CSH62(*rif*)	HfrH	*thi rif*	Rif^r Str^s
CSH54	F^-	Δ(*lacpro*) *supF trp pyrF his strA thi*	Pro^- Rif^s Str^r
CSH55	F^-	*ara* Δ(*lacpro*) *supE nalA thi*	Pro^- Nal^r

Method

Day 1

Part A

Subculture fresh overnights of the HfrH and the F^- in rich broth. Dilute the donor 1:50 and the recipient 1:20. Aerate at 37° until the donor is at a density of $2\text{–}3 \times 10^8$ cells/ml (approximately $2\frac{1}{2}$ hours). Prepare a mating mixture either by mixing 2 ml of donor and 2 ml of recipient in a 100- or 125-ml flask or by mixing 5 drops of each in a large test tube. Allow to stand without shaking in a 37° waterbath for 3 hours.

Part B

Before plating on selective medium it is necessary to grow the recipients for several generations in the absence of rifampicin. Add 8 ml LB broth if the mating was done in the flask, and 2 ml LB broth if the mating was done in a large test tube with the specified volumes. The broth should contain enough streptomycin so that the final concentration is 100–150 μg/ml. Aerate at 37° for $2\frac{1}{2}$ hours.

Part C

Plate 0.1 ml of a 10^0, 10^{-1}, 10^{-2}, and a 10^{-3} dilution on glucose minimal plates with rifampicin, streptomycin, uracil, tryptophan, and histidine. Incubate at 37°.

Day 3

Pick and purify 30 $Pro^+Str^rRif^r$ recombinants onto the selection plates.

Day 4, 5

The recombinants will be tested both for parental markers and for the ability to donate the Pro^+ marker in an Hfr cross. Inoculate 10 different colonies (if possible more should be tested) into 5 ml LB broth and grow with aeration at 37° until a density of $2\text{–}3 \times 10^8$ cells/ml is reached. Prepare a mating mixture with each of these strains by mixing 5 drops with 5 drops of an exponentially growing culture of a Pro^-Nal^r recipient (CSH55) in a large test tube. Incubate at 37° in a waterbath for 60 minutes and then plate out a 10^{-2} dilution of each donor candidate and the recipient alone as controls. Plate 0.1 ml of a 10^{-2} dilution of each mating mixture onto glucose minimal plates with nalidixic acid.

From each donor candidate culture streak with a loop onto glucose minimal plates with: tryptophan, histidine, and uracil; tryptophan and histidine; tryptophan and uracil; and histidine and uracil. Streak 6 per plate. These are tests for Trp^+, His^+ and $PyrF^+$, respectively. Incubate at 37°.

Day 5, 6

Examine the plates from the day before. Record the number of strains which became Hfr as a result of the original cross and which retained the parental markers.

Materials

Day 1 **Part A**

fresh overnights of CSH54 and CSH62 Rif^r
2 test tubes with 5 ml LB broth
2 5-ml and 2 pasteur pipettes
100- or 125-ml flask
waterbath at 37°

Part B

8 ml LB broth containing 150 μg/ml streptomycin
1 10-ml pipette

Part C

5 0.1-ml and 2 1-ml pipettes
4 glucose minimal plates with rifampicin (100 μg/ml), streptomycin (100 μg/ml), tryptophan, uracil, and histidine
3 dilution tubes

Day 3

5 plates identical to those in Part C above
30 round wooden toothpicks

Day 4, 5 **(per 10 strains tested)**

10 test tubes with 5 ml LB broth
exponentially growing culture of CSH55
10 test tubes
42 0.1-ml and 12 pasteur pipettes
21 glucose minimal plates with nalidixic acid (20 μg/ml)
2 each of the following plates: glucose minimal plates supplemented with
- (A) uracil and tryptophan (to test for His^+)
- (B) uracil and histidine (to test for Trp^+)
- (C) histidine and tryptophan (to test for $PyrF^+$)
- (D) uracil, histidine, and tryptophan (control)

21 dilution tubes
waterbath at 37°

B. Conversion of an F^- Strain to Hfr Cavalli

In the following experiment we will cross an HfrC against an F^- which is $Pro^-Trp^-Str^r$. Selection is for a late and an early marker. The $Pro^+Trp^+Str^r$ recombinants are then tested for the ability to donate Pro^+ to a Pro^- strain. Streptomycin is used as the counterselective agent in the first cross, and nalidixic acid in the second. It is also possible to use a selection for the late marker alone (Trp^+ in this case).

Strains

Number	Sex	Genotype	Important Properties
CSH61	HfrC	*trpR thi*	Pro^+ Trp^+ Str^s
CSH53	F^-	*ara* Δ(*lacpro*) *strA thi tonB trp* (φ80d*lacI*)	Trp^- Pro^- Str^r Nal^s
CSH55	F^-	*ara* Δ(*lacpro*) *supE nalA thi*	Pro^- Nal^r

Method

Day 1 Subculture a fresh overnight of the HfrC strain CSH61 into fresh LB broth and aerate at 37° until a density of 2–3 × 10^8 cells/ml is reached. Prepare a mating mixture by mixing equal volumes of the donor with an exponentially growing culture of a Trp^-Str^r female (in this case CSH53) and incubate without interruption at 37° for 3 hours. Plate dilutions (0.1 ml of a 10^{-1}, 10^{-2}, and 10^{-3} dilution) and also 0.1 ml direct on glucose minimal plates with streptomycin. Also plate 1 drop of both the donor and recipient alone as a control. Incubate at 37°.

Day 3 Pick and purify $Pro^+Trp^+Str^r$ colonies onto glucose minimal plates and incubate at 37°.

Day 4, 5 Test the $Pro^+Trp^+Str^r$ colonies for sensitivity to male phage and also the ability to donate the Pro^+ marker in an Hfr cross. Test as many colonies as possible, but each group should test at least 15. For each strain to be tested, inoculate a colony into 5 ml of LB broth and grow at 37° with aeration until a density of 2–3 × 10^8 cells/ml is reached. Also subculture CSH55 ($F^-Pro^-Nal^r$) by diluting a fresh overnight 1:20 into LB broth. Prepare mating mixtures by mixing equal volumes of both donor and recipient and gently shaking at 37° for 60 minutes. At the end of this time plate 0.1 ml of a 10^{-2} dilution onto a glucose minimal plate with nalidixic acid. Incubate at 37° for 24–36 hours. Test for male phage sensitivity by the method described in Experiment 9. Cross streak from each donor culture directly before mating for this test.

Day 6, 7 Examine the plates. All of the Trp^+ recombinants from the cross which became Hfr's should be able to donate Pro^+ to a Pro^- recipient. What percentage of the Trp^+ colonies were Hfr's?

Materials

Day 1

overnight of CSH61 and of a $Trp^-Pro^-Str^r$ strain (CSH53)
2 tubes with 5 ml LB broth, empty test tube
waterbath at 37°
3 dilution tubes

5 0.1-ml, 2 1-ml, and 6 pasteur pipettes
6 glucose minimal plates with streptomycin

Day 3

8 glucose minimal plates with streptomycin
50 round wooden toothpicks

Day 5

(per 15 strains tested)

17 test tubes with 5 ml LB broth each
fresh overnight of CSH55
15 glucose minimal plates containing nalidixic acid (20 μg/ml)
15 dilution tubes
31 0.1-ml pipettes or 31 pasteur pipettes, 1 1-ml pipette
shaking waterbath at 37° or rotor (30 rpm) in 37° room
lysate of male-specific phage
4 LB or R plates

EXPERIMENT 37

F′ Factor Formation; Use of Rec^- Strains to Isolate F′ Factors Carrying Different Regions of the *E. coli* Chromosome

How do F′ factors arise in a bacterial population? Their derivation can be envisioned from Hfr strains, in which the F factor has previously integrated into the bacterial chromosome. In some of these cells the F factor comes out of the chromosome, occasionally incorporating some of the nearby bacterial genes into its own circular structure. The model in Figure 37A depicts this event as a rare genetic exchange occurring between regions of the bacterial chromosome on both sides of the F factor. This results in the formation of both a circular F′ factor and a chromosomal deletion. This has been demonstrated for the episome F13 shown in Figure 37A (Scaife, 1967).

The non-homologous exchange can also occur between a region of the chromosome and part of the F factor. This produces a different type of F′ factor (one carrying markers initially on only one side of the integrated F factor rather than one carrying markers on both sides). In addition, this event leaves some of the F factor still incorporated into the chromosome. After curing, another F factor can be introduced, which effects chromosome mobilization at a higher rate than otherwise, due to the homology with the residual F factor genetic material (Adelberg and Burns, 1960).

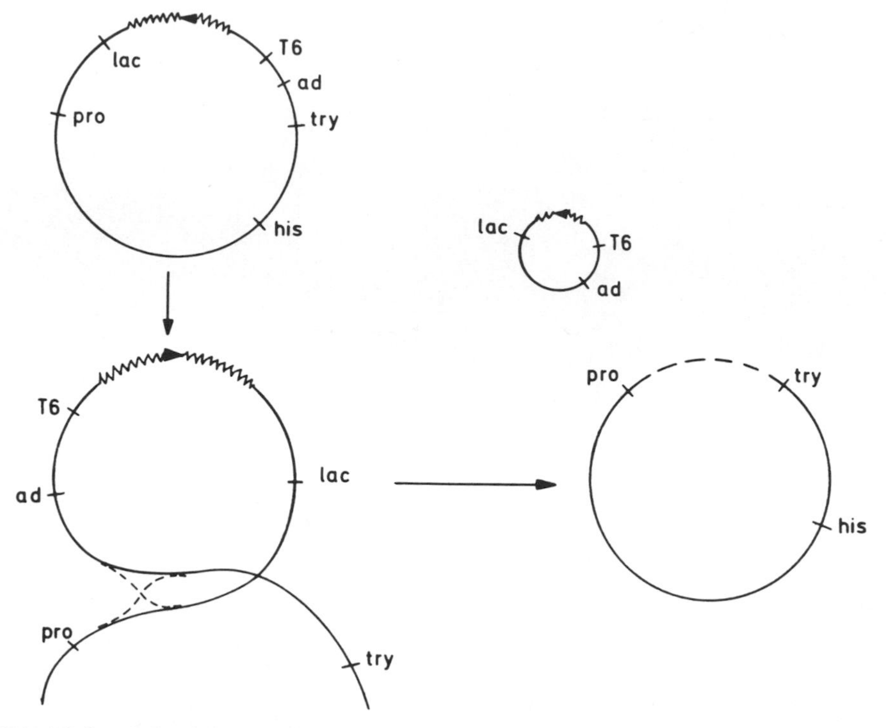

Figure 37A. F′ factor formation. These genetic elements are not drawn to scale. (Reprinted with permission from Broda et al., 1964.)

Isolation of F′ Factors

If an Hfr × F^- mating mixture is interrupted after 60 minutes and the donors killed by streptomycin, no transfer of terminal markers should occur. If we select for recipients which have received a terminal marker in such a cross, rather than finding recombinants, we obtain cells which contain an F′ factor carrying that marker. This is because F′ factors present in the original Hfr population are

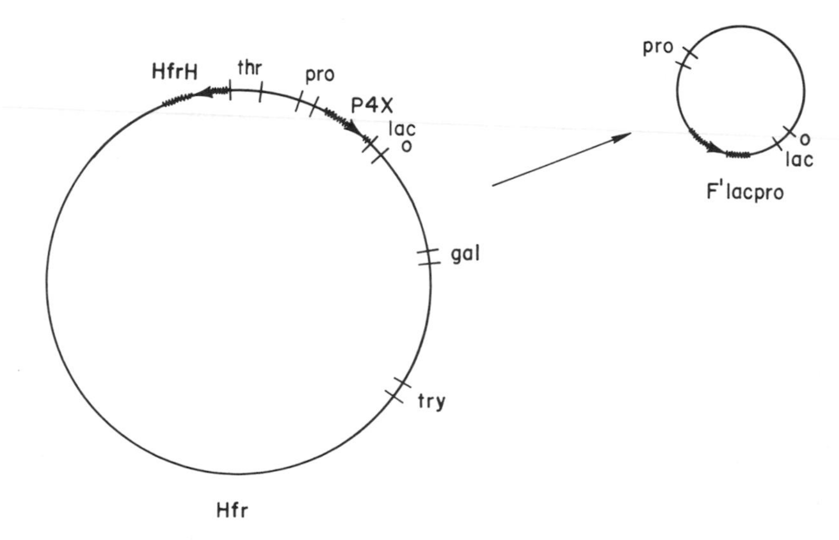

Figure 37B. Formation of F′*lac*.

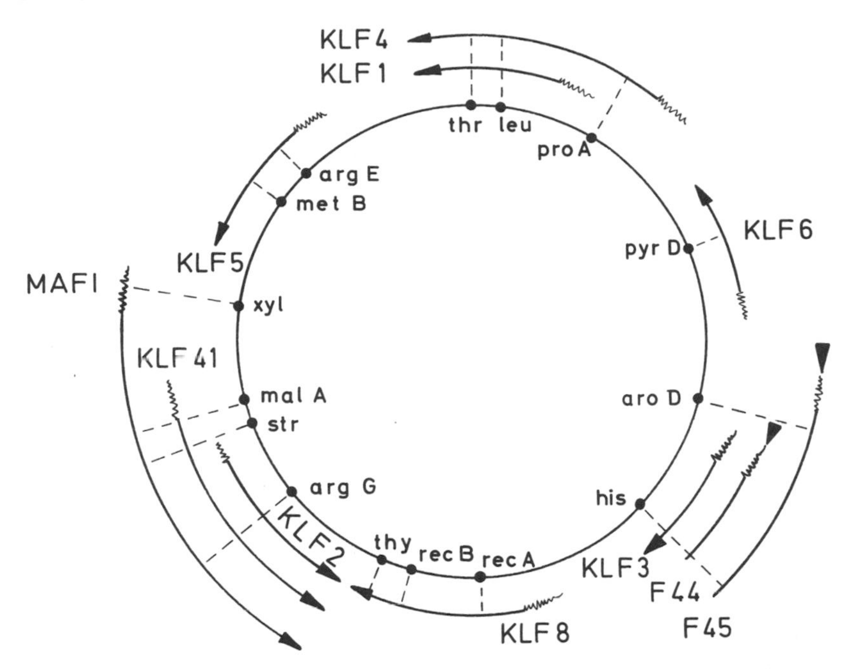

Figure 37C. F′ factors isolated from Rec⁻ recipients. (Redrawn from Low, 1968.) For a more complete list of F′ factors in *E. coli*, see Appendix.

transferred early (Jacob and Adelberg, 1959). Figure 37B shows two Hfr strains, P4X and HfrH. From P4X we could select for a *lacpro* episome by mating for a short time against a Lac^-Pro^- strain, since the *lac* region is transferred as a terminal marker by this strain.

Suppose, however, we wish to obtain an F′ carrying *thr* and *leu* from an HfrH. Since these markers are transferred early by this Hfr, the above method cannot be used. However, it is possible to isolate such an F′ factor by making use of the fact that a $Thr^-Leu^-Rec^-$ (*recA*) strain cannot form Thr^+Leu^+ recombinants but is able to carry an F′thr^+leu^+. Thus, merodiploids for the *thr-leu* region can be obtained by crossing an HfrH Str^s strain with an $F^-Thr^-Leu^-Str^r$ strain and selecting for cells which grow without threonine and leucine in the presence of streptomycin (Low, 1968). Figure 37C shows a set of F′ factors obtained by this method. It should be pointed out, however, that not all Hfr's give rise to F′ factors with detectable frequencies.

In the following experiment we use a Rec^- (*recA*) recipient to aid in the selection of F′ factors from HfrH (CSH62). Other Hfr strains are also provided to enable the selection of F′ factors carrying different regions of the *E. coli* chromosome. These are depicted in Figure IIE. After crossing HfrH (Str^s) against CSH58 ($F^-Thr^-Leu^-Rec^-Str^r$), the $Thr^+Leu^+Str^r$ colonies are tested for the presence of an F′thr^+leu^+ by donation to a Leu^- recipient. Nalidixic acid is used as a counter-selective agent in this case.

References

ADELBERG, E. A. and S. N. BURNS. 1956. A variant sex factor in *Escherichia coli*. *Genetics 44:* 497.

ADELBERG, E. A. and S. N. BURNS. 1960. Genetic variation in the sex factor of *Escherichia coli*. *J. Bacteriol. 79:* 321.

BRODA, P., J. R. BECKWITH and J. SCAIFE. 1964. The characterization of a new type of F-prime factor in *Escherichia coli* K12. *Genet. Res. 5:* 489.

CAMPBELL, A. M. 1969. *Episomes.* Harper and Row.

DAVIS, R. W., M. SIMON and N. DAVIDSON. 1971. Electron microscope heteroduplex methods for mapping regions of base sequence homology in nucleic acids. *In* (L. Grossman and K. Moldave, ed.). Methods in Enzymol. XXI D: 413.

JACOB, F. and E. A. ADELBERG. 1959. Transfert de caractères génétiques par incorporation au facteur sexuel d'*Escherichia coli. Compt. Rend. Acad. Sci. 249:* 189.

LOW, B. 1968. Formation of merodiploids in matings with a class of Rec$^-$ recipient strains of *Escherichia coli* K12. *Proc. Nat. Acad. Sci. 60:* 160.

SCAIFE, J. 1966. F-prime factor formation in *E. coli* K12. *Genet. Res. 8:* 189.

SCAIFE, J. 1967. Episomes. *Ann. Rev. Microbiol. 21:* 601.

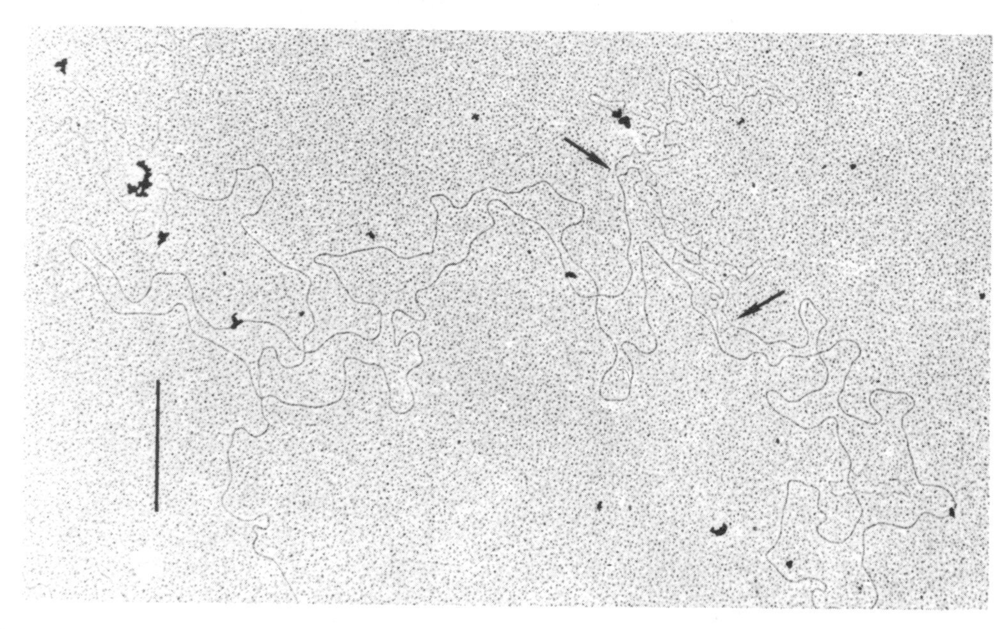

Figure 37D. Heteroduplex between an F′*gal* (F′8 *gal*) and an unsubstituted F factor, formed by the re-annealing of two single strands originating from two DNAs with partially similar base sequence. The heteroduplex is mounted by the formamide basic protein technique, as described in Davis et al., 1970, in which single strands appear as extended filaments. In this heteroduplex the larger single strand contains the portion of the bacterial chromosome carrying the *gal* genes, but not the λ attachment site. During the formation of the F′ from the parent Hfr a portion of the F factor was probably deleted. The piece of F deleted in the F′, but contained in F, forms the shorter side of the single-stranded loop (see arrow). (Photo courtesy of Phil Sharpe.)

Strains

Number	Sex	Genotype	Important Properties
CSH58	F$^-$	*ara thr leu proA lac gal trp his nalA recA1 thyA strA xyl mtl argE thi sup*	Thr$^-$ Leu$^-$ Rec$^-$ Str^r Nal^s Pro$^-$ Trp$^-$ His$^-$ Arg$^-$
CSH62	HfrH	*thi*	Str^s
CSH65	F$^-$	*leu lac nalA strA thi*	Nal^r Str^r Leu$^-$

Method

Day 1 Subculture 0.1 ml of a fresh overnight of both the donor (CSH62) and recipient (CSH58) into 5 ml of LB broth and aerate at 37° until both are at a density of approximately 2×10^8 cells/ml. Prepare a mating mixture by mixing 0.2 ml of the donor with 2 ml of the recipient in a 100- or 125-ml flask in a 37° waterbath. Shake gently for 60 minutes at 37°.

Interrupt the mating by using the apparatus described in the Appendix. This is to prevent the *recA*$^+$ gene from being transferred during mating. Chill the cultures by placing on ice. Plate 0.1 ml of the mating mixture direct, and also a 10^{-1} and a 10^{-2} dilution onto glucose minimal plates with proline, histidine, arginine, tryptophan, and streptomycin. Also streak the donor and recipient alone onto this type of plate as a control.

Day 3 Pick Leu$^+$Thr$^+$Str^r colonies from the cross. Purify 20–25 onto the same type of plate, streaking for single colonies.

Day 5 Inoculate 10 purified colonies (each arising from a different "recombinant") into 10 test tubes containing 5 ml LB broth each. Aerate at 37° until a density of $2–3 \times 10^8$ cells/ml is reached. Also subculture a fresh overnight of CSH65 into 2 test tubes with 5 ml LB broth each.

Prepare a mating mixture by mixing 0.5 ml of CSH65 (growing exponentially) with 0.5 ml of each of the test cultures in a test tube and place on a 30–33 rpm rotor in a 37° room for 30 minutes. Plate out a 10^{-2} dilution (0.1 ml) onto a glucose minimal plate with nalidixic acid. It is also feasible to simply streak each mating mixture with a loop onto a sector of the selective plate. Also streak each donor and the recipient (CSH65) as a control. The nalidixic acid is not absolutely necessary, since each donor is an auxotroph.

As an alternative procedure, it is possible to prepare a master plate from the purified colonies and then replicate, after 8 hours growth, onto a lawn of the recipient, CSH65, on a glucose minimal plate. In this case, nalidixic acid should not be used since it stops mating fairly quickly. Although it does not interfere with the transfer of small episomes, such as F′*lac*, it may with larger substituted F factors.

Day 6, 7 Observe the plates. Some of the cells which became Leu$^+$Thr$^+$ in the initial cross did so because they received an episome which carried the *thr-leu* region. These should be able to donate this episome to a Leu$^-$ recipient. Strain CSH65 is Leu$^-$Nal^r, while the CSH58 cells which became Thr$^+$Leu$^+$ are Nal^s. Some Thr$^+$Leu$^+$ cells cannot donate an episome. Apparently these cells have received an F′ factor which is now defective in its ability to transfer.

Materials

Day 1

overnight of CSH58 and CSH62
2 test tubes with 5 ml LB broth
shaking waterbath at 37°
100- or 125-ml flask
ice bucket
4 glucose minimal plates with proline, histidine, arginine, tryptophan, and streptomycin
2 dilution tubes
3 13 × 100 mm test tubes with 3 ml F-top agar each, at 45°
5 0.1-ml, 2 1-ml, 1 5- or 10-ml, and 2 pasteur pipettes

Day 3

3 glucose minimal plates with proline, histidine, arginine, tryptophan, and streptomycin
20–25 round wooden toothpicks

Day 5

(per 10 colonies tested)

12 test tubes with 5 ml LB broth each
fresh overnight of CSH65
12 glucose minimal plates with nalidixic acid
10 dilution tubes
20 0.1-ml (or pasteur) pipettes, 11 1-ml and 1 5-ml pipettes
10 test tubes
rotor in a 37° room or water bath at 37°

EXPERIMENT 38

Use of Homogenotization in Strain Construction

Often it is desirable to transfer a **specific** marker from one strain to another. Hfr crosses and P1 transduction are the usual means of achieving this. The former method requires that our strain be an Hfr, and also that we have a linked marker available in the recipient. P1 transduction (Experiment 28) is a feasible method in this case only if the strains involved contain the appropriate closely linked markers. Suppose we wish to transfer the marker z^-_{U118} from strain A to strain B. If strain A is $z^-_{U118}pro^+$, and strain B lac^+proC^-, we can use P1 phage grown on strain A and transduce strain B to Pro$^+$. Among the Pro$^+$ transductants of strain B will be cells which are now $z^-_{U118}pro^+$ (see Experiment 28). If a nearby selective marker is not present, however, P1 transduction cannot be used without other genetic constructions.

F′ Factor as Intermediate

A different method makes use of an F′ factor as an intermediate in this type of construction. The z^-_{U118} marker can be crossed into an F′*lac*, and this episome can then be transferred to a new strain. A second recombination event can then occur which transfers the marker onto the new chromosome. Figure 38A diagrams

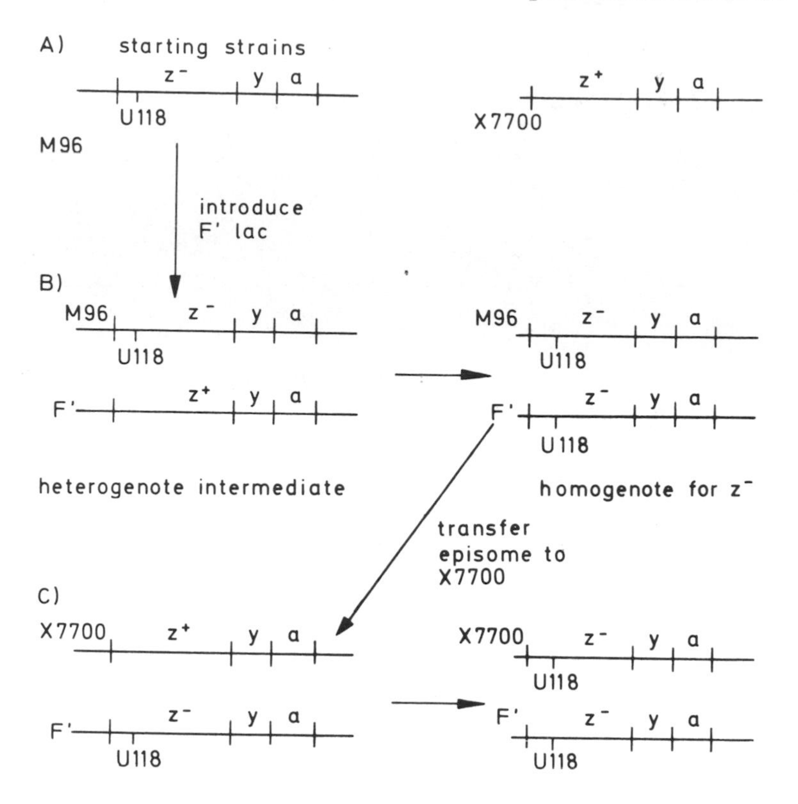

Figure 38A. Conversion of X7700 to X7700z^-_{U118}.

(A) two strains, CSH46 (M96) and CSH51 (X7700). CSH46 is streptomycin-sensitive and carries the z^- polar mutation U118. CSH51 is streptomycin-resistant and is z^+. We wish to transfer the z^- marker (U118) from strain CSH46 to strain CSH51 to produce X7700 z^-_{U118}, using an episome as an intermediate in the construction. The introduction of an F'*lacpro* episome into CSH46 creates a heterogenote for the *lac* region (B). Since the combination z^+/z^- is Lac$^+$, this heterogenote forms Lac$^+$ (white) colonies on lactose tetrazolium plates. Frequent recombination occurs between the diploid regions in these heterogenotes, and segregant Lac$^-$ colonies, of the genotype z^-/z^- can usually be observed at 0.2% to 5% of the population. Purified Lac$^-$ colonies have the U118 marker on the episome. The episome can now be transferred to the new strain, CSH51. This creates a new heterodiploid, shown in (C). By plating single colonies on lactose tetrazolium plates, z^-/z^- red colonies can be observed and purified. These should be the "homogenote" shown in the right half of (C). Curing of the episome with acridine orange completes the construction. Homogenotization can be used for almost all markers. For systems in which indicator plates can be employed, it is invaluable.

As an optional exercise, we suggest doing the same procedure using the *recA*$^-$ derivative of CSH51. Since homogenotization requires the recombination enzymes, it should be prevented in this strain.

Strains

Number	Sex	Genotype	Important Properties
CSH28	F′*lac*$^+$ *proA*$^+$,*B*$^+$	Δ(*lacpro*) *supF trp pyrF his strA thi*	Pyr$^-$ Trp$^-$ His$^-$
CSH46	F$^-$	*ara* Δ(*lacpro*) *thi* (λCI857St68d*lacIlacZ*)	Lac$^-$ (z^-_{U118}) Pro$^-$ Str^s, temperature-sensitive; must grow below 34°
CSH51	F$^-$	*ara* Δ(*lacpro*) *strA thi* (φ80d*lac*)	Lac$^+$ Pro$^-$ Str^r

Method

Day 1 Subculture an overnight of strain CSH28 into broth and after approximately 2 hours, when the culture is at 2–3 × 10^8 cells/ml, mix 0.5 ml with 0.5 ml of an exponentially growing culture of CSH46. Place in a waterbath at 30–33° for 1 hour with gentle shaking or preferably on a rotor at 30–33°. Plate 0.1 ml of a 10^{-2}, 10^{-3}, and 10^{-4} dilution onto glucose minimal plates and incubate for 36–48 hours at 30–33°. CSH46 carries its *lac* region on a temperature-inducible prophage. It is therefore necessary to grow this strain at temperatures below 34°.

Day 3 Pick single colonies off the selection plate and streak for single colonies on lactose tetrazolium plates. These colonies are CSH46(z^-) with an F′*lac*$^+$*pro*$^+$ and should be white.

Day 4 Approximately 1/200 colonies will be red (z^-_{U118}/z^-_{U118}). Pick and repurify on lactose tetrazolium plates several of these colonies. If whole red colonies cannot be seen, pick and purify red sectors of colonies.

Day 5 Inoculate overnights from 6 of the Lac$^-$ colonies.

Day 6 Subculture the overnights of the CSH46 z^-_{U118}/z^-_{U118} homogenotes into broth and when the cultures have reached 2–3 × 10^8 cells/ml, transfer the episome into CSH51 by doing spot matings on glucose minimal plates with streptomycin. Incubate at 37°.

Day 7 Purify the new heterogenotes (CSH51/F′*lacpro* z^-_{U118}) by picking from the center of the spots and streaking for single colonies onto a glucose minimal streptomycin plate. (Some of the supposed homogenotes used as donors in Day 5 are really CSH46 cells which have lost the episome, since these are also Lac$^-$. These will not donate a *lacpro* episome to CSH51.)

Day 9 Streak for homogenotes on lactose tetrazolium plates at 37°.

Day 10 Observe homogenotes of the genotype CSH51 z^-/F′ z^-.

Materials

Day 1

overnights of CSH46 (grown at 30°) and strain CSH28
3 glucose minimal plates
1 test tube
3 dilution tubes
2 test tubes with 5 ml broth
5 0.1-ml, 3 1-ml, and 2 pasteur pipettes
rotor or shaking waterbath at 30–33°
incubator at 30°

Day 3

5 lactose tetrazolium plates
5 round wooden toothpicks
incubator at 30–33°

Day 4

3 lactose tetrazolium plates
18 round wooden toothpicks
incubator at 30–33°

Day 5

6 test tubes with 5 ml broth each
rotor or shaking waterbath at 30–33°

Day 6

overnights of 6 homogenotes of CSH46; overnight of CSH51
13 pasteur pipettes
glucose minimal streptomycin plate

Day 7

6 round wooden toothpicks
glucose minimal streptomycin plate

Day 9

3 lactose tetrazolium plates
3 round wooden toothpicks

UNIT V

TRANSDUCTION AND THE ISOLATION OF SPECIALIZED TRANSDUCING PHAGE LINES

INTRODUCTION TO UNIT V

Advantages of Specialized Transducing Phage

More and more research today is directed towards constructing specialized transducing phage. In such phage, specific bacterial genes, for instance *lac* or *gal*, are fused to phage DNA. Why are these phage so desirable? The basic advantage is that the bacterial DNA in a transducing phage is replicated and packaged as if it were phage DNA. This makes possible many useful experiments:

A. Gene Enrichment

Since the relative concentration of a bacterial gene carried by a transducing phage is about 100 times greater in a phage lysate than it is in a bacterial culture, such phage are a useful source of specific DNA. Frequently, the DNA isolated by this method is pure enough to allow binding studies of regulatory proteins (Gilbert and Müller-Hill, 1967) and to provide a template for *in vitro* mRNA synthesis and subsequent hybridization (see Unit VI). Also, the sensitivity of cell-free protein synthesis is greatly increased by the use of DNA templates enriched for the genes under study (Unit IX; Zubay et al., 1970; de Crombrugghe et al., 1971).

B. Transduction

Lysogenization with transducing phage is a convenient method for creating stable partial diploids. An effective way to introduce genes from one organism into another is by transduction or transformation, which requires DNA rich in a particular gene (see review by Beckwith, 1970).

C. Mutagenesis

Gene-specific mutagenesis can be carried out *in vitro*, allowing high mutant yields without affecting the potential host strain. Recent techniques for selecting deletion mutants of phage and transducing phage make this particularly desirable (see Experiment 39).

D. Translocation

Use of transducing phage lines allows transposition of the bacterial genes on the phage to different regions of the chromosome (Experiments 40 and 41). This greatly facilitates the isolation of deletions, and also makes possible operon fusion studies (Experiment 42).

E. Enzyme Overproduction

A lysogen whose prophage incorporates a gene of interest can be induced to yield a culture synthesizing large amounts of the gene product. This is facilitated by the use of phage mutations allowing heat induction and preventing lysis (Müller-Hill et al., 1968). Pulse-chase experiments are also made easier by the large increase in specific activity after heat induction (Platt et al., 1970).

Putting Genes on Transducing Phage

How do we construct such phage? We are able to exploit the fact that temperate phage occasionally make errors during excision from the host chromosome which result in the formation of phage carrying nearby regions of the chromosome (Kayajanian and Campbell, 1966). The frequency of these events becomes undetectable as the marker of choice is positioned further away from the lysogenized phage. Therefore, in practice the isolation of a transducing phage is a two-step process. The first problem is to position the particular bacterial gene under investigation near to the prophage or prophage attachment site. Several approaches to this problem are demonstrated in Experiments 40 and 41.

Having accomplished this, the next step is to select for a transducing phage carrying the desired markers. The frequency of transducing particles for a nearby marker in an induced lysate can be on the order of 10^{-5} or 10^{-6} per induced cell, although for more distant markers the frequency is much lower. Therefore, powerful selection techniques are often required. This problem is considered in Experiment 39.

References

BECKWITH, J. R. 1970. *Lac:* The genetic system. *The Lactose Operon*, p. 5, Cold Spring Harbor Laboratory.

CAMPBELL, A. M. 1969. *Episomes.* Harper and Row. Offers detailed discussion of transducing phage.

DECROMBRUGGHE, B., B. CHEN, W. ANDERSON, P. NISSLEY, M. GOTTESMAN, I. PASTAN and R. PERLMAN. 1971. The essential elements for controlled *lac* transcription. *Nature New Biol. 231:* 139.

GILBERT, W. and B. MÜLLER-HILL. 1967. The *lac* operator is DNA. *Proc. Nat. Acad. Sci. 58:* 2415.

HAYES, W. 1968. *The Genetics of Bacteria and Their Viruses*, 2nd Edition, p. 620. John Wiley and Sons, Inc., New York. Detailed discussion of transducing phage.

KAYAJANIAN, G. and A. CAMPBELL. 1966. The relationship between heritable physical and genetic properties of selected gal^- and gal^+ transducing λdg. *Virology 30:* 482.

MÜLLER-HILL, B., L. CRAPO and W. GILBERT. 1968. Mutants that make more *lac* repressor. *Proc. Nat Acad. Sci. 59:* 1259.

PLATT, T., J. H. MILLER and K. WEBER. 1970. *In vivo* degradation of mutant *lac* repressor. *Nature 228:* 1154.

ZUBAY, G., D. A. CHAMBERS and L. C. CHEONG. 1970. Cell-free studies on the regulation of the *lac* operon. *The Lactose Operon*, p. 375. Cold Spring Harbor Laboratory.

EXPERIMENT 39

Construction of Specialized Transducing Phage Lines

The following section describes the steps for isolating specialized transducing phage lines. Although the particular phage used here is $\phi 80h$, the methods employed in these experiments are quite general.

Because we are isolating transducing phage lines carrying the *trp* genes in this experiment, we have chosen to use $\phi 80$, since the $\phi 80$ attachment site is located near the *trp* region on the bacterial chromosomes. The *h* mutation enables $\phi 80$ to infect the *tonB* strains which are subsequently used for mapping the *trp* region carried by the transducing particles.

LFT Lysates

When a lysogen is induced, the prophage is excised from the chromosome. Rarely, an error is made in excision which results in the addition of some bacterial genes to the phage and the loss of some phage genes (Figure 39A). If the loss of the phage genes prevents the phage from replicating and producing progeny phage particles in a sensitive host, the phage is termed **defective** (abbreviated *d*). Most defective phage can grow normally if the missing functions are supplied by a wild-type phage, termed **helper.** Thus, in an induced lysate of $\phi 80h$, there will be

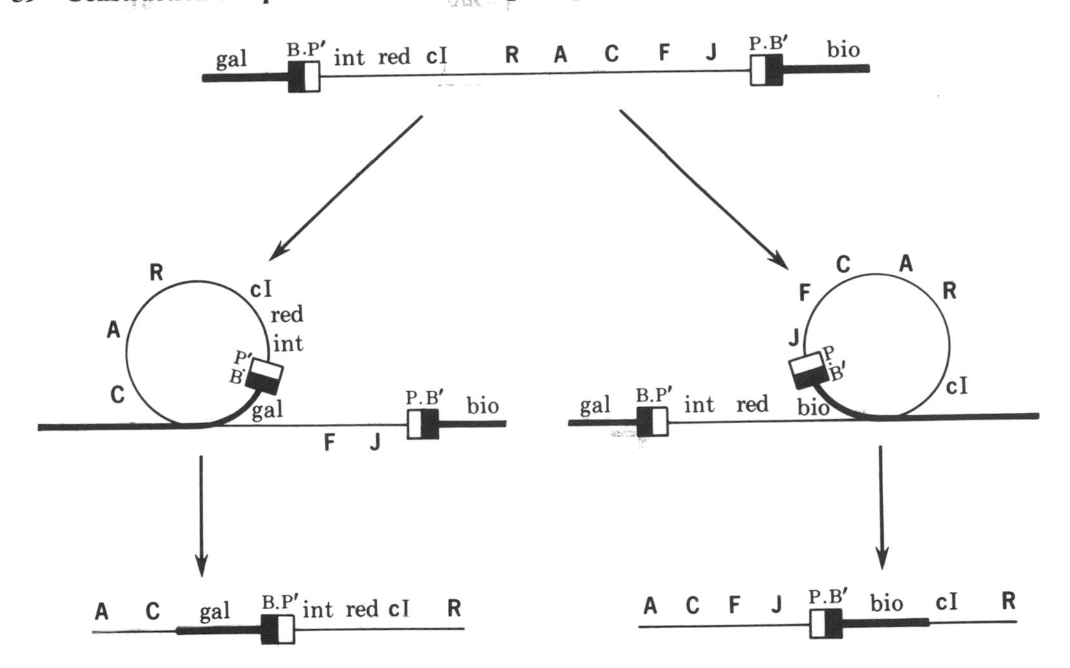

Figure 39A. The origin of transducing phage lines. The light lines represent the phage and prophage chromosomes, and the heavy lines the bacterial chromosome. Recombination is assumed to take place at the bottom of the circular loops. The formation of a λ*gal* (left) and a λ*bio* (right) are shown. The *att* of the phage chromosome is called **P.P′** and that of the bacterial chromosome **B.B′**. (Reproduced with permission from Gottesman and Weisberg, 1971.)

a small percentage of phage carrying various segments of the nearby *trp* operon (Figure 39B). If we use this lysate to infect a Trp^- strain, a small fraction of the recipients will be transduced to Trp^+. Since the frequency of Trp^+ cells is low, the lysate is termed an LFT (low frequency transducing). Typically, each transduced cell will carry a ϕ80*htrp* prophage in addition to the original mutation in the *trp* gene of the recipient. Such transductants are therefore diploid for *trp*. Since the LFT lysate usually contains large numbers of non-transducing ϕ80*h*, many of the transductants will be double lysogens, carrying also a non-transducing prophage.

HFT Lysates

If a single colony, which is a double lysogen containing both a ϕ80*htrp* prophage and a wild-type ϕ80*h* prophage as a helper, is grown up and induced, the resulting lysate will now contain a high proportion of ϕ80*htrp* phage, since each induced cell should liberate both transducing and non-transducing phage particles. This lysate will transduce a Trp^- strain with a high frequency, and is termed an HFT (high frequency transducing). Each HFT is homogeneous, since it is a phage line which resulted from a specific transducing particle. An LFT lysate, on the other hand, contains a collection of different transducing particles. If the Trp^- recipient we use carries a mutation in *trpA*, the *trp* cistron closest to the ϕ80 attachment site, then the ϕ80*htrp* required to transduce it to Trp^+ need have only that *trpA* cistron intact and functioning. Thus, by using a $trpA^-$ recipient, we can find a wide range of *trp* transducing phage lines ($trpE^+$, D^+, C^+, B^+, A^+; $trpD^+$, C^+, B^+, A^+; $trpC^+$, B^+, A^+; $trpB^+$, A^+; and $trpA^+$) each of which has been formed by a different excision error.

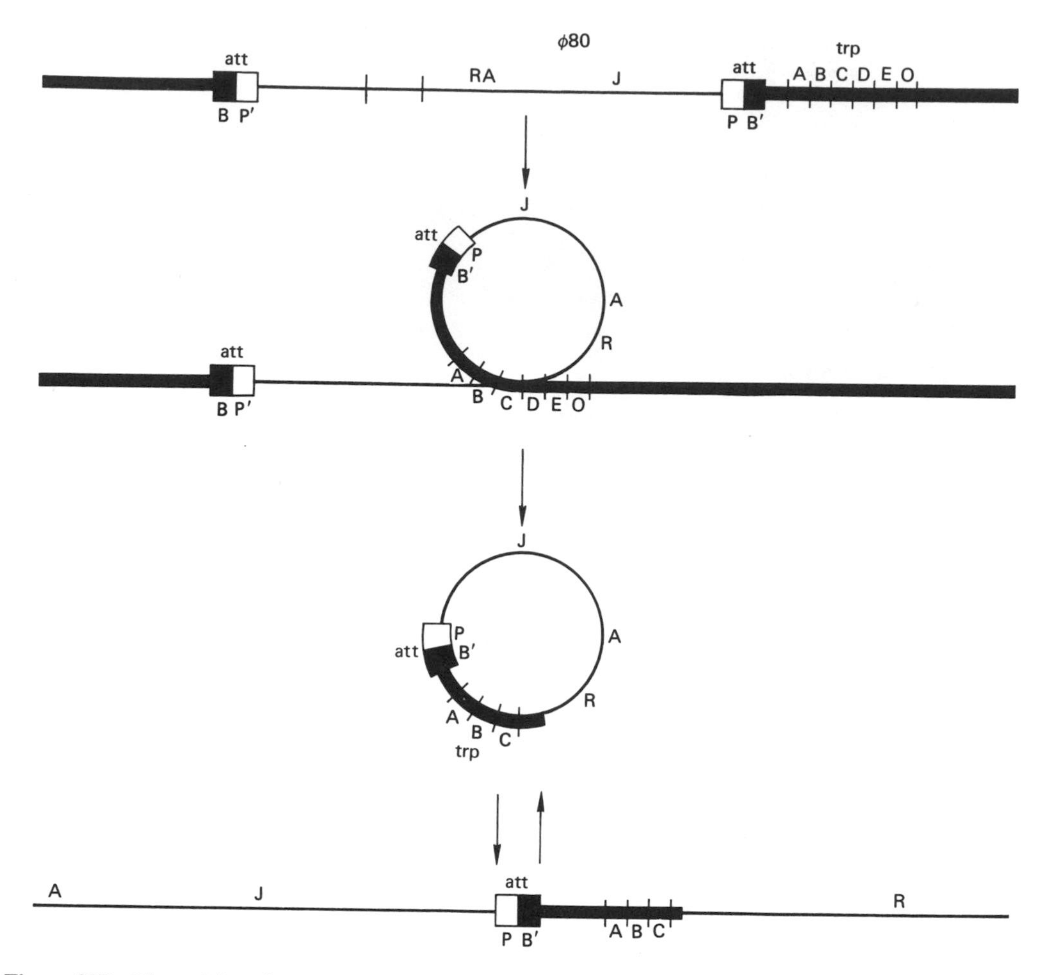

Figure 39B. The origin of a *trp* transducing phage line. The same conventions as in Figure 39A are used. Here recombination at the bottom of the circular loops results in the formation of a defective ϕ80 phage carrying part of the *trp* operon.

Not every transductant from the LFT will yield an HFT. Some of the transductants are haploid segregants in which the trp^+ gene of the transducing phage has replaced the trp^- gene of the host by recombination, and the phage genes have subsequently been lost. A proportion of the transductants are single lysogens. If the transducing particle is defective, such single lysogens will not liberate phage particles upon induction unless helper phage is added.

Plaque-forming Transducing Phage Lines

Some transducing phage lines retain enough phage genes to be able to propagate without added helper. These are termed plaque-forming (p) phage, and are easier to work with than defective lines. In addition to ϕ80ptrp^+ phage, λpbio^+, λplac^+, and λphut^+ (histidine utilization) phage have been isolated.

For nearby markers on the right side of the prophage (see Figure 39A), plaque-forming transducing phage can be isolated by the following powerful selection technique. It is known that λ cannot make plaques on P2 lysogens. However, λ mutants occurring at a frequency of 10^{-6} do form plaques and are found to have lost the function of several genes at the left end of the prophage (Lindahl et al.; Smith). Since a frequent mutation which leads to a loss of several functions is a deletion, and since terminal prophage deletions of λ are readily formed by the excision errors which result in the generation of a transducing particle, λ phage mutants which form plaques on P2 phage are mostly transducing phage. In fact, at least 40% of the λ phage which formed plaques on a P2 lysogen were shown to transduce part of the *hut* operon (Smith). This method offers, therefore, a 10^6-fold enrichment for plaque-forming transducing phage, and is particularly valuable for isolating transducing phage lines carrying bacterial markers which cannot be selected for.

In vitro Mutagenesis

One advantage of having bacterial genes in phage particles is the ease with which one can isolate deletion mutants *in vitro*. Because the rate of phage inactivation at low ionic strength is dependent on the amount of DNA in the λ phage head, selection for deletion mutants can be achieved by the isolation of phage resistant to EDTA or citrate treatment (Parkinson and Huskey). This method has recently been used to isolate *trp* deletions from a λh80*trp* transducing particle (Burdon; see addendum to this experiment).

References

BURDON, M. G. 1970. A potential method for selecting genetic deletions in *Escherichia coli*. *Mol. Gen. Genet. 108:* 288.

CAMPBELL, A. 1962. Episomes. *Adv. Genet. 11:* 101.—Original formulation of model of integration of circular episomes into the host chromosome.

GOTTESMAN, M. E. and R. A. WEISBERG. 1971. Prophage insertion and excision. *The Bacteriophage Lambda*, p. 113. Cold Spring Harbor Laboratory.—Recent review of prophage integration and excision and transducing phage formation.

KAYAJANIAN, G. 1970. *Gal* transduction by phage λ: On the origin and nature of LFT transducing genomes. *Mol. Gen. Genet. 108:* 338.—This and the following reference give detailed characterization of λ transducing phage.

KAYAJANIAN, G. and A. CAMPBELL. 1966. The relationship between heritable physical and genetic properties of selected gal^- and gal^+ transducing λ*dg*. *Virology 30:* 482.

LINDAHL, G., G. SIRONI, H. BIALY and R. CALENDAR. 1970. Bacteriophage lambda: Abortive infection of bacteria lysogenic for phage P2. *Proc. Nat. Acad. Sci. 66:* 587.

MANLY, K. F., E. R. SIGNER and C. M. RADDING. 1969. Nonessential functions of bacteriophage λ. *Virology 37:* 177.—Study of deletions in the λ genome.

MATSUSHIRO, A. 1963. Specialized transduction of tryptophan markers in *Escherichia coli* K12 by bacteriophage φ80. *Virology 19:* 475.—First description of φ80 *trp* transduction.

PARKINSON, J. S. and R. J. HUSKEY. 1971. Deletion mutants of bacteriophage lambda. I. Isolation and initial characterization. *J. Mol. Biol. 56:* 369.

SIGNER, E. R. 1968. Lysogeny: The integration problem. *Ann. Rev. Microbiol. 22:* 451.—Comprehensive review of temperature phage integration into the host chromosome.

SMITH, G. R. 1971. Specialized transduction of the Salmonella *hut* operons by coliphage λ: Deletion analysis of the *hut* operons employing λ*phut*. *Virology 45:* 208.

Strains

Number	Sex	Genotype	Important Properties
CSH23	F′*lac*⁺ *proA*⁺,*B*⁺	Δ(*lacpro*) *supE spc thi*	Trp⁺
CSH27	F⁻	*trpA33 thi*	Trp⁻
CSH29	F⁻	*trpB thi*	Trp⁻
CSH30	F⁻	*trpC thi*	Trp⁻
CSH31	F⁻	*trpD thi*	Trp⁻
CSH32	F⁻	*trpE thi*	Trp⁻

Method

Preparation of a $\phi 80htrp$ Phage Line

Day 1

Preparation of a lysogen

Place several drops of an overnight culture of strain CSH23 in a small test tube and add 2.5 ml of H-top agar (from a bottle kept at 45°). Mix and plate onto the surface of a plate containing H agar. Let dry; then apply a drop of a lysate of $\phi 80h$. Incubate at 37° overnight after the spot dries. Keep the plate face up. The lysates of $\phi 80h$ will be supplied by the instructor.

Day 2

Part A

Examine the plate. There should be a lawn of bacteria growing around the spot. The spot should be clear with a turbid center. This is due to bacteria which have become lysogenic and are growing in the center of the spot. Inoculate a tube containing 10 ml of broth with material from the center of the spot, and aerate at 37°.

Part B. Preparation of lysate

When the culture reaches a density of $2\text{–}3 \times 10^8$, spin down and resuspend in 5 ml 0.1 M magnesium sulfate. Pour into an open petri dish and irradiate with UV to induce the $\phi 80h$ (see Experiment 2). Immediately pour into a tube covered with aluminum foil containing 5 ml LB broth and shake vigorously at 37° for 2.5 hours. Add 5 drops of chloroform and shake on the vortex for 30 seconds. Let stand for 10 minutes and then spin down the debris. The supernatant is a lysate of $\phi 80h$ which must be titered.

Part C. Titering the lysate

Make serial dilutions of the lysate. In each of 3 small test tubes add 0.1 ml of a 10^{-7}, 10^{-8}, and 10^{-9} dilution, respectively. Add 3 drops of an overnight culture of strain CSH23 and incubate at 37° for 8 minutes. Plate out onto H plates by adding 2.5 ml of H-top agar kept at 45°. Let dry and incubate overnight at 37°.

Day 3

Part A

Count the $\phi 80h$ plaques and determine the titer of the lysate (the number of phage per ml).

Part B. Preparation of LFT from $\phi 80h$ lysate

Spin down an overnight culture of a *trpA* auxotroph (CSH27) grown up in LB broth, wash, and resuspend in buffer. Divide the contents into 3 small test tubes. To one, add enough phage to give an moi of 10, and to a second, an moi of 1. Do not add any phage to the third tube. Incubate all 3 tubes in a 37° water bath for 20 minutes.* Note the volume of each tube. Plate out 1 ml and 0.1 ml

* The procedure is the same for λ except that it is necessary to preadsorb in 0.01 M $MgSO_4$.

of each mixture directly. Also plate 0.1 ml of a 10^{-1} dilution of each mixture. Use F-top agar. Incubate at 37° for 48 hours.

Day 5

Score the Trp^+ transductants. What was the frequency of transduction compared to the reversion frequency of the control? Pick and purify about 30 transductant colonies onto glucose minimal plates and incubate at 37°.

Day 7

Part A. Preparation of lysates with high frequency transducing titer (HFT's)

Inoculate 10 ml of LB broth in a large test tube with one of the purified transductants. Inoculate heavily from the plate. Do this for at least 10 of the Trp^+ colonies. Shake at 37° until a density of $2\text{–}3 \times 10^8$/ml is reached and then prepare UV-induced lysates in the same manner as above in the preparation of LFT's.

Part B. Testing lysates for HFT's

A *trpA* auxotroph should be used to screen the 10 lysates. Spread about 5 drops of an overnight culture of a *trpA* mutant (CSH27) onto a glucose minimal plate. Allow to dry, and then spot a drop of each lysate onto a sector of the plate. Spot 5 different spots onto each of 2 plates. Do likewise for 2 glucose minimal plates onto which no bacteria were spread, as a control. Allow to dry, and then incubate at 37° for 24 hours. In the center of each plate spot the original ϕ80*h* LFT. Store the lysates in the cold.

Day 8

Examine the plates. The results of the spot test should tell you which of the 10 independently prepared lysates are HFT's. On these spots there should be confluent growth, compared to a spot which did not have an HFT, such as the ϕ80*h* control. One can retest more carefully by incubating, as done on Day 3, with part of a resuspended overnight culture of a *trpA* strain and plating out dilutions to determine the frequency of transduction.

Any lysates which are HFT's for the *A* segment of the *trp* operon should be tested to determine whether the transducing phage also carries other *trp* genes. To do this, the HFT lysates are spotted onto a lawn of different *trp* auxotrophs in exactly the same manner as done for the *trpA* mutant. Onto each of 4 glucose minimal plates spread a few drops of an overnight culture of Trp^- strains, containing a mutation in one of the first 4 *trp* genes (CSH29–32). Incubate at 37° overnight (actually 24 hours). Score the results. Which phage contained parts of the *trp* operon?

Additional Methods

Using a lambda ϕ80p*trp* transducing phage, Burdon has used the following technique to isolate *trp* deletions: A lysate of the phage (10^8/ml) is incubated in EDTA (0.01 M, pH 7) for 60 minutes at 37° and then plated directly on a ϕ80-sensitive strain. The survival rate of plaque-forming phage under these conditions is 10^{-5}. In this experiment as many as 10% of the surviving phage carry *trp* deletions of varying length. Further details on this method and its use are in Burdon and also Parkinson and Huskey.

Materials

Day 1

1 ml overnight culture of strain CSH23
1 H plate
H-top agar in 45° water bath
2 pasteur pipettes
1 small test tube
1 5-ml pipette

Day 2

Part A

1 large test tube with 10 ml of LB

Part B

ice bucket
2 plastic centrifuge tubes to spin down 10 ml
desk-top centrifuge
5 ml 0.1 M magnesium sulfate
petri dish, preferably glass
UV lamp, turned on 60 minutes prior to experiment
dark tube, or regular with aluminum foil, containing 5 ml LB broth
2 5-ml pipettes
shaker bath set at 37°
chloroform, vortex

Part C

1 ml overnight culture of strain CSH23
6 dilution tubes
3 small test tubes
7.5 ml H-top agar at 45°
7 0.1-ml, 2 1-ml, 2 5-ml or 1 10-ml pipettes, and 1 pasteur pipette
3 H plates
waterbath at 37°

Day 3

5 ml overnight culture of CSH27 in LB broth
2 small centrifuge tubes
desk-top centrifuge
9 small test tubes
10 ml phosphate buffer, pH 7.0
$\phi80h$ lysate (at least 2×10^{10} pfu/ml)
water bath at 37°
9 0.1-ml (or pasteur), 5 1-ml, 3 5-ml, and 3 10-ml pipettes
9 glucose minimal plates
3 dilution tubes with 0.9 ml buffer each
25 ml F-top agar

Day 5

5 glucose minimal plates
30 sterile round wooden toothpicks

Day 7

Part A

10 large (25 mm) test tubes with 10 ml LB broth
10 large centrifuge tubes to spin down 10 ml
10 petri dishes, preferably glass
UV lamp, turned on 60 minutes prior to experiment
10 large test tubes with 5 ml LB each, covered with aluminum foil
50 ml 0.1 M $MgSO_4$
10 5-ml or 10-ml pipettes
shaking waterbath at 37°
10 large centrifuge tubes
chloroform
10 test tubes

Part B

overnight culture of *trpA* mutant (CSH27)
12 pasteur pipettes
4 glucose minimal plates

Day 8

overnight cultures of each of the 4 *trp* auxotrophs (CSH29–32)
4 glucose minimal plates
8 pasteur pipettes

INTRODUCTION TO EXPERIMENTS 40–41

Positioning Genes near Prophage Attachment Sites

Because the isolation of specialized transducing phage for a specific marker depends on the nearness of that marker to a particular prophage, it is often necessary to move genes closer to a prophage attachment locus, or vice versa. Three different approaches to this problem have been used successfully. In one, **episomes carrying bacterial genes are integrated at different sites on the chromosome**. Selection for integration of an F′*lac* episome into the *tonB* locus, near the ϕ80 attachment site, is demonstrated in Experiment 40. This is a general technique for relocating chromosomal markers which are carried on F′ factors. The transposition to the *tonB* site enables the isolation of ϕ80 transducing phage for the markers in question. Phage lines carrying *lac*, *ara*, and *his* have been isolated in this fashion. By looking for loss of galactose sensitivity in a $galE^-$ background (see Experiment 19) episome integrations into the *gal* operon near to the λ attachment site can also be selected for (Figure 40A), and from these transpositions λ*lac* transducing phage have been isolated (Shapiro et al., 1969; Ippen et al., 1971).

Another technique involves the **integration of a prophage into different sites on the chromosome**. Thus, in a strain in which the normal attachment site for λ is deleted, λ is found to integrate at many different sites (Experiment 41). A wide assortment of transducing phage have been isolated with this method.

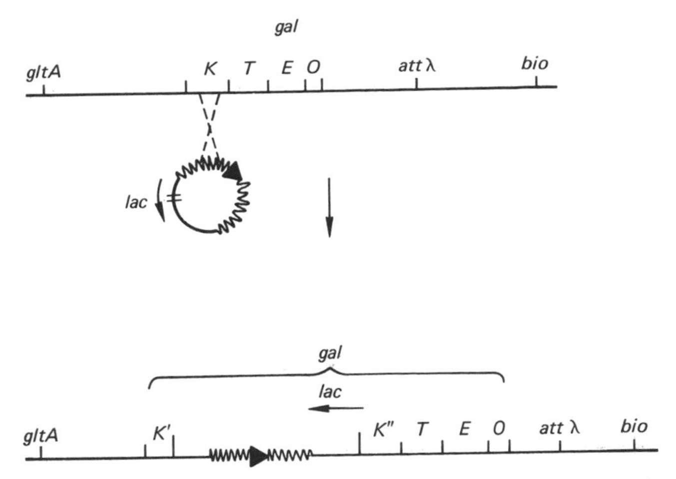

Figure 40A. Integration of F′*lac* into the *gal* operon. By transposing the *lac* region to the *gal* region in this manner, *lac* is now positioned near the λ attachment site. For methods of translocating bacterial genes, see Experiment 40. (Reprinted with permission from Ippen et al., 1971.)

In a third method, reviewed at the end of this unit, **selection for a large fused episome** is made, starting from two different episomes, one containing the prophage attachment site and the other the marker under study.

References

IPPEN, K., J. SHAPIRO and J. BECKWITH. 1971. Transposition of the *lac* region to the *gal* region of the *Escherichia coli* chromosome: Isolation of λ*lac* transducing bacteriophages. *J. Bacteriol. 108:* 5.

SHAPIRO, J., L. MACHATTIE, L. ERON, G. IHLER, K. IPPEN and J. BECKWITH. 1969. Isolation of pure *lac* operon DNA. *Nature 224:* 768.

EXPERIMENT 40

Transpositions of the *lac* Region

A. General Transpositions

F′*lac* episomes integrate with high frequency into a wild-type chromosome. This integration occurs because of a recombination event between the homologous *lac* regions on the chromosome and the episome. When a large deletion covering the entire *lac* region is introduced into the chromosome, the homology between the episome and the chromosome is eliminated. In this case the frequency of integration of an F′*lac* episome into the chromosome is reduced by at least 100-fold. Now, however, instead of integrating at the *lac* region (which is deleted here), the episome integrates at various places in the *E. coli* chromosome. Integrations of this type have been selected for by using strains with an F′*lac* and a large chromosomal deletion covering the *lac* region, and growing the cells in the presence of acridine, selecting for Lac^+ colonies. Since the episome cannot replicate in the presence of acridine, only those cells in which the episome has integrated into the chromosome will maintain the *lac* genes and be able to grow on lactose.

Instead of acridine, a temperature-sensitive ($F'_{ts}lac$) episome can be used. This episome is lost from the population at high temperature (42.5°) unless it has integrated into the bacterial chromosome. Cells which remain Lac^+ at high

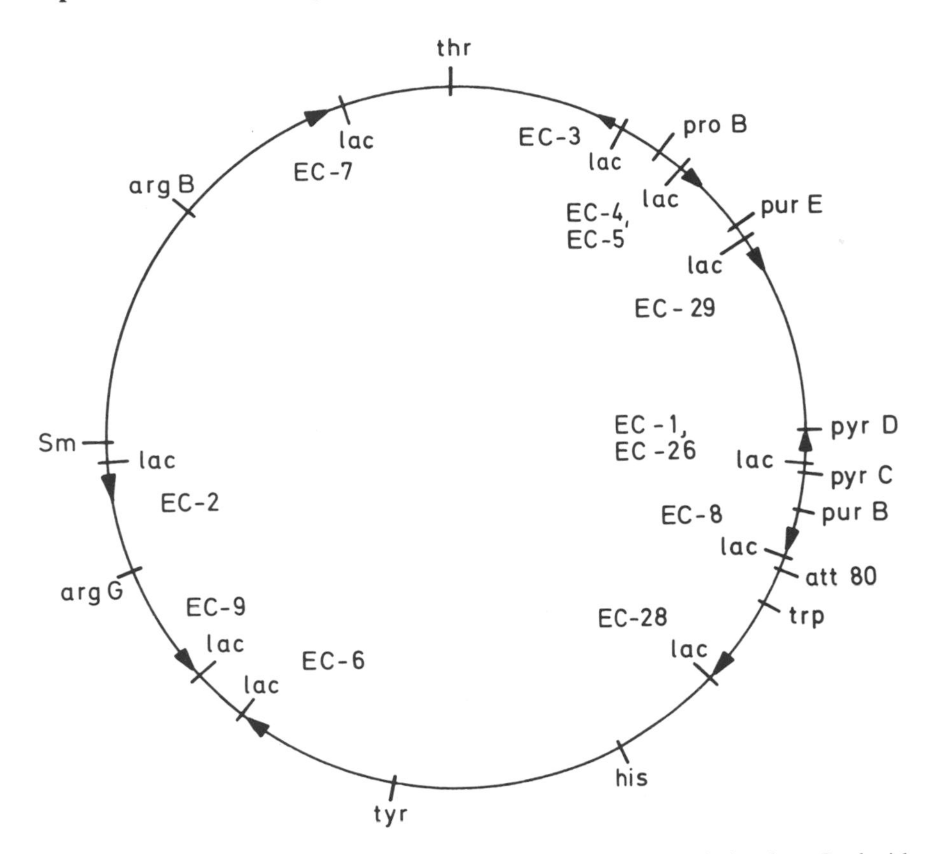

Figure 40B. Hfr strains from the integration of F'_{ts}*lac*. (Reprinted with permission from Beckwith et al., 1966.)

temperature must have had an integration of the F'_{ts}*lac* episome somewhere in the chromosome. In these strains the *lac* region has now been **transposed** to a new locus on the *E. coli* genetic map. These strains will be Hfr's, with their origin near the point of integration. Interrupted matings can be used to demonstrate both the location of the origin of the new Hfr's and the direction of transfer.

Experiment 40A demonstrates the isolation of new Hfr strains by integration of an F'_{ts}*lac* into the *E. coli* chromosome. The selection is for Lac^+ at 42.5°. We then attempt to determine the location of the transposed *lac* operon by interrupted matings. Figure 40B shows some of the Hfr strains derived by this method (Beckwith et al.).

B. Directed Transpositions

When an F′*lac* integrates into the chromosome at a site other than the normal *lac* region, it is likely that the integration occurs in the middle of a gene. In this case, the gene is inactivated. For instance, integration within the gene for T1 phage receptor sites (*tonB*) would lead to the loss of the ability to synthesize T1 receptors, and the cell would become T1 resistant. Using a double selection (integration of the episome and T1 resistance),* we can specifically select for integrations at the *tonB*

* We select mutants resistant to $\phi 80v$ and colicin V,B lysates in this case. T1 itself is never used in a bacterial genetics laboratory for this purpose.

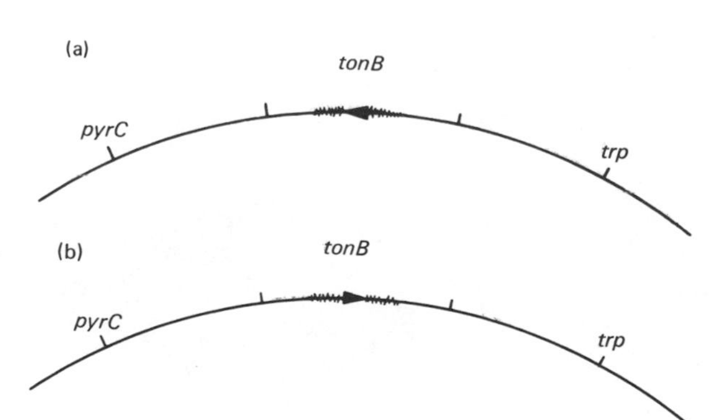

Figure 40C. Integration of F′*lac* into the *tonB* locus. (Reprinted with permission from Beckwith et al., 1966.)

locus. This type of double selection is a powerful tool and has led to the isolation of Hfr's arising from F'_{ts}*lac* integrations at the following specific sites: *tonA*, *tsx*, *gal*, *tonB*, and *thyA*.

Experiment 40B is designed to select strains in which the F'_{ts}*lac* has integrated at a specific locus, *tonB*. Candidates are tested by interrupted mating experiments to determine whether the origin of the selected Hfr is at the *tonB* locus. The Lac^+ colonies at 42–43° which are T1 resistant will be tested for early transfer of *pyrC* and *trp*. These two loci are close to the *tonB* locus, but on either side. An Hfr with its origin at *tonB* will donate one of these two markers, but not the other, with high frequency in a 30-minute interrupted mating. Recall that the orientation of the inserted F factor determines the direction of transfer (Figure 40C).

References

Beckwith, J. R., E. R. Signer and W. Epstein. 1966. Transposition of the *lac* region of *E. coli*. *Cold Spring Harbor Symp. Quant. Biol. 31:* 393.

Cuzin, F. and F. Jacob. 1964. Délétions chromosomiques et intégration d'un épisome sexuel *F-lac*$^+$ chez *Escherichia coli* K12. *Compt. Rend. Acad. Sci. 258:* 1350.

Ippen, K., J. Shapiro and J. Beckwith. 1971. Transposition of the *lac* region to the *gal* region of the *Escherichia coli* chromosome: Isolation of λ*lac* transducing bacteriophages. *J. Bacteriol. 108:* 5.

Scaife, J. G. and A. P. Pekhov. 1964. Deletion of chromosomal markers in association with F-prime factor formation in *Escherichia coli* K12. *Genet. Res. 5:* 495.

Shapiro, J., L. MacHattie, L. Eron, G. Ihler, K. Ippen and J. Beckwith. 1969. Isolation of pure *lac* operon DNA. *Nature 224:* 768.

EXPERIMENT 40A

Selection of F′*lac* Integrations

Strains

Number	Sex	Genotype	Important Properties
CSH24	$F'_{ts}lac^+$	Δ(*lacpro*)*supE thi*	temperature sensitive for the Lac^+ character
CSH57	F^-	*ara leu lacY purE gal trp his argG malA strA xyl mtl ilv metA* or *B thi*	Ilv^- Met^- Ara^- Leu^- Lac^- Pur^- Gal^- Trp^- His^- Arg^- Str^r

Method

Day 1 **Isolation of transposed strains**

Plate dilutions of a fresh overnight culture of CSH24 (ECO) onto lactose tetrazolium plates which have been pre-warmed at

42–43°. Try to obtain 500–1000 colonies per plate, and use at least 10 plates. Incubate at 43° overnight. The temperature sensitivity of the F'_{ts}*lac* is not complete, and it is important that the incubators are carefully regulated.

Day 2

Observe the lawn of colonies under the dissecting scope. Almost all colonies should be red. Pick and purify with a fine wire white (Lac^+) colonies, or white streaks or sectors of colonies. Streak these onto lactose tetrazolium plates and incubate at 43° overnight.

Alternative procedure: Instead of using lactose tetrazolium plates, use lactose minimal plates with proline (200 μg/ml). The leakiness of the F'_{ts}*lac* will complicate this procedure, however.

Day 3

Lac^+ colonies, those which are white on lactose tetrazolium plates at 43°, are likely to be insertions. These will be tested for the order of transfer of chromosomal markers to determine the point of integration of the episome. The transposition strains have a range of stabilities with respect to maintaining the F'_{ts}*lac* in the chromosome. The exact reason for this is not clear. It is advisable, however, to use the most stable of those isolated in the following experiment. Stability can be monitored by continued streaking on lactose indicator plates at 43°. Strains which segregate the least number of Lac^- colonies on indicator plates should be used.

Testing Transposition Strains for Origin of Transfer

Day 4

To test for the origin of transfer of a transposition Hfr, we will perform a 30-minute interrupted mating against a multiply marked recipient, CSH57. Dilutions are then plated on minimal plates with streptomycin, containing different combinations of required nutrients. After the origins have been approximately located, other markers can be used for a more precise placement.

Inoculate 5 ml of LB broth with a single Lac^+ colony from Day 3 of this experiment. When the density reaches 2×10^8 cells/ml, mix 0.2 ml of this with 2 ml of an exponentially growing culture of CSH57 in a 100- or 125-ml erlenmeyer flask and shake gently in a 37° waterbath for 30 minutes. At the end of this time vortex a 10^{-2} dilution of the mating mixture to disrupt the mating pairs. Plate out 0.1 ml of this dilution onto minimal plates selecting for:

Recombinant phenotype	Marker (map location)	Plate type
Leu^+	*leu* (1 min)	C
Ade^+	*purE* (13 min)	E
Trp^+	*trp* (25 min)	G
His^+	*his* (39 min)	H
Arg^+	*argG* (61 min)	A
Met^+	*metA* or *B* (77–78 min)	B

The plates are prepared exactly as in Unit II (see Table IIC) except that proline is added to all plates (recall that CSH24 is Pro^- and thus we wish to avoid complications caused by crossing in the *pro* region from CSH24).

Plate 0.1 ml of a 10^{-2} dilution of both the donor and recipient onto each type of plate as a control.

Day 5, 6

Examine the plates and record the number of colonies on each type of selection. From the relative frequencies of observed recombinants, it should be possible to determine the approximate location of the origin of the Hfr, and possibly the direction of transfer. To determine these more precisely, repeat this experiment using other strains provided with this manual. Also, it is possible to do interrupted matings at different time points (5, 10, 15, 25, etc.) with CSH57 for the marker which was transferred with the highest frequency.

Materials

Day 1

overnight of CSH24 grown at 30°
15 lactose tetrazolium plates
3 dilution tubes
5 pasteur pipettes and 1 1-ml pipette
incubator at 43°

Day 2

wire needle
4 lactose tetrazolium plates
incubator at 43°

Day 4

colony of Lac^+ transposition strain derived from CSH24
exponential culture of CSH57
test tube with 5 ml broth
shaking waterbath at 37°
100- or 125-ml erlenmeyer flask
3 dilution tubes containing 1 ml LB and 9 ml buffer
50 ml F-top agar at 45°
18 small test tubes
3 minimal plates of the following types (see Table IIC, Unit II):
C, E, G, H, B, A, all with proline as an extra supplement
6 0.1-ml, 1 1-ml, 1 5-ml, and 6 10-ml pipettes

EXPERIMENT 40B

Transpositions at the *tonB* Locus

Strains

Number	Sex	Genotype	Important Properties
CSH24	$F'_{ts}lac^+$	Δ(*lacpro*) *supE thi*	temperature sensitive for the Lac^+ character

Method

Day 1 **Selection for transposition strains**

Treat 10 overnight cultures of CSH24 for T1 resistance, exactly as described in Experiment 42. Plate the entire contents of each preadsorption mixture onto 4 plates (2 lactose tetrazolium plates* and 2 lactose MacConkey plates) and incubate at 42–43°.

* Strains which are *tonB* are reported to indicate better on tetrazolium plates if $FeSO_4$ (3.3×10^{-6} M) is included in the medium.

Day 2 **Purification**

Examine the plates. Only mutants which are *tonB* should form colonies. Some of these will be Lac^+ due to an insertion of the F'_{ts}*lac* into the *tonB* locus. These will appear white on lactose tetrazolium plates and red on lactose MacConkey. *TonB* strains in general indicate poorly on tetrazolium plates. However, these plates have been used in this type of selection. Purify up to 6 Lac^+ colonies from each culture. If the lactose tetrazolium plates are too ambiguous, pick deep red colonies from the MacConkey plates. Incubate at 42–43°.

Day 3, 4 **Mapping**

Pick several pure Lac^+ colonies from each original culture and inoculate each into 3 ml of LB broth. Aerate at 37° until a density of 2×10^8 cells/ml is reached. Prepare a mating mixture with X181a (CSH59) by mixing 0.5 ml with 5 ml of CSH59 in a 100-, 125-, or 250-ml erlenmeyer flask. Place in a 37° waterbath and shake gently. Interrupt the mating after 30 minutes by agitating with a vortex a 10^{-2} dilution of the mating mixture for 30 seconds, and plate 0.1 ml onto plates selecting for Ura^+ (*pyrC*) and Trp^+. Use dilution tubes with 1 part LB broth per 9 ml buffer, and plate in 2.5 ml F-top agar. Incubate at 37°.

Also plate 0.1 ml of a 10^{-2} dilution of CSH59 onto each type of selection plate as a control.

Day 5, 6

Examine the plates. Hfr's with their points of origin at *tonB* will transfer either *trp* early and *pyrC* late or vice-versa (see Figure 40C). Therefore, a strain in which *lac* has been transposed in this manner to the *tonB* locus will give many recombinants with CSH59 on one of the selection plates but not on the other.

Materials

Day 1 **(per group)**

10 overnights of CSH24 grown at 30°
10 small centrifuge tubes
desk-top centrifuge
waterbath at 37°
5 ml lysate of $\phi 80v$
5 ml lysate of colicin V, B (see Experiment 42)
20 lactose tetrazolium plates
20 lactose MacConkey plates
2 1-ml and 10 pasteur pipettes
incubator at 42–43°

Day 2

10 lactose tetrazolium plates or 10 lactose MacConkey plates
60 round wooden toothpicks
incubator at 42–43°

Day 3, 4

(per colony tested)

1 test tube with 3 ml LB broth
overnight of CSH59
empty test tube
2 small test tubes
100-, 125-, or 250-ml erlenmeyer flask
waterbath at 37° (preferably a shaking waterbath)
vortex
1 dilution tube with 1 ml LB broth and 9 ml buffer
1 glucose minimal plate with tryptophan and streptomycin
1 glucose minimal plate with uracil and streptomycin
5 ml F-top agar at 45°
3 0.1-ml, 1 1-ml, and 2 5-ml pipettes

(for the control)

1 glucose minimal plate with tryptophan and streptomycin
1 glucose minimal plate with uracil and streptomycin
1 dilution tube
2 0.1-ml and 1 5-ml pipettes
5 ml F-top agar at 45°
2 small test tubes

EXPERIMENT 41A

Insertion of λ DNA into New Sites on the Chromosome

The chromosomes of phage such as λ integrate into *E. coli* at specific sites on the DNA termed attachment sites or *att*. A specific phage function (Int) mediates recombination between an attachment site on the λ chromosome and a site on the bacterial chromosome (near *gal* for phage λ).

New Prophage Locations

If the *E. coli* chromosome is deleted for the λ attachment site, the frequency of integration of λ DNA is lowered approximately 200-fold. Although still requiring the *int* gene, integration now occurs at different sites on the bacterial chromosome. Shimada and coworkers have determined the prophage locations in the resulting abnormal lysogens and find that preferential sites for insertion exist, although the integration of λ DNA into many different locations can be detected (Figure 41A), including sites within genes. Thus, in the same manner that we selected a specific transposition of F′*lac* in Experiment 40B, we can select for specific integrations of λ DNA using a host which is deleted for the normal attachment site. The λ prophage can then be induced and from such lysates transducing phage can be isolated, using methods described in Experiment 39. With this technique many new transducing phage have been isolated.

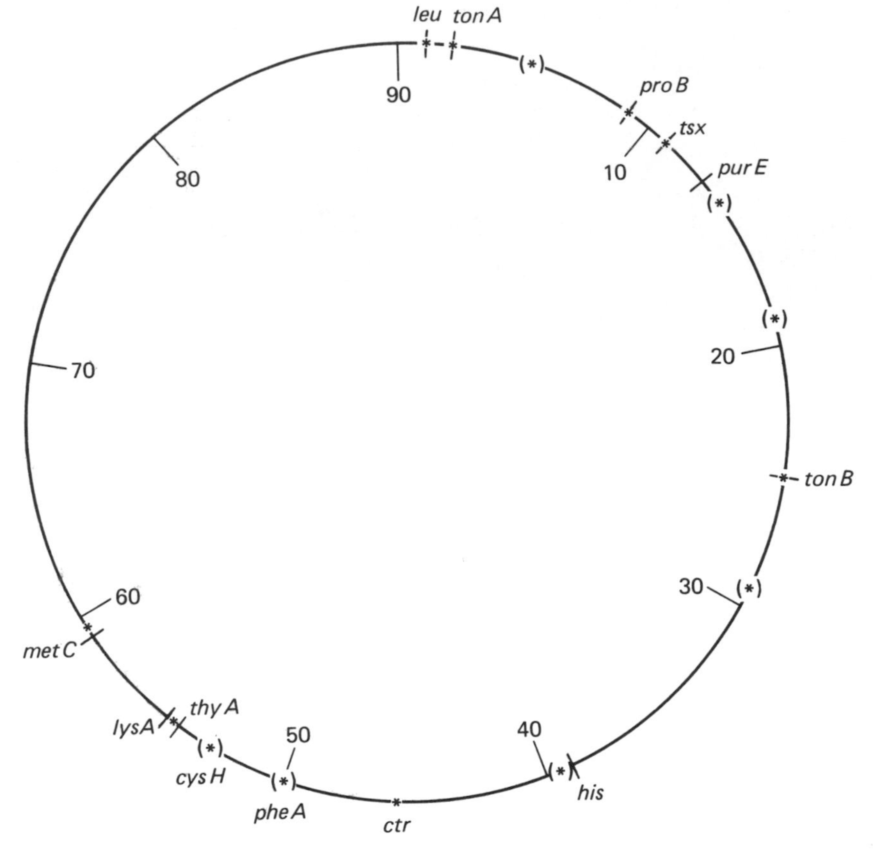

Figure 41A. λ prophage integrated at different sites on the *E. coli* chromosome. (Courtesy of K. Shimada, R. A. Weisberg and M. E. Gottesman.)

In the following experiment we select λ lysogens in a strain deleted for the λ attachment site. The new sites of λ integration are then mapped by interrupted matings.

References

Jacob, F. and E. L. Wollman. 1961. *Sexuality and the Genetics of Bacteria.* Academic Press, New York.

Shimada, K., R. A. Weisberg and M. E. Gottesman. 1972. Prophage lambda at unusual chromosomal locations. I. Location of the secondary attachment sites and properties of the lysogens. *J. Mol. Biol. 63:* 483.

Strains

Number	Sex	Genotype	Important Properties
CSH71	HfrH	Δ(*gal att*λ *bio uvrB*) *thi*	Str^s, deleted for λ attachment site
CSH25	F^-	*supF thi*	$Su3^+$
CSH54	F^-	Δ(*lacpro*) *supF trp pyrF his strA thi malA*	Lac^- Pro^- $Su3^+$ Trp^- Pyr^- His^- λ^r Str^r

Number	Sex	Genotype	Important Properties
CSH57	F^-	*ara leu lacY purE gal trp his argG malA strA xyl mtl ilv metA* or *B thi*	Ara^- Leu^- Lac^- Pur^- Gal^- Trp^- His^- Arg^- Mal^- λ^r Str^r Xyl^- Mtl^- Ilv^- Met^-
CSH62	HfrH	*thi*	Str^s
CSH65	F^-	*leu lac nalA strA thi*	Lac^- Str^r

Method

The phage used in this experiment is λ*CI857S7*. Strain CSH45 is a lysogen carrying this prophage. Lysates are prepared either by UV induction as described in Experiment 2, or by heat induction as described in Experiment 43. The *S7* mutation, which prevents lysis but not DNA synthesis after induction, is suppressed by *supF* (Su3). λ*CI857S7* forms plaques on strain CSH25, which carries *supF* (Su3). Therefore, this strain should be used to titer lysates of this phage.

Day 1

Selection for rare lysogens

Spin down a fresh overnight culture, grown in tryptone broth, of strain CSH71, and also CSH62 (a control). Wash, resuspend in the same volume of 0.01 M $MgSO_4$, and shake for 60 minutes at 37°. At this point withdraw 0.1 ml from each culture and titer by plating 0.1 ml of a 10^{-5} dilution on EMBO plates. Preadsorb λ*CI857S7* at a multiplicity of approximately 10:1 (phage:bacterium) and incubate for 10 minutes at 33°.

Spread 0.1 ml of a 10^{-2}, 10^{-3}, and a 10^{-4} dilution of CSH71, and 0.1 ml of a 10^{-4} and a 10^{-5} dilution of CSH62 on EMBO plates. The plates should be seeded with approximately 10^9 phage that contain mutations preventing integration and the establishment of immunity. We could use λb2*C*, or λ*intC*, for this purpose, although the selection is cleaner with the phage supplied with the strain kit, λ*Ch*80del9.

The survivors of this selection are either mutants resistant to the λ*Ch*80del9 phage (which in this case would be φ80-resistant, since we are using a phage with a φ80 host range) or else are lysogens, carrying λ*CI857S7* as a prophage. Incubate all plates at 33°.

Day 2

After approximately 20 hours, true survivors of this selection form pink colonies on EMBO plates. These should be tested for either resistance to φ80*v* at 33° by cross streaking, or for growth at 42°. Mutants resistant to φ80*v* or strains which grow at 42° are not lysogens but resistant mutants with lesions at either the *tonA* or *tonB* locus.

Pick single colonies and stab into a sector of 2 rich plates, incubating the first at 42° and the second at 33° after streaking for single colonies. By streaking 8 per plate and streaking for single colonies, pure colonies of verified lysogens are ready on Day 3 for further testing. Since strain CSH71 grows poorly on some indicator plates at 42°, particularly tetrazolium plates, it is recommended that LB plates be used for this step.

Determine the titer of each culture from the plates prepared in Day 1, before infection.

Day 3

Examine the plates. Compute the number of true lysogens per ml from each original culture, and using the viable cell count from Day 2, determine the relative efficiency of integration of λ in each of the two strains.

Grid a master LB plate with single colonies from each lysogen and incubate overnight at 33°. This can be replicated on Day 4 onto minimal medium to test for auxotrophs. The master plates can also be used to test the donor properties of the lysogens by replica plating onto a lawn of CSH65 selecting for Leu^+ (use glucose minimal plates with streptomycin). Some of the lysogens isolated by this procedure are found to have lost their Hfr properties.

Mapping Prophage Locations

In order to map the prophage locations in a large collection of different lysogens, it is convenient to use **zygotic induction**. We can measure the time of appearance of infective centers in interrupted mating experiments, or else follow the interference with the inheritance of donor markers by a sensitive recipient (see Introduction to Unit II and Jacob and Wollman, p. 103).

Method I

The genetic effects of zygotic induction are seen as a sharp reduction in the inheritance of markers closest to the prophage. After a long mating, the ratio of recombinants for two different markers will remain roughly constant compared with a control cross in which no zygotic induction occurs, unless one of the markers is near or distal to the prophage integration site (Figure 41B). Using CSH57 as a multiply marked recipient, we can use both CSH71 (as a control) and the λ lysogens derived from CSH71 as donors in an uninterrupted mating as described in Experiment 7.

Relative to the control, a sharp drop in the number of recombinants should be seen for those markers near to or beyond the site of the translocated prophage.

Prepare mating mixtures of each strain to be tested, exactly as described in Experiment 7 except that all donors must be grown at 33–34°. Carry out the mating at 33–34° for 100–120 minutes with gentle shaking. Plate 0.1 ml of a 10^{-2} dilution on each selective plate (types A to I; see Unit II). Incubate all plates at 33°. CSH57 is λ^r, which eliminates the complications that arise from infection of the recipient by free λ phage.

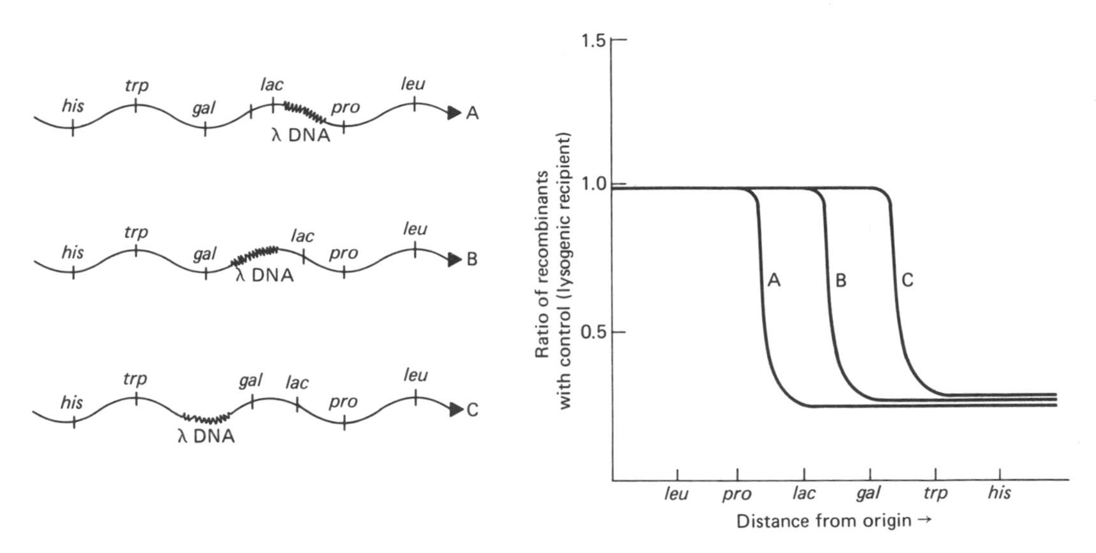

Figure 41B. Effect of zygotic induction on the inheritance of donor markers. Three HfrH strains with λ prophage at different sites are depicted above. In matings with sensitive recipients (those having no λ immunity), a reduction in the number of recombinants relative to a control cross with a lysogenic recipient is seen for those markers distal to the site of integration of the prophage. Therefore, in Hfr (**A**) the *lac*, *gal* and other regions transferred later than the λ DNA are inherited at a reduced frequency, as shown by the graph. However, in Hfr (**C**) *lac* is inherited with a normal frequency, while *trp* and *his* are still interfered with. The region of the map where the number of recombinants relative to the control begins to fall represents the approximate location of the prophage. For markers located very near the prophage, as much as a 100-fold reduction in the number of recombinants has been reported. However, the reduction is not as sharp for distal markers located further away from the site of integration of the prophage.

Method II

In order to score for infectious centers (Figure 41C), two strains carrying Su3 are required, since $\lambda CI857S7$ requires Su3 to be able to lyse the cell and form plaques. The first strain, CSH54, carries several nutritional markers and is λ^r. Because chromosome transfer is slower at 33° than at 37°, internal controls are required to allow the comparison of times of appearance of infectious centers with times of entry of known markers.

The second strain, used as an indicator for infectious centers, is CSH25 (λ^s). In order to use these strains, either Nal^r derivatives of both CSH25 and CSH54 must be constructed (Experiment 32), or preferably a Str^r derivative of CSH25 should be made. If scoring for infectious centers is employed to map the prophage locations, then prepare mating mixtures of CSH54 or CSH54(Nal^r) and of each λ lysogen derived from CSH71. Mix exponential cultures in a ratio of 5 F$^-$:1 Hfr. The CSH71 derivatives should be grown at 33–34°. Place 2–5 ml of the mating mixture in a 100- or 125-ml erlenmeyer flask and incubate at 33–34° with gentle shaking.

At 10 minute intervals pipette 0.1 ml into 10 ml of 0.1 M phosphate buffer, pH 7.0 and vortex vigorously for 30–60 seconds. Plate 0.1 ml in 2.5 ml of H-top agar together with 10^8 bacteria from a fresh overnight culture of strain CSH25(Str^r or Nal^r) on tryptone plates or H plates containing 100–200 μg/ml streptomycin or 20 μg/ml nalidixic acid.

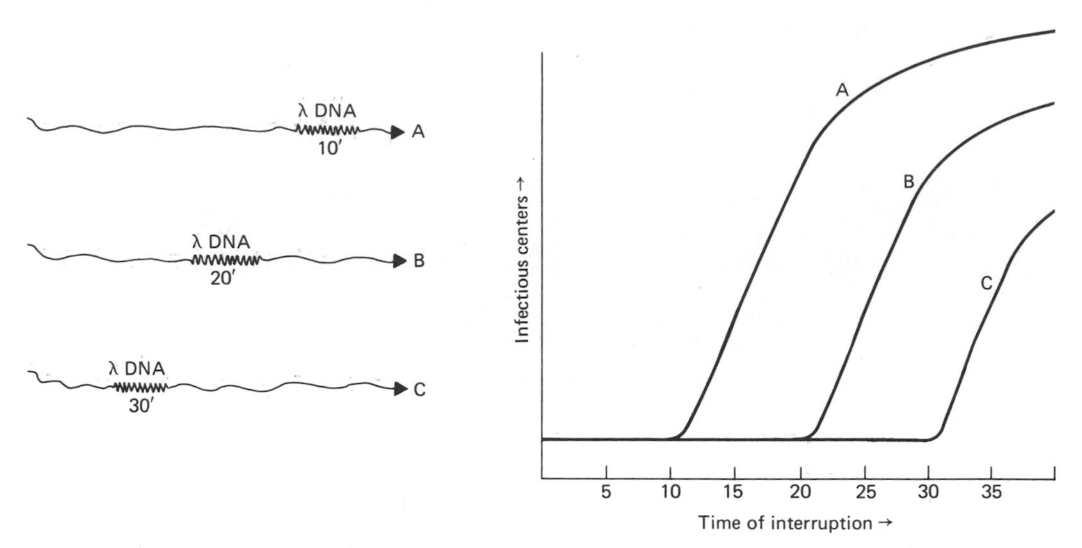

Figure 41C. Appearance of infectious centers after zygotic induction. Three Hfr strains are depicted above, each with λ DNA integrated at a different site. The λ prophage is transferred to a sensitive recipient after 10, 20, and 30 minutes of mating, respectively. To measure the number of infectious centers, the mating is interrupted at various times, and dilutions of each mixture are plated with 10^8 bacteria from a sensitive strain, to allow plaque formation. The recipient strain itself should be resistant to infection by λ to prevent interference from free phage particles in cultures of the donor lysogen.

Also plate 0.1 ml of the diluted (10^{-2}) mating mixture at each time interval on medium selecting for Pro^+, Trp^+, and His^+ recombinants. Incubate all plates at 33°.

Day 4, 5 Replicate the master plate prepared on Day 3 onto a glucose minimal plate supplemented with biotin, and then onto an LB plate. Incubate both at 33°. From the results of the mating experiments with CSH57, or by scoring infectious centers produced after zygotic induction, estimate the approximate location of each translocated prophage. Can you separate all of the lysogens into 4–5 groups according to approximate map position of the prophage?

Materials

Day 1

- fresh overnight cultures of CSH71 and CSH62 grown in tryptone broth
- 10 ml phosphate buffer, pH 7.0 (0.1 M)
- 10 ml 0.1 M $MgSO_4$
- 2 centrifuge tubes
- 2 empty test tubes
- rotary shaker at 37°
- 12 dilution tubes
- 17 0.1-ml, 7 1-ml, and 2 10-ml pipettes
- lysate of λ*CI857S7*, and also λ*Ch*80del9 at approximately 10^{10} pfu/ml
- waterbath at 33°
- 7 EMBO plates (see Experiment 2)
- incubator at 33°

Day 2

(per 40 lysogens tested)

10 LB plates
incubators at 30–33° and 40–42°
40 round wooden toothpicks

Day 3

1 LB plate and 40 toothpicks to test for auxotrophs

(for each lysogen to be used as a donor with CSH57: Method I)

exponential culture of CSH71 and of a λ lysogen derived from CSH71 grown at 33°; exponential culture of CSH57
waterbath at 33–34°
2 100- or 125-ml erlenmeyer flasks
4 0.1-ml, 2 1-ml, and 2 5-ml pipettes
2 plates of each type A to I (Unit II)
2 dilution tubes
incubator at 33°

(for each lysogen to be tested for infectious centers after zygotic induction taking 10 points: Method II)

exponential cultures of a λ lysogen of CSH71 grown at 33°, and of CSH54 or CSH54(Nal^r); fresh overnight culture of CSH25 (Str^r) or of CSH25(Nal^r)
100- or 125-ml erlenmeyer flask
20 0.1-ml, 1 1-ml, 11 5-ml, and 10 pasteur pipettes
vortex mixer
10 dilution tubes with 10 ml of phosphate buffer (0.1 M), pH 7.0
25 ml of H-top agar, 10 small test tubes
10 H plates or tryptone plates
10 glucose minimal plates containing streptomycin, biotin (0.1 μg/ml), tryptophan, uracil, and histidine
10 glucose minimal plates containing streptomycin, biotin, proline, uracil, and histidine
10 glucose minimal plates containing streptomycin, biotin, proline, tryptophan, and uracil
incubator at 33°

Day 4

1 glucose minimal plate containing biotin (0.1 μg/ml)
1 LB plate
incubator at 33°
replica velvet, replicating block

Additional Exercises

1. Combine the selection for rare lysogens with a selection for a drug resistance or auxotrophic marker. One technique which has been used to select lysogens in which the prophage is integrated within a necessary gene is to pick a large number of colonies which come up in the selection and inoculate a single flask containing fresh broth, and then carry out a penicillin enrichment for the desired marker. Can you devise a procedure which is less cumbersome?

2. Although most of the cells from a culture of a *λCI857S7* lysogen do not form colonies at 40–42°, it is possible to detect rare survivors. Many of these are strains in which both phage genes and near-by bacterial genes are deleted. Some of the strains carrying these deletions will be auxotrophs. The determination of the particular lesion responsible for the nutritional requirement, among strains in which the location of the prophage is approximately known, can lead to a precise location of the site of integration. Use this method to attempt to map the prophage integration sites more carefully.

EXPERIMENT 41B

Bringing Two Genes into Close Proximity by Isolation of a Fused Episome

Under normal conditions, two F factors cannot replicate in the same *E. coli* cell. Selection for cells which maintain two F′ factors has led to the isolation of strains harboring F′ factors which have the properties of "fused" F′ factors (Figure 41D), a large episome arising from the fusion of two smaller ones (Press et al.). This finding can be used to link genes which are normally far apart on the chromosome, by positioning them on the same fused episome. If one of the episomes contains a prophage attachment site, such as $\phi 80att$, then we can position different genes near this site, and greatly increase the chances of obtaining specialized transducing phage for the desired marker.

For this procedure we would cross an F′*trp* episome containing $\phi 80att$ into a recipient carrying an F′ with the marker of interest, let us say x. The recipient is converted to a phenocopy F^- (see Experiment 10). The recipient chromosome is $x^- trp^-$. The selection is for X^+Trp^+, and we are thus selecting for the maintenance of both the $F'trp^+$ $\phi 80att$, and the $F'x^+$ in the recipient. To prevent recombination between the episome and the host chromosome, the recipient must be $recA^-$. To eliminate the transfer of the $recA^+$ allele from the donor, the donor should also be $recA^-$. Using drug resistance markers (described in Experiment 32), we select against the donor in this cross (see Figure 41E).

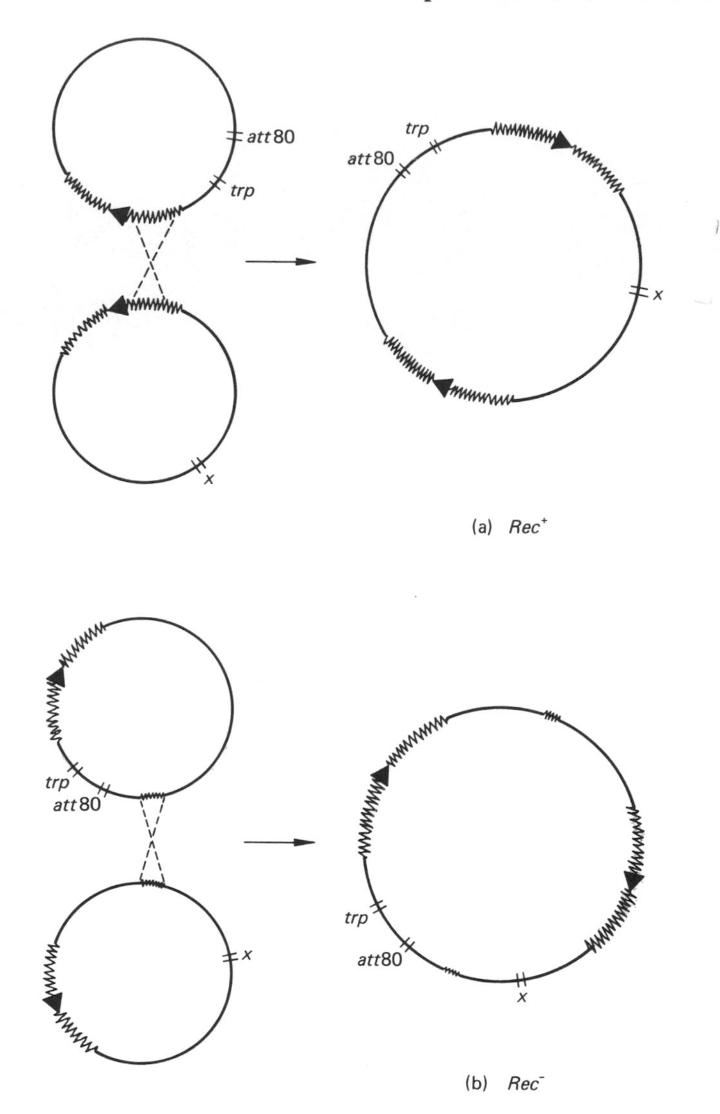

Figure 41D. Fusion of two F′ factors. This diagram depicts two F′ factors, one carrying the *trp* and *att*80 regions, and the other a genetic region denoted as *x*. The type of fusion resulting from homologous pairing and recombination in a Rec^+ strain is shown in part (**a**). In this case, the marker *x* is located on the opposite side of the new F′ factor from *att*80. However, the non-homologous recombination which is shown in part (**b**) results in the formation of a fused F′ factor in which *x* is near *att*80. To greatly reduce the type of fusion events depicted in (**a**), a Rec^- ($recA^-$) strain is employed.

The colonies which are X^+Trp^+ in such a cross are of two types. One class are revertants to either Trp^+ or X^+ which have kept one of the episomes. The second class are due to fused episomes, many of which have ϕ80*att* and *x* in close proximity. Subsequent lysogenization with ϕ80 and then induction of the prophage allow the screening for ϕ80*x* transducing particles. This method has been successfully used to isolate ϕ80*arg* and ϕ80*met* transducing phage (Press et al.).

We supply in the accompanying strain kit the strain CSH72 which carries the $F'_{ts}trp^+$ *att*80 factor originally described by Press et al., 1971. This makes possible

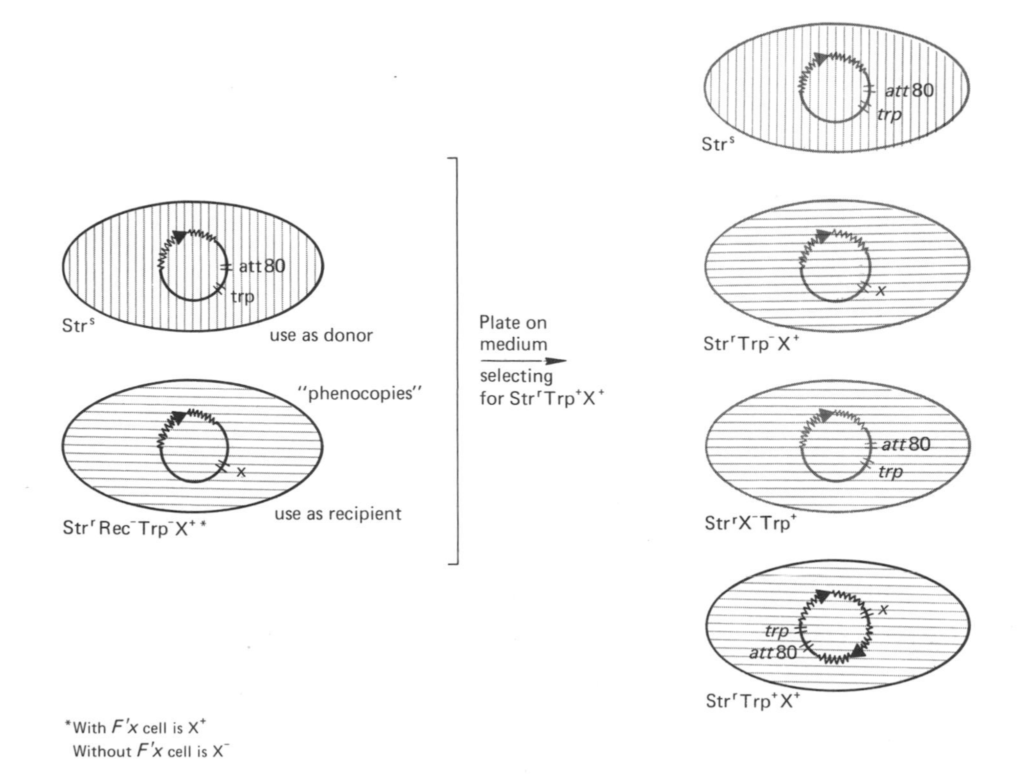

Figure 41E. Selection for strains carrying a fused F′ factor. This figure depicts the manner in which strains carrying a fused F′ factor with *trp* and *x* are selected. A strain which is Rec^- Trp^- X^- Str^r is used. An F′ factor carrying *trp* and *att*80 can be introduced into this strain, converting it to Trp^+, or else as shown here, an F′ factor carrying *x* is introduced. This converts the strain to X^+. A phenocopy cross is then carried out with a strain carrying the other F′ factor, in this case one carrying *trp* and *att*80. Selection is for Trp^+ X^+. Streptomycin is used to counterselect against the donor. In order to grow on this medium the cell must retain parts of both F′ factors. Here the Rec^+ Str^s background is represented by vertical shading, and the Rec^- Str^r background by horizontal shading.

experiments selecting for fused episomes using different F′ factors, for instance those isolated in Experiment 37, and the subsequent isolation of a variety of transducing phage lines.

Reference

Press, R., N. Glansdorff, P. Miner, J. de Vries, R. Kadner and W. K. Maas. 1971. Isolation of transducing particles of ϕ80 bacteriophage that carry different regions of the *Escherichia coli* genome. *Proc. Nat Acad. Sci. 68:* 795.

EXPERIMENT 42A

Isolation of Fusion Strains for the *lac* and *trp* Operons

Construction of Transposed Strains

The *lac* and *trp* regions in *E. coli* are normally separated by approximately 12% of the bacterial chromosome. As demonstrated in Experiment 40, it is possible to select transposition strains in which the *lac* genes are now situated very close to the *trp* operon. Using methods described in Experiment 39 specialized transducing phage can be isolated which carry near-by bacterial genes. In this manner ϕ80d*lac* phage have been isolated. Lysogenizing a *lac* deletion strain with a ϕ80d*lac* phage produces strains such as X7700 (CSH51), pictured in the top part of Figure 42A. Since the ϕ80 attachment site is located near *trp* and *tonB*, the *lac* operon in these strains is again transposed. In this particular ϕ80d*lac*, the *lac* region replaces late genes of ϕ80. When this phage is inserted at the ϕ80 attachment site (att80), its *lac* operon is oriented in the same direction as the *trp* operon.

Fusion Strains

The *tonB* locus, situated between *lac* and *trp* in transposed strains such as CSH51, controls the sensitivity of *E. coli* to bacteriophage T1,* ϕ80 and colV,B.

* T1 phage are never used in a bacterial genetics lab since they are air stable and kill bacterial strains on a wide scale.

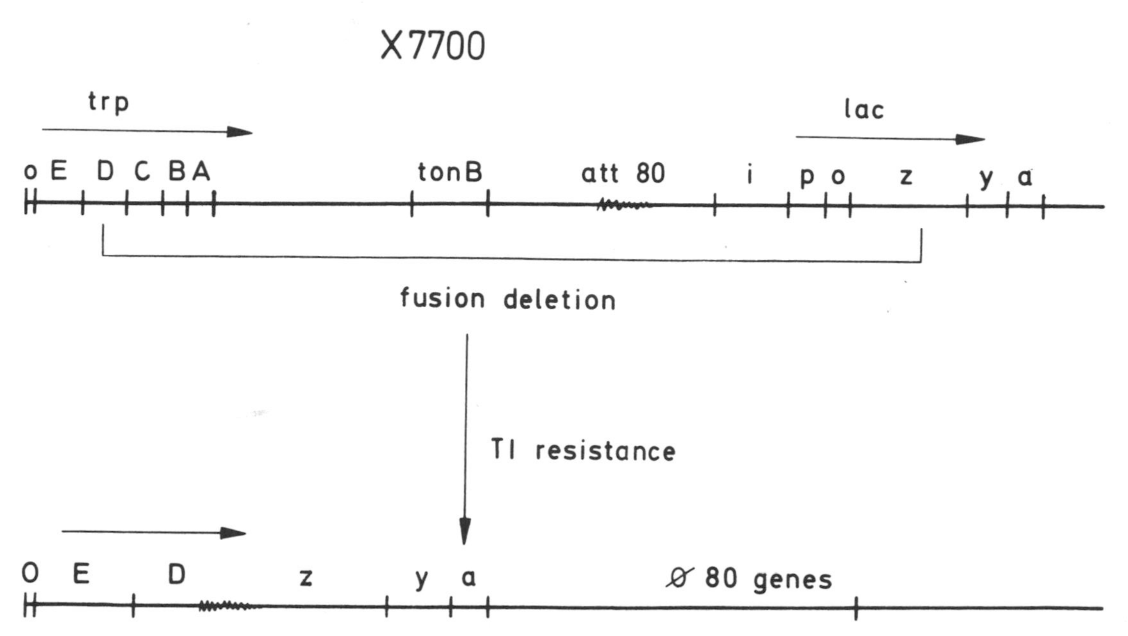

Figure 42A. Fusion of the *lac* operon to the *trp* operon.

Mutants resistant to ϕ80 and colicin, isolated from CSH51, often contain deletions of the *tonB* locus which extend into the *trp* region at one extremity or into the *lac* region at the other. Some of these deletions end within the *trp* operon and also within the *lac* operon. In these cases the remaining genes of the *lac* operon are fused to the *trp* operon (see bottom part of Figure 42A). Table 42A shows levels of transacetylase measured in such fusion strains. In the parent strain, X7700 (CSH51), transacetylase is synthesized at the same rate regardless of whether the *trp* region is repressed or derepressed. However, two fusion derivatives, 291 and 293, can be seen to produce transacetylase completely under *trp* control. The use of strains in which various genes have been transposed near to *tonB* has greatly facilitated the isolation of important deletion mutants. In addition to deletions extending into *z* and *y*, strains such as X7700 have also been used to select deletions extending into the *i* gene. Deletions such as these are a great aid to mapping studies and studies of operon fusion.

Table 42A. Synthesis of Anthranilate Synthetase and Transacetylase in *trp-lac* Fusions

Strain	Anthranilate synthetase		Transacetylase	
	R^+	R^-	R^+	R^-
X7700	0.003	0.470	0.720	0.760
291	0.004	1.020	0.002	0.410
293	0.006	0.490	0.001	0.130

Strains 291 and 293 carry *trp-lac* fusions. The levels of enzymes specified by the *trpE* gene (anthranilate synthetase) and *lacA* gene (transacetylase) are assayed in both a *trpR*$^+$ and a *trpR*$^-$ (constitutive for the *trp* enzymes) background. Strain X7700 (CSH51), the non-fusion parent strain, is also shown.

Design of Experiment

This experiment will involve the isolation of strains which are resistant to colV,B lysates and ϕ80*v*. These are *tonB* lesions, many of which are deletions. Since we use CSH51, many of the deletions will extend into or past *lac*. By recombination tests with different Lac$^-$ strains, the *lac* region will be mapped. All Lac$^-$ TonB strains will then be tested for their *trp* character and the extent of the deletion into the *trp* operon. Strains which appear to have deletions ending within the *trp* operon at one end, and within the *lac* operon at the other, will be tested for functional fusion of the two operons. If the *trp* operon is derepressed in these strains, the remaining genes of the *lac* operon should also be derepressed. This can be accomplished by crossing in a mutation in the *trp* regulatory gene (*trpR*), by means of an Hfr cross. After constructing *trpR*$^-$ derivatives of the deletion strains, we will compare the levels of the *lac* permease with that of the *trpR*$^+$ parent.

At 42° *E. coli* requires the *lac* permease to grow on melibiose. A level of permease corresponding to approximately 8% of the fully induced wild-type amount is needed for growth on this sugar. Strains with the *lac* permease under the control of the *trp* operon which are derepressed (by being *trpR*$^-$) have sufficient levels of permease to be able to grow on melibiose at 42°. Some *trpR*$^+$ fusion strains will be able to grow on melibiose if the deletion is in phase. Also, some out of phase deletions will generate severe polarity and even *trpR*$^-$ derivatives of these fusion strains will not grow on melibiose at 42°.

Selection of *tonB* Strains

Survivors of cultures treated with phage T1 are found to be of two types. About 95% are normal size colonies which have a mutation that maps near the *leu* locus. The remaining 5% form small colonies which have a mutation located near the *trp* region. The former site is called the *tonA* locus, and the latter *tonB*. We are interested only in the *tonB* colonies. It is difficult to work directly with T1 phage in a bacterial genetics laboratory, since the phage is air stable and can destroy bacterial stocks on a devastating scale. The survivors of cultures treated with lysates of ϕ80 virulent (ϕ80*v*) are mutant at either the *tonB* site or a second site, *tonA*. Similarly, the survivors of a culture treated with lysates of colicinV,B are mutant at either the *tonB* locus or a different second locus. If we treat sensitive strains with lysates of both colicin and ϕ80*v*, the only survivors are those which are mutant at the *tonB* locus. A majority of spontaneous mutations at this site are deletions, and some will cut into the *lac* operon in CSH51. In this experiment the survivors of a colicinV,B + ϕ80*v* treatment will be directly plated onto lactose MacConkey plates. On these plates Lac$^+$ colonies appear red and Lac$^-$ colonies white. Thus, we can pick TonB Lac$^-$ deletion mutants in a single step.

Mapping of the *lac* End of the *tonB* Deletions

The basic method for mapping the *lac* end of the T1-resistant deletions will be by spot tests for recombination with episomes carrying mutations in the *lac* region. The *lac*$^-$ mutation X90 is an ochre mutation late in the *z* gene. All Lac$^-$ isolates will be screened for recombination with X90. Those TonB Lac$^-$ strains which recombine with this marker will be saved and tested further (Figure 42B).

The donor strains carrying the *lac*$^-$ mutations on an episome are Str^s. The recipients, in this case all derivatives of CSH51, are Str^r. Therefore, if we carry out this spot mating (see Experiment 5) on a selection plate containing lactose as a carbon source and streptomycin, only those mated recipients which have had a recombination event between the *lac* regions of the chromosome and the episome will grow. As a control for Lac^+ revertants on the episome, a *lac* deletion recipient can be used as a control. In this manner we can test many strains at one time.

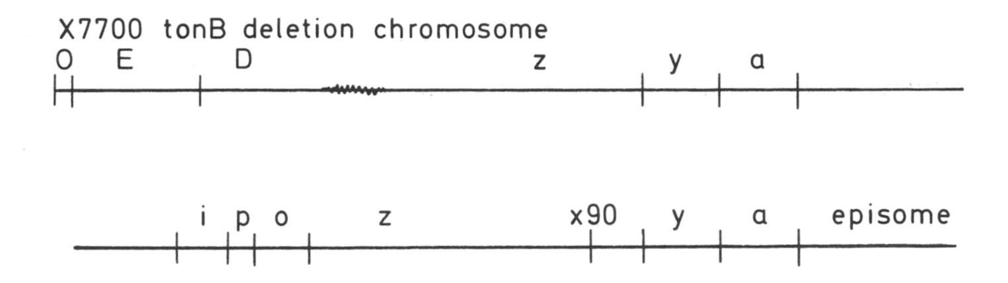

A) Deletion does not go past x90. Recombination can occur, yielding Lac^+ colonies.

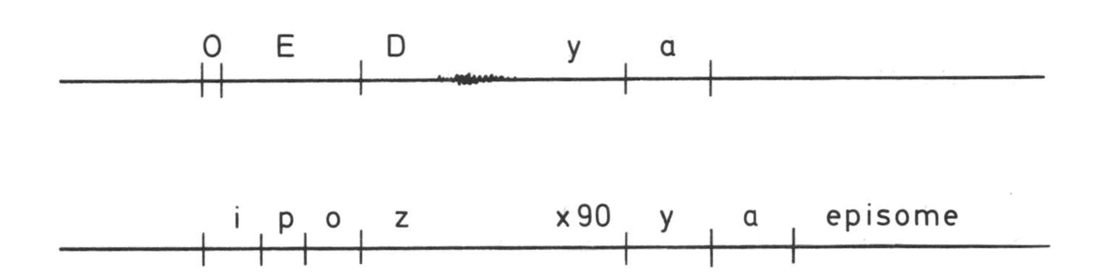

B) Deletion goes past x90. No Lac^+ recombinants can arise.

Figure 42B. Recombination test for the *lac* region.

Mapping the *trp* End of *tonB* Deletion Strains

A. Nutrient tests

1. Tryptophan requirement: This is determined by replicating about 50 colonies onto minimal agar plates with and without tryptophan. The plates should contain proline, since the starting strain is Pro^-.

2. Growth on indole: Strains with deletions cutting into *trpA* grow slowly in the absence of tryptophan if indole is supplied. By replication tests one can determine whether any *trp* deletions end in the *trpA* gene.

B. Complementation

The basic principle here is to construct a stable, partial diploid for the *trp* region, and ask whether or not there is at least one intact copy of each of the five cistrons needed for growth in the absence of tryptophan.

1. Specialized transducing phage, $\phi 80hdtrp$: A transducing phage which carries all or part of the *trp* operon can be constructed (Experiment 39). One phage carries the last four *trp* cistrons (*D*, *C*, *B*, *A*) intact and functioning. When a drop of a lysate of this phage is spotted onto a drop of a culture of a Trp^- strain, a high percentage of the Trp^- cells will become lysogenic for the transducing phage. If the recipient possessed an intact *trpE* cistron, then the cistron would complement the four phage-supplied cistrons. TonB strains are resistant to $\phi 80$. However, phage mutants have been isolated which overcome this barrier. These are termed host range (*h*) mutants. The specialized transducing phage used in this experiment was derived from a $\phi 80h$ mutant. Lysates are prepared by UV treatment of strain CSH48 (see Experiment 2).

2. Episomes: Diploids can be constructed with a series of episomes which contain mutations in each of the *trp* cistrons. The diploids are then examined for the Trp character.

References

BECKWITH, J. R., E. R. SIGNER and W. EPSTEIN. 1966. Transposition of the *lac* region of *E. coli*. *Cold Spring Harbor Symp. Quant. Biol. 31:* 393.

JACOB, F., A. ULLMANN and J. MONOD. 1965. Délétions fusionnant l'opéron lactose et un opéron purine chez *Escherichia coli*. *J. Mol. Biol. 13:* 704.

MILLER, J. H., W. S. REZNIKOFF, A. E. SILVERSTONE, K. IPPEN, E. R. SIGNER and J. R. BECKWITH. 1970. Fusions of the *lac* and *trp* regions of the *Escherichia coli* chromosome. *J. Bacteriol. 104:* 1273.

WANG, C. C. and A. NEWTON. 1969. Iron transport in *Escherichia coli:* Relationship between chromium sensitivity and high iron requirement in mutants of *Escherichia coli*. *J. Bacteriol. 98:* 1135.

Note: TonB strains are especially sensitive to the trace amounts of Cr^{+++} present in agar, and require citrate for growth on solid medium. The standard minimal medium formula ($1 \times A$) has sufficient citrate to enable TonB strains to grow. Otherwise, medium should contain 5×10^{-3} M sodium citrate (see Wang and Newton).

Strains

Number	Sex	Genotype	Important Properties
CSH51	F^-	*ara* Δ(*lacpro*) *strA thi* (ϕ80d*lac*$^+$)	Str^r, *lac* region located near *trp* operon
CSH21	*F'lacZ proA*$^+$,*B*$^+$	Δ(*lacpro*) *supE thi*	Str^s, donates F'*lacpro* with late z^- mutation X90
CSH50	F^-	*ara* Δ(*lacpro*) *strA thi*	Lac^- Pro^- Str^r

Method

Day 1 Inoculate 10 tubes containing 5 ml each of rich broth with a single colony of CSH51. Use a different colony for each tube. Since each colony originated from a single cell, this will insure that mutants arising from different cultures are independent. Allow the cultures to grow to saturation at 37° with good aeration (overnight).

Day 2

Treatment for T1 resistance

Part A. Pre-adsorption

Spin down the cells from each tube in a desk-top centrifuge. Resuspend the pellet in 0.4 ml colicin lysate + 0.4 ml $\phi 80v$ lysate (allow to stand in a 37° waterbath for 30 minutes). The $\phi 80v$ lysate should be 2×10^{11}/ml. The lysates will be supplied by the instructor. A method for preparation of the respective lysates is presented in Experiment 2.

Part B. Plating for survivors

Plate the entire contents of each tube onto 2 lactose MacConkey plates. If the plates are allowed to stand partially open and face down at 37° for 2 hours prior to use, it will be easier to spread this volume of cells. Incubate the plates at 37° for 20 hours.

Day 3

Part A. Scoring results

After 16 hours examine the plates. Record the number of survivors from each culture, and the number of Lac^- colonies. Recall that on lactose MacConkey, Lac^+ colonies are red, and Lac^- colonies are white.

Part B. Purification

Using a toothpick, pick 4 Lac^- colonies from each plate (a total of 8 for each original culture) and stab onto a section of a fresh lactose MacConkey plate. Streak for single colonies. Use 8 sections per plate.

Day 4

Preparation of master plate for replica plating

Prepare a master plate by picking from the purification plates onto a rich (LB) plate. Use sterile round toothpicks and stab a single colony onto each square. Grid 40–50 colonies per plate. Incubate overnight at 37°.

Day 5

Replica plating

Using sterile velvet pads, replicate each master plate first onto a minimal glucose plate (containing citrate and proline) and then onto a glucose plate containing tryptophan (and citrate and proline). Incubate all plates, including the master plate, at 37° overnight.

Day 6

Inoculation of Trp^- strains

Score the results of the nutrient test. Record the percentage of Trp^- colonies among the $T1^r$ *lac* deletion strains. Make an overnight in 2–3 ml LB or NB broth of each Trp^- strain. Shake overnight at 37°. Inoculate from the master plate or the minimal plate (with tryptophan).

Day 7 **Spot test for *lac* marker**

Place a drop of a saturated culture of each Trp$^-$ strain onto a sector of a lactose minimal plate (containing streptomycin, tryptophan, citrate, and proline). When this dries, place a drop of a growing culture of strain CSH21 on each spot. Strain CSH21 contains an episome which has the z^- mutation X90. After the plate dries, incubate at 37° for 48 hours. Use a fresh culture of the episome donor at about $3–5 \times 10^8$ cells/ml. Prepare this by subculturing a stock culture 2 hours before use, aerating at 37°.

As a control, use strain CSH50. This has a complete deletion of the *lac* region and should give a negative result in the spot test with strain CSH21.

Day 9 **Lysate: φ80*hd*trp (*D, C, B, A*)—test for *trpE***

1. Score the results of the cross against X90. Set aside all of the strains which gave a positive result.

2. Spot test for the *trpE* cistron. Spread 4 drops of an overnight culture of the test strain on a glucose minimal plate containing proline and citrate. After the plate dries, apply a drop of a lysate of *φ80hdtrp*. A procedure for the preparation of these lysates is provided in Experiment 2. This lysate contains phage with the last 4 *trp* cistrons. Put a drop of the lysate onto a section of the plate. Apply a drop of broth to a different section of the plate as a control. Incubate at 37° for 24–32 hours. In this manner, test all Trp$^-$ strains which recombine with X90. Trp$^-$ strains which do not recombine with X90 can also be tested.

Day 10 Record the results of the *trp* spot test. This will tell you which strains have at least the *trpE* gene intact, and thus which strains end within the *trp* operon.

Notes: There are several types of survivors that can leak through this selection and complicate the results. These are:

(A) Mucoids. Mucoids have an altered cell wall and are thus immune to many phage. These grow much faster than *tonB* cells, particularly at low temperatures, and a single mucoid colony can overtake an entire plate. The continued appearance of mucoids here indicates that the colicin lysate is not potent enough.

(B) Survivors at other loci: i.e., *tonA*. These can be recognized by their larger colony size. As a test to be absolutely sure that the survivors are really *tonB*, spot test against the lysates of colicin and *φ80v*. Plate out 10^8 bacteria in H-top agar (2.5 ml) onto an H plate. Onto $\frac{1}{2}$ of the plate, put a drop of a *φ80v* lysate, and onto the other half, a drop of colicin lysate. Incubate at 37° for 20 hours. The spot should be visible as a large clearing if the strain was sensitive to that lysate, but not visible at all if the strain was resistant to only one of the lysates.

(C) Use of Xg. Instead of plating on MacConkey plates, one can plate on indoxylgalactoside (Xg) plates. These plates will indicate for deletions cutting into the *i* gene, since constitutive colonies turn blue.

(D) Dangers of nutrient test by replica plating. Be careful about interpreting results of nutrient tests. A common error is to transfer too much material onto the minimal test plate. This can give a false appearance of growth after 24 hours. Another danger is cross-feeding. This occurs when a Trp^+ colony (for example) excretes enough tryptophan into the medium immediately surrounding it to allow neighboring Trp^- cells to grow.

Materials (per group)

Day 1

single colonies of CSH51
10 test tubes with 5 ml of LB broth each
10 spaces in a 37° shaker bath or on a rotor at 37°

Day 2

4 ml of a colicin lysate
4 ml of a $\phi 80v$ lysate
10 plastic centrifuge tubes or small test tubes
desk-top centrifuge
vortex mixer
37° water bath
20 lactose MacConkey plates (pre-dried)
2 5-ml and 10 pasteur pipettes

Day 3

10 lactose MacConkey plates
80 sterile toothpicks

Day 4

80 sterile toothpicks
2 rich plates (LB)
graph paper grid (see Experiment 4)

Day 5

2 sterile velvet pads, replicating block
2 glucose minimal plates with proline and citrate (0.005 M)
2 glucose minimal plates with proline, tryptophan, and citrate

Day 6

35 test tubes with 2.5 ml of LB broth each

Day 7

6 lactose minimal plates with proline, tryptophan, citrate, and streptomycin
exponential culture of strain CSH21
overnight culture of strain CSH50
overnight cultures from Day 6
37 pasteur pipettes

Day 9

(per 10 strains tested)

10 glucose minimal plates with proline and citrate
12 pasteur pipettes
lysate of $\phi 80hdtrp$
fresh overnight cultures of 10 Lac^- Trp^- strains to be tested

EXPERIMENT 42B

Construction of *trpR*$^-$ Derivatives of *trp-lac* Fusion Strains

The purpose of this exercise is to demonstrate that when two operons are fused, the second operon (or the part that remains) is now under the control of the first. We can show this by taking all the strains in which the *trp* operon is fused into the *z* gene of the *lac* operon, and proving that the permease (*y* gene product) is now under *trp* control.

At 42° the *lac* permease is required for growth on melibiose. The level of permease required is at least 8% of the fully induced level. Thus, strains in which *trp* is fused into the *z* gene will not grow on melibiose at 42° when the *trp* operon is repressed, since the level of permease produced would not be high enough. However, when these strains are derepressed for the *trp* operon, the levels of *lac* permease should now be sufficient. We can bring about derepression of the *trp* operon by crossing in a *trpR*$^-$ allele. This mutation, which is closely linked to the arabinose region of the *E. coli* chromosome, results in the absence of a normal *trp* repressor, causing constitutive synthesis of the *trp* enzymes. With the particular allele which we are using, the rate of synthesis of the *trp* enzymes increases 50-fold (under our conditions of growth) for the first two *trp* enzymes (specified by *trpE* and *trpD*), and 15-fold for the last three enzymes (specified by *trpC*, *trpB*, and *trpA*).

Design of Experiment

Since the starting strain (CSH51) was Ara^-, all of the fusion strains derived from this strain are also Ara^-. We will use an Hfr Cavalli which donates ara^+ and $trpR^-$ within the first 30 minutes of mating. Mating pairs will be broken up by agitation, and the mixture plated out on arabinose minimal plates with streptomycin. After purification, the colonies will be scored for the $trpR^-$ character.

Scoring for the $trpR^-$ is done by testing for 5-methyltryptophan resistance. This compound inhibits the first enzyme (anthranilate synthetase) of the *trp* operon. Inhibition can be overcome by the constitutive synthesis of high levels of this enzyme; thus, $trpR^-$ cells will grow normally in the presence of this compound, while $trpR^+$ cells will not.

Strains

Number	Sex	Genotype	Important Properties
CSH61	HfrC	*trpR thi*	Ara^+ $TrpR^-$ Str^s
CSH33	F′*colV*$^+$,*B*$^+$ *trp*$^+$ *cysB*$^+$	*thi*	Str^s

Method

Day 1 **Selection for Ara^+ recombinants**

Grow the Hfr (CSH61) in LB broth to a density of 2–3 × 10^8/ml at 37°. To 2 drops of the recipient (a density of about 5 × 10^8/ml) in the bottom of a large test tube, add 8–9 drops of the Hfr. Place in a 37° bath for 60 minutes.* At the end of this time add 0.1 ml of the mating mixture to 10 ml of dilution buffer containing 1 ml of LB broth (a 10^{-2} dilution). Also plate out a 10^{-3} and a 10^{-4} dilution (each dilution tube should contain 1 ml of LB broth per 10 ml of dilution buffer). To plate out the dilutions, add 0.1 ml to a small tube, and then add F-top agar and spread over an arabinose minimal plate containing proline, tryptophan, citrate, and streptomycin. Incubate at 37° for 48 hours, and purify the recombinants on arabinose minimal plates.

In a similar manner plate a 10^{-2} dilution of the donor and the recipient alone as a control.

Day 3 **Scoring for proline**

A proportion of the Ara^+ recombinants will receive the normal *lacpro* region donated by the Hfr. Test for Pro^+ by picking each Ara^+ recombinant with a toothpick and stabbing first into an arabinose minimal plate without proline and then into an arabinose minimal plate with proline. Pick 50–100 colonies and grid two sets of plates in this manner. The minimal plates should also contain citrate, tryptophan, and streptomycin. This step also serves as a further purification of the Ara^+ recombinants.

* Shake gently.

Day 5

Scoring for *trpR*$^-$

In order to score for the *trpR*$^-$ marker, the strains must be converted to Trp$^+$ strains. This must be done because if tryptophan is put into the media (to enable Trp$^-$ strains to grow), it will be concentrated by the cells, interfering with the 5-methyltryptophan test. The strains can be converted to a Trp$^+$ form by crossing in an F′*trp* episome. Since the donor is streptomycin sensitive, this can easily be done in a spot test. Use CSH33 as the donor. Grow up 10 Ara$^+$ Pro$^-$ recombinants from the preceding cross and spot on the donor, as done in previous spot tests. Use glucose minimal plates with streptomycin and proline.

Test for melibiose growth

To save time, streak directly from the cultures which were grown up for the cross with CSH33 onto melibiose minimal plates, along with the starting T1-resistant derivative. Plate at 42° for 36–48 hours.

Day 6

Test for *trpR*$^-$

Purify the diploids prepared in Day 5 by picking from the spots directly onto 5-methyltryptophan plates. After 18 hours, the 5-methyltryptophan-resistant derivatives will have grown up, while the sensitive strains will not. The plates contain 100 μg/ml of 5-methyltryptophan. Pick first onto 5-MT plates and then onto a glucose minimal plate (each with proline) for comparison.

Day 7

Observe which streaks have grown on melibiose, and compare with results from the 5-methyltryptophan test.

Optional Exercise

Assay fusion strains in the R$^+$ and R$^-$ state for transacetylase (Experiment 53).

Materials

Day 1

overnight culture of CSH61
overnight culture of TonB Lac$^-$ deletion strain derived from CSH51
2 test tubes
10 ml LB broth
waterbath at 37°
5 arabinose minimal plates with tryptophan, proline, and streptomycin
12.5 ml F-top agar at 45°
9 0.1-ml, 1 1-ml, 2 10-ml, and 4 pasteur pipettes
5 dilution tubes containing 1 ml LB broth and 9 ml buffer

Day 3

100 round wooden toothpicks

2 arabinose minimal plates with tryptophan, proline, and streptomycin*
2 arabinose minimal plates with tryptophan and streptomycin

Day 5

10 test tubes with 5 ml LB broth each
2 glucose minimal plates with proline and streptomycin
10 pasteur pipettes
exponential culture of CSH33
2 melibiose minimal plates containing tryptophan and proline
42° incubator

Day 6

2 glucose minimal plates with proline
2 5-methyltryptophan plates (100 μg/ml) with proline; 5-methyltryptophan is autoclaved with the salts
10 round wooden toothpicks

* If minimal medium other than 1 × A is used, 0.005 M citrate must be present.

UNIT VI

TRANSFORMATION WITH λh80d*lac* DNA AND MEASUREMENT OF *lac* MESSENGER RNA

EXPERIMENT 43

Preparation of λh80d*lac* Phage and Transformation with λh80d*lac* DNA

DNA from phages such as λ can under certain conditions infect *E. coli* cells. In order to allow infection of bacteria by λDNA, it is necessary to use intact phage particles as an additional "helper," although spheroplasts can be infected by λDNA alone. If the DNA is prepared from transducing phage carrying bacterial genes, such as *lac* or *gal*, then transformation of different markers can occur (Kaiser and Hogness).

In the following experiment a lysate containing both helper and *lac* transducing phage is prepared. The phage are separated from the bacterial cytoplasm by centrifugation through a CsCl density gradient. A CsCl equilibrium density gradient is then used to separate the defective phage from the normal phage. This particular defective phage has a higher density than the wild-type phage and bands in the lower position. Transducing phage DNA can then be isolated after phenol extraction. This DNA is then used to transduce a Lac$^-$ strain to Lac$^+$, in the presence of helper phage.

In the prophage state λ directs the synthesis of a repressor protein which inhibits the expression and replication of λ phage. The λ structural gene for this repressor is termed the *CI* gene. Mutations, for instance *CI857*, have been isolated in this gene which result in the production of a heat-labile repressor. It has been possible to construct phage which contain some genes from ϕ80 and some genes from λ. One such "hybrid" phage, called λh80, contains the immunity region of λ and the late genes (controlling late functions in phage development) of ϕ80.

When the *CI857* mutation is crossed into this phage, the new hydrid is designated λh80*CI857*. If we heat the bacteria which are lysogenic for this last phage, then the phage repressor is inactivated and the prophage is induced. The bacteria carrying λh80 would normally lyse after phage induction. The phage which we will use, however, carries an additional mutation (*St68*) which prevents lysis. The bacteria which have been heat-induced contain many whole phage particles. They can be centrifuged and disrupted with chloroform. To avoid adsorption of phage to the debris, we use bacteria which have a mutation which renders them resistant to ϕ80 adsorption.

References

KAISER, A. D. and D. S. HOGNESS. 1960. The transformation of *Escherichia coli* with deoxyribonucleic acid isolated from bacteriophage λdg. *J. Mol. Biol. 2:* 392.

KAISER, A. D. and R. INMAN. 1965. Cohesion and the biological activity of bacteriophage lambda DNA. *J. Mol. Biol. 13:* 78.

Strains

Number	Sex	Genotype
CSH43	F$^-$	*tonA* Δ(*lac*) *thi* (λh80*C*I857*St*68)
CSH44	F$^-$	*tonA* Δ(*lac*) (λh80*C*I857*St*68) *thi* (λh80*C*I857*St*68d*lac*$^+$)
CSH45	F$^-$	Δ(*lac*) *thi* (λ*C*I857S7) *trpR*

Method

Day 1

Heat induction: 20 ml of overnight cultures of strains CSH43 and 44 (grown at 32° in LB) will be used as inoculum for 2 2-liter flasks which contain 200 ml LB each. After shaking the cultures for ½ hour at 32°, transfer them to a 43–45° waterbath and shake for 20 minutes. (It is important not to go above 45°, otherwise the phage yield will drop sharply.) Shake the cultures for 3 hours at 37°. Centrifuge the cultures (5000 × *g* for 20 minutes), and resuspend the pellets in 10 ml SM buffer. After adding 1 ml chloroform, shake the suspension for 10 minutes at 37°. Add 10 μl DNase solution (1 mg DNase/ml H_2O). Shake slowly at 37° for 10 minutes. Store overnight at 4°C.

Day 2

CsCl block gradient: Repeat the DNase treatment with fresh DNase. Centrifuge the suspension (15 minutes at 10,000 rpm). Transfer the supernatant with a pasteur pipette to a test tube.

Prepare cesium chloride block gradients as follows: Use the large nitrocellulose tubes for the SW25 rotor of the Spinco L1. Add first 2.0 ml of a CsCl solution with specific density 1.6. Carefully add 1.5 ml CsCl solution with the specific density 1.5. Then add 1.5 ml CsCl solution with the density 1.4. Now slowly add the phage suspension. If there is not enough phage suspension, fill the tubes up to the top with SM buffer. Two students should use one tube. Centrifuge in the SW25 rotor in the Spinco centrifuge for 3 hours at 23,000 rpm. Take the tubes carefully out of the rotor. Put some tape underneath them. Very near to the bottom you should see a blue band of phage. Stick a hypodermic needle in.

Collect the first 2 ml of phage in a small nitrocellulose tube (for the SW39 rotor of Spinco L1), and fill the nitrocellulose tube with CsCl solution of density 1.5. Make an equilibrium run in the SW39 rotor for 22 hours (25,000 rpm at 4°).

Day 3

Transduction with λh80d*lac* phage (Lederberg Test)

Preparation of the indicator bacteria: 1 ml of an overnight culture of strain CSH45 is used by each group to prepare the recipient. The 1 ml is pipetted into 20 ml H medium. Incubate at 30° to an OD_{550} of 1 (which takes about 5 hours). Centrifuge 6 ml of this bacterial culture (20 minutes at low speed); resuspend the bacteria in 3 ml medium I.

CsCl gradient: After the equilibrium run you should see sharp bands in the tubes. The lysates from strain CSH43 form one band and the lysates from strain CSH44 form two bands, of which the lower contains the dlac^+ transducing phage. Collect the phage as you did in the block gradients. Add 4 ml SM to the fraction which contains the phage. Dialyze the phage suspension 2 times against 200 ml SM in the cold room. The suspension which contains the defective *lac* transducing phage will be tested biologically by the following method: Make a dilution series of 1 to 100 dilutions (0.05 ml + 5 ml) down to 10^{-10}. To a small tube containing 0.1 ml indicator bacteria (strain CSH45), add 0.1 ml of each dilution. Let the tube stand for 10 minutes at room temperature, then add 2 ml soft agar and pour onto EMB lactose plates. Measure all phage suspensions as follows: Add to a cuvette 0.1 ml phage and 0.9 ml SM; mix, and measure the OD_{260} in the Zeiss spectrophotometer against SM. One OD unit corresponds to 4×10^{11} phage/ml. The number of viable phage will be appreciably lower.

Day 4

DNA Preparation

Each group should pipette 1 ml of the λh80d*lac* phage suspensions in a screw-cap tube, add 1 ml neutralized phenol (If phenol is spilled on the skin, wash immediately with large amounts of ethanol; otherwise severe burns will result.), and put the tube for 30 minutes on a roller operating at room temperature. Cool the tube in an ice bath and spin at low speed for 5 minutes in the cold. Remove the low (phenol) phase with a pasteur pipette. Repeat the phenol treatment twice. The second time the tube should be on the roller 15 minutes, the third time 5 minutes. The water phase (which contains the DNA) is transferred to dialysis sacs. Wear disposable gloves! Dialyze 4 times against 100 ml DNA buffer. Each dialysis should take 2 hours and should be done in the cold.

Evaluation of the Transduction Experiment

Inspect the EMB plates for Lac^+ colonies. They are dark on a light-colored lawn of Lac^- bacteria.

Number of Lac^+ colonies
Control
10^{-2}
10^{-4}
10^{-6}
10^{-8}
10^{-10}

To how many transducing phage/ml do the data correspond?

Day 5

Prepare a new batch of indicator bacteria as for the Lederberg test. It is absolutely necessary to take freshly prepared cells. This can be done by first growing up an overnight at 30°C. Inoculate 20 ml of H medium with a drop of bacteria from a saturated culture; or inoculate 20 ml of H medium with 1 ml of a saturated culture in the same medium and incubate at 30° to an OD_{550} of 1.0.

Transformation with d*lac* DNA (Kaiser-Hogness Test): Place 3.5 ml indicator bacteria in a screw-cap tube. Remove 0.5 ml of the indicator bacteria to a second tube (these bacteria will not be infected with helper). Incubate both tubes 10 minutes in the 30° water bath, then transfer to an ice bath for 5 minutes. To the 3 ml bacteria add 3 ml helper phage. The helper phage suspension has to be previously adjusted to an OD_{260} of 0.05 with medium I. Incubate for 10 minutes at 30° and then place in an ice bath for 5 minutes. The helper-infected bacteria are now centrifuged (20 minutes at low speed in the cold). Then resuspend in 3 ml TCM medium and store again in the ice bath. Treat the control bacteria without helper in the same way. Dilute the DNA solution in 1 to 10 dilutions (0.5 ml + 4.5 ml [TCM]) down to 10^{-6} in TCM. The bacteria and the phage DNA are added according to the scheme on the facing page. The tubes are incubated for 40 minutes in a 30° bath. Keep the soft agar in a 45° bath. Then, pipette 2 ml of soft agar into each tube. Mix well and pour onto EMB lactose plates. Incubate at 32° when the soft agar has solidified.

Tube	0.2 ml each	0.1 ml each
1	Bacteria without helper	TCM
2	Bacteria without helper	DNA, undiluted
3	Bacteria helper infected	TCM
4	Bacteria helper infected	DNA, undiluted
5	Bacteria helper infected	DNA, 10^{-1} dilution
6	Bacteria helper infected	DNA, 10^{-2} dilution
7	Bacteria helper infected	DNA, 10^{-3} dilution
8	Bacteria helper infected	DNA, 10^{-4} dilution
9	Bacteria helper infected	DNA, 10^{-5} dilution
10	Bacteria helper infected	DNA, 10^{-6} dilution

Day 6 and 7 Inspect the EMB plates for Lac^+ transformants. They grow as dark colonies on a light-colored lawn of Lac^- bacteria. What is the number of Lac^+ transformants/ml DNA solution?
What is the number of Lac^+ transductants/ml phage suspension?
What is the efficiency of transformation?

Buffers

SM Buffer:
0.1 M NaCl
10^{-3} M $MgSO_4$
0.02 M Tris-HCl, pH 7.5
0.01% gelatin

H Medium:
0.1 M potassium phosphate buffer, pH 7.0
0.15 M $(NH_4)_2SO_4$
10^{-3} M $MgSO_4$
1.8×10^{-6} M $FeSO_4$
1 mg/ml glucose and 1 μg/ml B1 added after autoclaving

Medium I:
0.01 M Tris-HCl, pH 7.1
6×10^{-5} M $MgCl_2$
6×10^{-4} M potassium phosphate buffer, pH 7.0
5×10^{-4} M $(NH_4)_2SO_4$
4×10^{-10} M $FeSO_4$
1 mg/ml glucose and 1 μg/ml B1 added after autoclaving

DNA Buffer: 0.01 M Tris-HCl, pH 7.5
0.001 M EDTA

TCM Medium: 0.01 M Tris-HCl, pH 7.5
0.01 M $MgSO_4$
0.01 M $CaCl_2$

Materials

Day 1

20 ml of fresh overnight cultures of CSH43 and CSH44 grown at 30–33° in LB or 2YT medium
2 2-liter flasks containing 200 ml LB each (preferably baffle-bottom)
water bath at 50° and also at 42°
Sorvall or similar centrifuge with rotors, tubes, and beakers to spin down the 200 ml of each culture and the 20 ml debris
shaker at 37° to enable the incubation with aeration of 2 2-liter flasks
20 ml SM buffer
2 ml chloroform
20 μl DNase (1 mg/ml) in H_2O
1 10-μl and 2 10-ml pipettes

Day 2

20 μl DNase
Sorvall centrifuge as in Day 1
3 5-ml, 1 10-μl, and 2 pasteur pipettes
2 test tubes
1 Spinco L1 or a similar centrifuge with a SW25 and SW39 rotor (one Spinco accommodates six students at once) and the nitrocellulose tubes for each rotor (there are also rotors with six buckets SW50.1 and SW27)
CsCl solutions: (see also Experiment 45)

		required amount
density 1.4:	39% w/w in SM buffer	1.5 ml
density 1.5:	45% w/w in SM buffer	6.5 ml
density 1.6:	51% w/w in SM buffer	2 ml

scissors, tape, hypodermic syringe #20

Day 3

1 ml fresh overnight of CSH45 grown at 30° in H medium
20 ml H medium
3 ml medium I
2 small centrifuge tubes
450 ml SM buffer
2 pasteur pipettes, 12 0.1-ml, 8 1-ml and 5 10-ml pipettes
5 dilution tubes
6 small test tubes
12 ml soft agar: per liter, 4.5 g agar, 10 g Bacto tryptone, and 5 g NaCl
6 lactose EMB plates
Zeiss or a similar spectrophotometer to measure the OD at 260 mμ

Day 4

roller which can accommodate 6 small screw-cap tubes
6 small screw-cap tubes
6 ml phenol (redistilled) saturated with 0.1 M Tris-HCl, pH 7.5
ice bath
clinical centrifuge
24 pasteur pipettes, 9 1-ml pipettes
disposable gloves
dialysis tubing, Visking #8 (40 cm per student); see Unit VIII for notes on preparation of dialysis tubing
400 ml DNA buffer

Day 5

fresh overnight of CSH45 grown in H medium
20 ml H medium
2 test tubes with screw-top caps
water bath at 30° and 45°
ice bath
8 0.1-ml, 11 1-ml, and 6 10-ml pipettes
clinical centrifuge
40 ml TCM medium
7 test tubes with 4.5 ml TCM medium each
10 small test tubes
20 ml soft agar
10 lactose EMB plates
incubator at 30–33°

INTRODUCTION TO EXPERIMENTS 44–47

The use of specialized transducing phage allows the detailed study of the control and properties of messenger RNA in *E. coli*. Transducing phage provide an ideal source of genetically defined DNA which can be used to detect homologous RNA by the DNA-RNA hybridization assay. Thus, the *lac* transducing phage can be used to detect, quantitate and characterize *lac* messenger RNA. The DNA of the *lac* transducing phage contains genes of both bacterial and viral origin. The phage genes themselves have very little homology with the bacterial chromosome and so the only portion of bacterial RNA which will hybridize with the phage DNA will be the RNA corresponding to the bacterial genes carried by the phage.

The next four experiments describe the preparation of materials for a *lac* DNA-RNA hybridization assay and the details of the assay itself. In the first experiment radioactive RNA is prepared from both IPTG-induced and uninduced cultures. Experiment 45 describes the preparation of DNA from *lac* transducing phage for the hybridization assay. Finally in Experiments 46 and 47 the level of *lac* message from induced and uninduced cells is measured by the hybridization assay, and the size of the *lac* message is determined by sedimentation on a sucrose gradient.

References

Bøvre, K. and W. Szybalski. 1971. Multistep DNA-RNA hybridization techniques. *Methods in Enzymol. XXI-D:* 350.

Contesse, G., M. Crépin and F. Gros. 1970. Transcription of the lactose operon in *E. coli*. *The Lactose Operon*, p. 111. Cold Spring Harbor Laboratory. For gradient determination of size and hybridization curves.

Gillespie, D. 1968. The formation and detection of DNA-RNA hybrids. *Methods in Enzymol. XII-B:* 641. Discussion of all commonly used methods for detection of homology between DNA and RNA.

Goff, C. G. and E. G. Minkley. 1970. The RNA-polymerase sigma factor: A specificity determinant. *LePetite Colloq. Biol. Med. 1:* 124. For *in vivo* messenger extraction, DNA purification procedures, and separated strand hybridizations.

Naono, S., J. Rouvière and F. Gros. 1965. Preferential transcription of the lactose operon during diauxic growth of *Escherichia coli*. *Biochem. Biophys. Res. Commun.* 18: 664. For filter hybridization.

Roberts, J. W. 1969. Termination factor for RNA synthesis. *Nature 224:* 1168. For separated strand hybridizations and TCA precipitation.

Shapiro, J., L. MacHattie, L. Eron, G. Ihler, K. Ippen and J. Beckwith. 1969. Isolation of pure *lac* operon DNA. *Nature 224:* 768. Explains how the p*lac*5 phage was constructed, the separation of its strands and hybridization of *lac* mRNA.

Summers, W. C. 1970. A simple method for extraction of RNA from *E. coli* utilizing diethyl pyrocarbonate. *Anal. Biochem. 33:* 459.

Summers, W. C. and W. Szybalski. 1968. Size, number, and distribution of poly G binding sites on the separated DNA strands of coliphage T_7. *Biochim. Biophys. Acta 166:* 371. For separation of phage DNA strands.

Szybalski, W., H. Kubinski, Z. Hradecna and W. C. Summers. 1971. Analytical and preparative separation of the complementary DNA strands. *Methods in Enzymol. XXI-D:* 383. Methods for preparation of separated strands from different phages, especially λ and T7, with a thorough discussion of the principles involved.

EXPERIMENT 44

Isolation of RNA Labeled *in vivo* after IPTG Induction

The synthesis of *lac* messenger RNA in the cell may be stimulated by the addition of the gratuitous inducer IPTG. In the following experiment we extract labeled messenger RNA from parallel cultures of Lac^+ bacteria, grown with and without IPTG. ^3H-uridine is added for a short time to label the RNA, which is then extracted with hot phenol equilibrated with a sodium acetate buffer, pH 5.2, containing 0.5% SDS. This procedure lyses the cells, inactivates nucleases, extracts the proteins, and partitions the RNA in one step. The hot phenol and SDS destroy the cell membrane and denature most proteins. The phenol, at this pH and temperature (60°), also partitions the proteins and most of the DNA from the RNA.

Strains

Number	Sex	Genotype
CSH23 or	F′*lac*$^+$*proA*$^+$,*B*$^+$	Δ(*lacpro*) *supE spc thi*
CSH62	HfrH	*thi*

Method

Day 1

Preparation of cultures

Inoculate 4 ml of M9 glycerol medium (containing B1) with 0.1 ml of a fresh overnight culture of strain CSH23 or CSH62. Aerate at 37° for 2–3 hours until a density of approximately 2×10^8 cells/ml is reached ($OD_{550} \simeq 0.2$). Subculture 0.2 ml into 2 test tubes containing 2 ml each of M9 glycerol medium, one of which is supplemented with 5×10^{-4} M IPTG. Grow for approximately 3 hours at 37° until the density is again at 2×10^8 cells/ml (any OD_{550} from 0.1–0.3 is satisfactory).

Labeling the cells

To each tube add 100 μc of ^{3}H-uridine (specific activity approximately 25 c/mM). Shake the cells vigorously for one minute at 37°. This pulse time allows the labeled uridine to be incorporated into the labile *lac* message in significant amounts before large quantities of label appear in the stable RNA species.

Add 2 ml of ice-cold (−10°C) M9 medium containing 0.02 M NaN_3 (a poison) to each tube to prevent the further incorporation of label. Spin down the cells in the cold in a desk-top Sorvall for 5 minutes at $10{,}000 \times g$. Carefully decant the radioactive supernatant and resuspend the cells in 2 ml of ice-cold M9-azide medium, using a vortex. Spin down the cells as before. Decant the supernatant and resuspend the cells in 2 ml of a solution of 0.02 M sodium acetate, 0.5% SDS, and 0.001 M Na EDTA, pH 5.5.

Phenol extraction

Add 4 ml of redistilled phenol saturated with the same sodium acetate buffer. (The phenol is saturated by adding an equal volume of the buffer and mixing vigorously. The phenol forms the lower phase.)

Gently vortex the mixture and shake in a 60° water bath for 5 minutes. Transfer the mixture to an acid-washed conical centrifuge tube and spin in a clinical centrifuge at room temperature to separate the phases. Remove the upper aqueous phase with an acid-washed pasteur pipette. Do not suck up the phenol or any of the material at the interface. Re-extract the aqueous phase at room temperature with 4 ml of the saturated phenol and collect the aqueous phase as before in a clinical centrifuge tube.

Nucleic acid precipitation

Adjust the aqueous phase to 0.1 M KCl by adding approximately 50 μl of 4 M KCl and then precipitate the nucleic acids by adding 2 volumes of 95% ethanol (precooled to −20°C). Chill for 3 hours or longer at −20° to allow the precipitate to form. Spin down the precipitate for 30 minutes in the cold in a clinical centrifuge. Carefully pipette off as much of the ethanol as possible and resuspend the pellet in a volume of 200 μl of $2 \times$ SSC. Store in the cold.

Materials (per group)

Day 1

overnight culture of strain CSH23 or CSH62 grown in M9 glycerol medium supplemented with B1. Overnight cultures are prepared by inoculating a single colony from a glucose minimal plate.
20 ml M9 glycerol medium supplemented with B1 (see Appendix)
5 test tubes
1 ml 10^{-2} M IPTG
roller or bubbler at 37°
200 μc ^{3}H-uridine at 25 c/mM
20 ml ice-cold M9 medium
1 ml 0.1 M Na azide (to be handled with care!)
desk-top Sorvall
4 heavy-walled sterile 12-ml centrifuge tubes for Sorvall
8 plastic gloves
20 ml of 0.02 M sodium acetate, 0.001 M Na EDTA, 0.5 %SDS
20 ml redistilled phenol (may be stored frozen)
4 large culture tubes
vortex mixer
60° waterbath
37° waterbath
4 acid-washed conical test tubes
4 acid-washed pasteur pipettes
1 ml 4 M KCl
10 ml ethanol precooled to −20°C
5 0.1-ml, 5 1-ml, 8 5-ml and 4 pasteur pipettes

1 × SSC
0.15 M sodium chloride
0.015 M sodium citrate

2 × SSC
0.3 M sodium chloride
0.03 M sodium citrate

EXPERIMENT 45

Separation of the Strands of λp*lac* DNA

In this experiment we will grow and purify a *lac* transducing phage, λp*lac*5 *CI857S7* ($i^- z^+ y^-$). Strain CSH66 carries the λp*lac* as a prophage. Because the phage repressor is temperature sensitive, lysates can be prepared by heat induction. The phage also carries the *S7* mutation which allows overproduction of the phage by preventing the cells from lysing, as they do following induction of a wild-type phage. The phage are purified by banding first in a block gradient and then in an equilibrium gradient. The block gradient quickly separates the phage from nucleic acids which band below the phage and the proteins which band above.

If the filter method of hybridization is used, the DNA is extracted from the phage protein capsid by denaturing the proteins with SDS and then selectively precipitating them with KCl. Alternatively, the strands of the phage DNA can be separated by opening the phage and denaturing the DNA in one step and then allowing the denatured DNA to anneal to poly (U, G). Only the left half of one of the strands (termed the *r* strand) binds the poly (U, G). The DNA-poly (U, G) complex has a higher buoyant density than denatured DNA. We exploit this property in separating the DNA strands by banding them at different densities in a cesium chloride equilibrium gradient.

Strains

Number	Sex	Genotype
CSH66	F^-	Δ(*lac*) *thi* (λCI857S7p*lac*5$i^- z^+ y^-$)

Method

Day 1

Growth of phage

Inoculate 1 liter of 2YT in a 4-liter flask with 100 ml of strain CSH66 grown to 5 × 10^8 cells/ml (OD_{550} = 0.5). Prepare the inoculum the day before by inoculating 100 ml of 2YT in a 500-ml flask with a single lac^+ temperature-sensitive colony and aerating. Incubate at 34° overnight. (Test several lac^+ colonies for temperature sensitivity by streaking them on lactose indicator plates which are incubated at 34° and 42°. The only survivors at this high temperature should be a few cells which have lost the phage and are lac^-.)

Incubate the 1-liter culture with vigorous shaking at 34° for about 3 hours until the cells reach a concentration of 2.5 × 10^8 cells/ml ($OD_{550} \simeq 0.25$) (cultures at higher concentrations do not induce well). Put an alcohol-swabbed thermometer in the flask and heat it to 42° over a burner, swirling it all the time. Transfer the flask immediately to a 37° shaker. Incubate the culture at 37° for 4 hours. (If induction was successful, the cells by this time should be greatly elongated, which can be seen by looking at them under a phase contrast microscope.) Stop the growth by chilling the flask in ice.

Preparation of the lysate

Spin down the cells for 20 minutes at 9,000 rpm in a Sorvall GSA head. Resuspend them in 30 ml 0.01 M Tris pH 7.9, 0.1 M KCl, 10^{-4} M EDTA. Freeze them in an acetone dry-ice bath and then thaw. The mixture should then be very viscous. Add $MgCl_2$ to 10^{-2} M, then add 3 ml of 0.05 mg/ml of DNase I. Incubate with gentle shaking at 37° until the mixture becomes much less viscous (about 15 min). Chill and spin at 15,000 × *g* for 10 minutes in a desk-top Sorvall in the cold. Decant the supernatant and spin it again. Remove the supernatant which is the phage lysate. Discard the pellet. Store in the cold.

Day 2

Block gradient purification

Make 3 solutions of CsCl of densities 1.3, 1.5, and 1.7 by diluting a stock of saturated CsCl solution (65 g of CsCl plus 35 g H_2O) with phage buffer according to the recipe below.

Density	Saturated CsCl (room temperature)	Phage buffer
1.7	4.66 ml	2.00 ml
1.5	2.22 ml	1.78 ml
1.3	1.33 ml	2.11 ml

Phage buffer is 0.01 M $MgCl_2$, 0.01 M Tris pH 7.9, 0.01 M NaCl. The saturated CsCl solution should be filtered once through a Millipore filter before use.

For an SW25.1 or an SW27 Beckman rotor, form the step gradients by first adding 2 ml of CsCl density 1.3. Then carefully underlay it with 2 ml of the 1.5 solution and underlay the 1.5 solution with 1.5 ml of the 1.7 solution. Underlaying is accomplished by inserting a pasteur pipette filled with the solution through the upper layer(s) to the bottom of the tube and allowing the liquid to slowly displace the upper layer(s). You should be able to see the boundary between the three layers. (Mark the tube before you spin it so you know where the boundary layers are.) Then carefully layer the phage suspension over the block gradient. If you do not have enough lysate to fill the tube, layer phage buffer over the lysate until the tube is almost full ($\frac{1}{8}$ inch from the top).

Spin for 2 hours at 24,000 rpm, 8°. Collect the blue-white opalescent phage band, which should be in the 1.5 density band, into one tube by puncturing the tube just below the band with a 22-gauge needle. (You can see the band more clearly if you illuminate the tube from the side with a microscope lamp.) If you use nitrocellulose tubes, you must put a small patch of adhesive tape on the surface of the tube where you intend to puncture it, after roughening the surface with acetone on a paper towel; otherwise it will leak around the needle. Polyallomer tubes seal sufficiently without the patch. Avoid the pellet on the bottom and the bands above the phage. The phage may be stored in the cold until the next step.

Equilibrium banding: Adjust the phage-containing fractions to density 1.5 by adding either phage buffer or saturated CsCl. Bring the volume up to about 4 ml with cesium chloride of density 1.5. The refractive index should be (1.3813 ± 0.0002), that of H_2O being 1.3333. Band the phage by spinning 16–18 hours at 34,000 rpm in an angle 40 rotor at 8°.

Each angle tube should be filled about halfway with the phage solution. Fill the rest of the tube with mineral oil. Seal the tubes with a vacuum cap. The phage should form a band in the middle of the cesium gradient.

Day 3

Collect the phage as before, after removing the plug from the vacuum cap.

Extraction of phage DNA (to be used for the filter hybridization)

Dialyze the phage (about 2 ml) from the equilibrium banding against one liter of 0.05 M NaCl, 0.01 M Tris pH 7.9, 0.001 M Na EDTA for 2 hours in the cold. No K^+ ions should be present. After dialysis, adjust the phage concentration to 0.5–0.8 mg/ml (OD 1 equals 0.65 mg/ml at 260 mμ).

Place the dialyzed phage in a conical test tube. Add enough of a 10% solution of recrystallized SDS to bring the final concentration of SDS to 0.5%. Incubate the mixture with gentle mixing at 65° for about 5 minutes until the solution completely clears and becomes viscous. Then add $\frac{1}{8}$ of the volume of 4 M KCl to bring the solution to 0.5 M KCl. Chill the mixture for 15 minutes in ice to allow the precipitate to form. The precipitate is potassium-SDS and the capsid proteins. Centrifuge the precipitate for 5 minutes at 4° at 5000 rpm.

Use an acid-washed pasteur pipette to remove the viscous upper solution of DNA from the pellet. Be careful not to bring up any of the white pellet.

Dialyze the supernatant in #8 dialysis tubing at 4° against one liter at a time of 0.1 M NaCl, 0.01 M Tris pH 7.9, 0.001 M EDTA. Dialyze overnight with several changes of buffer because dialysis of SDS is slow. For storage dialyze into the same buffer with 0.1 M KCl replacing the 0.1 M NaCl.

Day 4

Strand separation (for separated strand hybridization)

This procedure is sensitive to contaminating nucleases. Use acid-washed glassware throughout and wear plastic gloves when handling dialysis tubing.

Dialysis

Dialyze 400 μg of DNA equivalents of phage for 2 hours against 1 liter of unbuffered 10^{-3} M EDTA pH 7.5 in the cold to remove the CsCl (OD 1 at 260 mμ of phage suspension is equal to 50 μg/ml of DNA). Keep the phage cold when you begin the dialysis. Remove the phage from the dialysis bag and put them in an acid-washed 12-ml conical test tube. Read the OD_{260} again. Transfer 300 μg of DNA equivalents of the dialyzed phage suspension into an acid-washed conical test tube. Add 0.25 ml of 1% sarcosyl NL97 (made up fresh each time) to this tube.

Denaturation, and annealing with poly (U, G)

Add to the dialyzed phage an amount of poly (U, G) equal to the weight of the DNA (in this case 300 μg). The lot number of poly (U, G) used is important. Some lots do not work. Schwartz lot #6801 works well. Bring the volume to 2.5 ml with cold 10^{-3} M EDTA and adjust the pH of the solution to 8–9 by using pH paper and adding about 7 μl of 1 N NaOH. Heat the mixture to 90–95° for 3.5 minutes. Measure the temperature of the mixture by placing a thermometer in a parallel tube filled with the same amount of liquid.

Mark the time from the point at which the temperature reaches 90°. If the DNA is not heated sufficiently, it will not denature; if it is heated too much, it will break. After 3.5 minutes chill the tube in an icewater mixture and immediately squirt 1.7 ml of ice-cold 10^{-3} M EDTA into the tube. Both sarcosyl and heat denature the capsid proteins, and the heat denatures the DNA which is prevented from re-annealing by rapid chilling. The poly (U, G) then binds to one of the two denatured strands. Rapid cooling is essential to avoid partial renaturation.

CsCl gradient

Add 5.62 g of optical grade cesium chloride and adjust the refractive index to 1.4035 ± 0.0001 density with solid CsCl or 10^{-3} M EDTA. Place the liquid in an angle 40 polyallomer tube, overlay it with mineral oil, cap the tube with a vacuum seal and spin it at room temperature for 40 hours in a Spinco angle 40 head at 37,000 rpm.

Day 5

Fraction collection

To collect the fractions, pierce the bottom with a cut-off 22-gauge needle (the Luer-lock part of which has been cut off so the needle is straight) and drip the gradient into tubes. Use pliers to grip the needle as you puncture the tube. Thirteen drops per tube gives about 45 fractions and gives good resolution of the peaks. The drops are smaller at the end of the gradient because the sarcosyl lowers the surface tension.

Add 0.2 ml of 2 × SSC to each fraction and read the optical density of each fraction at 260 mμ. The pattern should look like Figure 45A. The high OD at the bottom of the gradient is poly(U, G). The first peak from the bottom is the right strand, the second is the left strand. If the light fraction contains significantly more DNA than the heavy peak, you have broken the strands of the DNA. Likely causes are impure phage, too much heating, or allowing the phage to warm up once they have been dialyzed against EDTA. If you have a broad band of DNA at the top of the gradient, the DNA was either not fully denatured or chilling was not fast enough and the DNA has re-annealed.

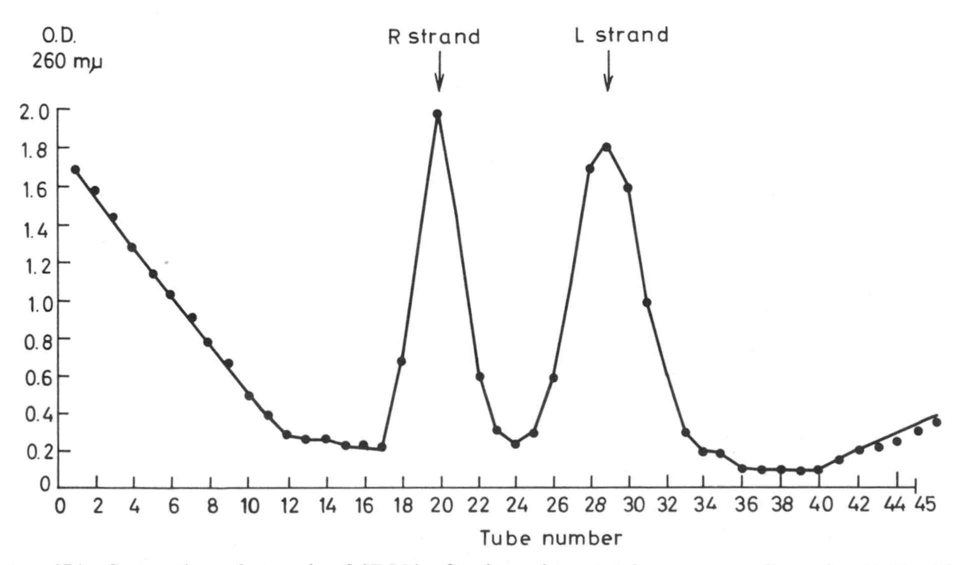

Figure 45A. Separation of strands of λDNA after heat denaturation, re-annealing of poly (U, G), and centrifugation in a CsCl gradient. λ*plac*5 DNA behaves in exactly the same way.

Pool the fractions containing the majority of each separated strand. Adjust to 0.25 M NaOH and incubate 4 hours at 37° to hydrolyze the bound poly (U, G). Dialyze overnight against 2 changes of 500 ml each of 2 × SSC.

Day 6 Heat for 4 hours at 67° to anneal any cross-contaminating strands. Store in the cold.

Materials (per group)

Day 1

100-ml culture of strain CSH66 grown to 5×10^8 cells/ml in 2YT
1 4-liter flask with one liter of sterile 2YT (a baffle-bottom flask if possible)
water baths or platforms at 34° and 37°
6 sterile bottles for the Sorvall GSA head
50 ml of 0.01 M Tris pH 7.9, 0.1 M KCl, 10^{-4} M EDTA
10 ml of 1 M $MgCl_2$
acetone dry-ice bath
DNase I, 1 mg/ml in 10^{-3} M HCl (Sigma)
desk-top Sorvall
2 sterile corex centrifuge tubes

Day 2

SW25.1 or SW27 Beckman centrifuge rotor
2 clean centrifuge tubes for the SW25.1 or SW27
22-gauge needle
microscope lamp (collimated beam)
1 angle 40 rotor
1 angle 40 centrifuge tube and vacuum cap
mineral oil
adhesive tape, acetone
CsCl, phage buffer (see Methods)

Day 3

microscope lamp
sterile tube for collecting phage
22-gauge needle
1 liter of cold 0.05 M NaCl, 0.01 M Tris pH 7.9, 0.001 M Na EDTA
1 Zeiss spectrophotomer
1 acid-washed conical test tube
5 ml 10% SDS (recrystallized)
1 65° waterbath
100 ml 4 M KCl
1 ice bucket
#8 dialysis tubing
1 clinical centrifuge (in a cold room)
1 acid-washed pasteur pipette
4 liters of 0.1 M NaCl, 0.1 M Tris pH 7.9, 0.001 M EDTA

Day 4

1 liter of 10^{-3} M EDTA pH 7.5 (cold)
plastic gloves
2 ml 1% sarcosyl NL97 (Geigy)
washed #8 dialysis tubing
2 acid-washed conical test tubes
400 μg poly (U, G)
1 ml 1 M NaOH
pH paper
stop watch
7 g optical grade CsCl
refractometer
2 angle 40 polyallomer centrifuge tubes with vacuum caps
mineral oil

Day 5

1 rack of 50 clean plating tubes
cut-off 22-gauge needle
37° waterbath
1 M NaOH
1 liter 2 × SSC (cold)
67° waterbath

Day 6

1 liter 2 × SSC
67° waterbath
#8 dialysis tubing

EXPERIMENT 46

Hybridization with *lac* mRNA

In this experiment the labeled mRNA extracted from cells induced and uninduced for the *lac* message (Experiment 44) is hybridized with λp*lac*5 DNA (Experiment 45). The uninduced messenger gives a suitable background control. Only the *lac* message should hybridize to the phage DNA for only it is homologous to the λp*lac*5 DNA.

Two different methods can be used for the hybridization. The first uses DNA immobilized on nitrocellulose filters; the second uses separated strands of phage DNA. The first method offers the advantage of low backgrounds but is time consuming and only a small portion of the RNA hybridizes to the filters. The method using separated strands is rapid, has high hybridization efficiency, but will give higher backgrounds.

Reference

CONTESSE, G., M. CRÉPIN and F. GROS. 1970. Transcription of the lactose operon in *E. coli*, p. 111. *The Lactose Operon*, Cold Spring Harbor Laboratory.

Method

The Filter Method

In this method denatured DNA is immobilized on nitrocellulose filters. The RNA is annealed to it, and the unhybridized RNA is then washed away and digested.

Day 1

Preparation of the filters

The DNA is bound to the filters after filtration with a solution containing the denatured DNA. The filters are then washed, dried, and the DNA fixed on the filter by baking.

The Millipore filters should be presoaked in 6 × SSC for at least 30 minutes prior to binding the DNA. Denatured DNA sticks to these filters. Denature 50 μg of λ*plac*5 DNA, prepared as described in Experiment 45, by adjusting it to 0.2 N NaOH with 1 N NaOH in an acid-washed conical test tube. Vortex the tube to mix in the base and immediately add the 10 ml of 6 × SSC. Prepare dilutions of the DNA with 6 × SSC so that the DNA to be bound to one filter is in a total volume of 2 ml. Use HAWP 24-mm Millipore filters.

Pass the 2 ml containing the DNA solution through the filter with gentle suction. Then wash each filter with 4 ml of 6 × SSC, also with gentle suction. Dry the filter over a wire mesh with heat lamps. Place the filters in a petri dish and bake them for 2 hours at 80° in a vacuum oven. The oven must be evacuated to keep the filters from charring. Filters prepared in this way are stable for a few months if stored in a vacuum desiccator.

For the hybridization experiment you will need filters with 0, 1, 2, 4, 7.5, and 10 μg of DNA each. Prepare 4 blank filters for use as controls by passing 6 ml of 6 × SSC through the filters and treating them as described above.

Hybridization

A constant amount of mRNA labeled *in vivo* is hybridized with filters containing increasing amounts of DNA until a saturation plateau is reached. The plateau in this case should represent about 5% of the labeled *lac* message which is but a fraction of the total message.

Place the filter containing the required amount of DNA in acid-washed vials for scintillation counting. Number the filters with a soft lead pencil before placing them in the vials. Add 800 μl of 2 × SSC to each vial and 50 μl of the labeled RNA. Do each tube in duplicate. A sample work sheet is given on the next page.

Seal each vial by wrapping it in Parafilm so it is air-tight to prevent evaporation of the liquid. Incubate the vials for 18–20 hours at 67° with gentle shaking. The temperature must be

Tube	DNA on filter (μg)	RNA added (μl) Induced	Uninduced
1	0	50	0
2	0	0	50
3	1	50	0
4	1	0	50
5	2	50	0
6	2	0	50
7	4	50	0
8	4	0	50
9	7	50	0
10	7	0	50
11	10	50	0
12	10	0	50

controlled accurately. If the temperature is too low, annealing kinetics are slow; if too high, the hybrids do not form.

Day 2

After the 67° incubation combine the filters in a large test tube. Wash them 5 times with 20 ml each time of 2 × SSC by mixing them for 30 seconds on a vortex mixer; discard the supernatants each time. Then add 5 ml of 2 × SSC containing 400 μg of RNase A (heat the RNase to 95°C for 2 minutes to inactivate traces of DNase before use) and incubate the filters at room temperature for 50 minutes, followed by a 10 minute incubation at 37°. Wash the filters 5 more times as before, dry, and count them in toluene scintillation fluid.

The curves should look like Figure 46A.

The control with no DNA on the filter will give a background which must be subtracted from all numbers.

The Separated Strand Method

In this method the labeled *lac* mRNA is annealed to the separated *l* strand of the λp*lac*5 DNA containing the sequence complementary to the message. Only the *lac* specific RNA will hybridize, for it is the only RNA complementary to the DNA. The hybridization can be done without immobilizing the DNA on filters because the separated strands cannot self-anneal.

Hybridization

A constant amount of RNA labeled *in vivo* is hybridized with increasing amounts of the *l* strand λp*lac*5 DNA prepared as described in Experiment 45. A range of from 0.01 to 0.2 μg per reaction is recommended. The hybridization is done in conical test tubes which have been acid-washed and baked to inactivate nucleases. The final reaction volume should be 200 μl. Each tube is done in duplicate. One set should contain no added DNA as a

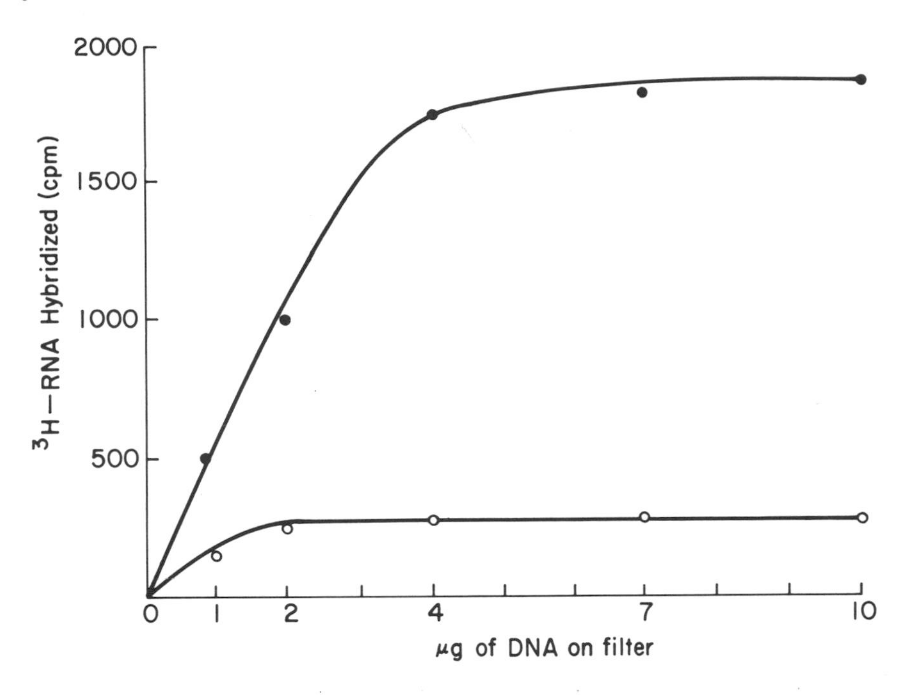

Figure 46A. Hybridization of RNA extracted from both IPTG-induced and uninduced cultures of *E. coli* to λp*lac*5 DNA immobilized on filters. —●— IPTG-induced; —○— uninduced.

control. Another control is the addition of a high concentration of the non-complementary strand to another set of tubes.

Day 1

Adjust the concentration of the separated strands to 20 μg/ml with 2 × SSC. (OD_{260} of 1.0 corresponds to about 36 mg/ml of single-stranded DNA.) Add the separated strands to the conical test tubes as indicated in the sample work sheet. Dilute 100 μl of the *in vivo* message prepared in Experiment 44 with 400 μl of 2 × SSC. Add 30 μl to each tube. Bring the final volume of each tube to 200 ± 5 μl with 2 × SSC.

A sample work sheet is shown on the next page. After all the components are added mix each tube lightly with a vortex mixer. Cap each tube separately and incubate for 3 hours at 67° ± 0.2°. Precise temperature regulation is essential.

After the 67° incubation, place the tubes in ice and add to each tube 750 μl of 2 × SSC containing 10 μg of RNase A heated to 95° for 2 minutes to inactivate DNase. Mix each tube thoroughly by vortexing at high speed for 20 seconds. Inadequate mixing gives irreproducible RNA digestion of unannealed RNA. The RNase will digest the unannealed RNA. Incubate the tubes for 30 minutes at 37°.

Separate strands of λ*plac*5 DNA at 20 μg/ml

Tube	DNA (μl)	2 × SSC (μl, ±5)	Labeled message (μl)	
			Induced	Uninduced
1	0	170	30	0
2	0	170	0	30
3	2	168	30	0
4	2	168	0	30
5	5	165	30	0
6	5	165	0	30
7	10	160	30	0
8	10	160	0	30
9	15	155	30	0
10	15	155	0	30
11	20	150	30	0
12	20	150	0	30
(13)	20 of non-complementary strand	150	30	0
(14)	20	150	0	30

After the 37° incubation chill the tubes in ice and add 5 ml of washing buffer (0.01 M Tris pH 7.5 and 0.5 M KCl) to each tube. Filter through Schleicher and Schuell B-6 nitrocellulose filters which have been presoaked for at least 15 minutes in washing buffer. Wash each filter with about 40 ml of washing buffer. After this wash, soak each filter for about one minute in washing buffer and dry under a heat lamp. Count the filters in toluene scintillation fluid. The control (DNA added) will give the background which must be subtracted from all numbers.

This method of collecting the hybrids works because single-stranded DNA binds to the filter, trapping the RNA hybridized to it.

Materials

Filter Method

Day 1

1 M NaOH
1 acid-washed conical test tube
16 HAWP 24-mm Millipore filters
Millipore filtration apparatus
500 ml 6 × SSC
8 baked conical test tubes
heat lamp
vacuum oven at 80°
vacuum desiccator
20 ml 2 × SSC
12 scintillation vials
Parafilm
67° shaking water bath

Day 2

400 μg RNase A (heat treated)
1 liter of 2 × SSC
large test tube
vortex mixer
37° water bath
scintillation counter and fluid (such as Omnifluor, New England Nuclear)
2 scintillation vials

Separated Strand Method

Day 1

14 acid-washed conical test tubes
50 ml 2 × SSC
2 liters washing buffer (0.01 M Tris pH 7.5, 0.5 M KCl)
Millipore filtration apparatus
20 SSB-6 filters
67° waterbath (Haake regulated)
37° waterbath
140 μg RNase A
scintillation counter and fluid
20 scintillation vials
heat lamp

A scintillation fluid similar to Omnifluor can be made by dissolving the following in 1 liter of toluene:
15.1 g PPO (2,5-diphenyloxazole)
0.38 g dimethyl POPOP (1,4-Bis-2-[4-methyl-5-phenyloxazolyl]-benzene)

EXPERIMENT 47

Determination of the Sedimentation Constant of *lac* mRNA

In this experiment we will measure the size of the *lac* messenger RNA by determining its sedimentation velocity in a sucrose gradient. The message used in this experiment is the same as that prepared in Experiment 44. You may also wish to prepare the stable RNA species as markers.

Reference

CONTESSE, G., M. CRÉPIN and F. GROS. 1970. Transcription of the lactose operon in *E. coli*, p. 111. *The Lactose Operon*, Cold Spring Harbor Laboratory.

Method

Day 1 **Preparation of *lac* messenger RNA and stable RNA**

Follow the protocol in Experiment 44 for the preparation of RNA from cells induced and uninduced for *lac*. Resuspend the ethanol-precipitated RNA in 0.20 ml of 2 × SSC.

To prepare the markers, grow and label the cells as in Experiment 44. To chase the label into stable species add 100 μl of 0.1 M cold uridine to the medium one minute after addition of the label and allow the cells to grow 15 more minutes. Extract and prepare the RNA as in Experiment 44.

Day 2

Preparation of gradients

Pour three gradients, one for the uninduced RNA, one for the induced sample, and one for the marker. Use an SB283 rotor for the International B-6 or SW41 rotor for the Beckman L-2. The gradients are 5–20% sucrose in 2 × SSC. Use Mann assayed ultra-pure RNase-free sucrose.

Introduce 6.1 ml of a 5% sucrose solution in 2 × SSC into the rear chamber (see Figure 47A) with the passage to the mixing chamber closed. Allow the small passage between the chambers to become filled with the diluted solution, then introduce 6.1 ml of a 20% sucrose solution in 2 × SSC into the mixing chamber. Start the stirring device (magnetic stirrer, reciprocating stirrer, etc.). Open the valve between the two chambers, and allow the sucrose solution to flow along the wall into the centrifuge tube. The exit tube of the gradient mixer must touch the wall of the centrifuge tube during the entire period of delivery of the gradient. Air bubbles in the passage between the two chambers can be forced out by gently pressing a piece of Parafilm over the top of the rear chamber. The gradients should be chilled for one hour before use. They are relatively stable for 6–12 hours in the cold.

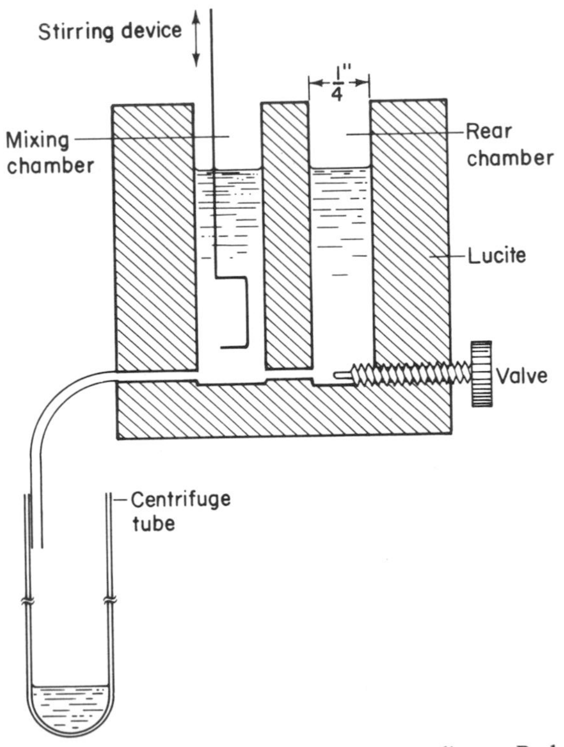

Figure 47A. Mixing chamber for production of linear sucrose gradients. Redrawn from Britten and Roberts, *Science 131*, 32 (1960).

Carefully layer the samples onto the gradients with a 250-μl pipette. Allow the sample to run down the inner edge of the tube just above the top of the gradient.

Place the tube in a chilled rotor and spin at 41,000 for 7 hours. If you want to spin longer, distance moved is proportional to ω^2. Spin at 4° with the diffusion pump on.

Collect the gradients into small heat-treated tubes by puncturing the bottom with a short 22-gauge Luer-lock needle. Nine drops per tube should give 19 tubes. Remember to poke out the plug in the needle with a fine wire before dripping the gradient.

Day 3

Hybridization

To determine the distribution of *lac* specific RNA, hybridize along the gradients with λ*plac*5 DNA. You may use the separated strand method or the filter method as described in Experiment 46. Use 100 μl aliquots from each tube in either case. Use 0.3 μg of DNA per tube of separated strands or 1 μg DNA per filter.

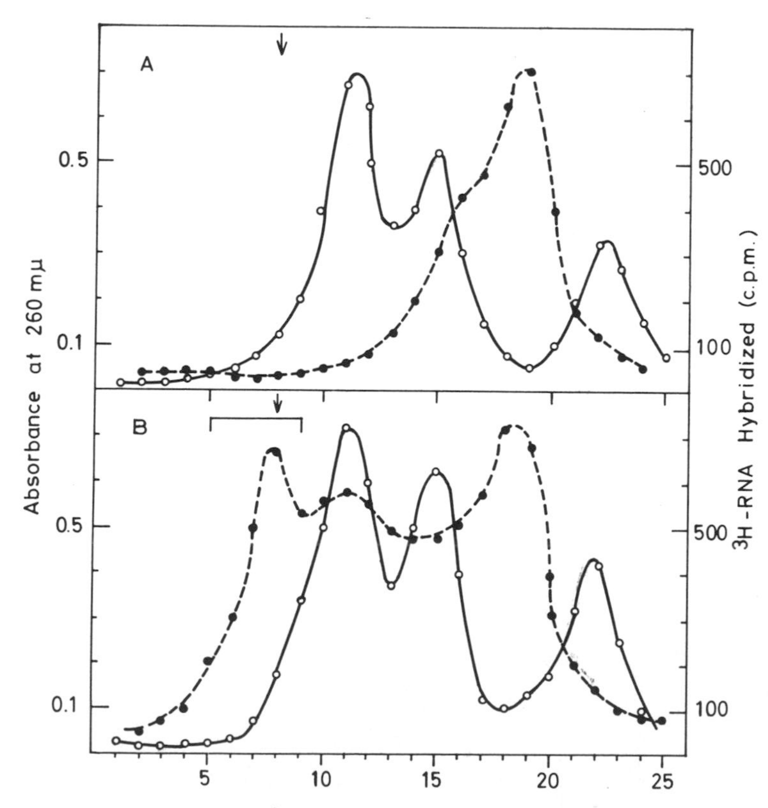

Figure 47B. Sedimentation pattern of ^{3}H-labeled *lac* mRNA from *lac*$^+$ strain. A portion (100 μl) of each fraction from a 5%–30% linear sucrose gradient was hybridized with 1 μg of ϕ80d*lac* DNA on a membrane filter.

— -●- — Amount of ^{3}H-RNA hybridized (cpm).
—○— Absorbance at 260 mμ.
↓ Determines the 30 S region in each gradient.

RNAs were prepared either from a non-induced culture (**A**) or from a culture induced by 5 × 10^{-4} M IPTG for 5 minutes (**B**). (Reprinted with permission from Contesse et al., 1970.)

Hybridization along the uninduced control gradient will give the distribution of RNA which is complementary to the bacterial DNA in the λ*plac*5 phage which is not *lac* DNA.

To determine the distribution of labeled RNA in the gradients measure out 50 μl aliquots from all three gradients in 2 ml of Aquasol. The 23S, 16S, and 4S peaks of the marker gradient will allow you to estimate the size of the *lac* specific RNA.

Your gradient should look like Figure 47B.

The *lac* specific RNA trailing the 30S peak is unfinished and degraded *lac* message. You may rerun the fast-sedimenting RNA extracted from the induced cells; it should run as a homogeneous peak at the original S value.

Materials

Day 1

As in Day 2 of hybridization experiment.

Day 2

(per student)

3 12-ml centrifuge tubes
1 International or Beckman centrifuge
1 20-ml gradient pouring device
1 stirring motor
100 ml each 5% sucrose in 2 × SSC
20% sucrose in 2 × SSC
(Mann assayed ultra-pure RNase-free sucrose)
1 250-μl pipette
1 ice bucket
3 racks sterile small test tubes, 25 tubes per rack
1 cut-off 22-gauge needle

Day 3

250 acid-washed tubes for hybridization (if single stranded technique)
Aquasol (New England Nuclear), or Bray's scintillator
60 Millipore filters DAWP 25-mm
Millipore filter apparatus
50 SSB-6 filters
67° waterbath
RNase A (heat treated)

UNIT VII

ASSAYS OF THE *lac* OPERON ENZYMES

INTRODUCTION TO UNIT VII

Most tests for bacterial proteins require the preparation of an extract of the bacteria. The methods of breaking open bacterial cells vary according to the type of assay. For certain enzymes it is sometimes sufficient to shake the bacteria with toluene. This disrupts the membrane, allowing small molecules to diffuse from the cell into the medium, while large enzymes such as β-galactosidase stay inside the punctured cell. Small substrates can readily enter toluene-treated cells. This method is used for the determination of β-galactosidase.

More concentrated extracts can be obtained by sonifying the cells or by grinding them with alumina. After the cells are broken, buffer is added, and the debris is spun down. The clear supernatant is a "crude extract" and contains most of the soluble cell protein at about 20–40 mg/ml. This method is used for the assay of *lac* repressor, which may be assayed directly in this fraction for binding either to inducer or to operator.

To assay *lac* permease activity, whole cells are used. The accumulation of a labeled galactoside inside the cells is measured by collecting the bacteria on a filter and counting in a gas flow or liquid scintillation counter.

EXPERIMENT 48

Assay of β-Galactosidase

β-Galactosidase is an enzyme which hydrolyzes β-D-galactosides. It can easily be measured with chromogenic substrates, colorless substrates which are hydrolyzed to yield colored products. An example is *o*-nitrophenyl-β-D-galactoside. This compound is colorless, but in the presence of β-galactosidase it is converted to galactose and *o*-nitrophenol. The *o*-nitrophenol is yellow and can be measured by its absorption at 420 mμ. If the *o*-nitrophenyl-β-D-galactoside (ONPG) concentration is high enough, the amount of *o*-nitrophenol produced is proportional to the amount of enzyme present and to the time the enzyme reacts with the ONPG. In order for the assay to be linear, the ONPG must be in excess. For best results, the amount of enzyme should be such that it takes between 15 minutes and 6 hours for a faint yellow color to develop. The reaction is stopped by adding a concentrated Na_2CO_3 solution, which shifts the pH to 11. At this pH β-galactosidase is inactive.

References

PARDEE, A. B., F. JACOB and J. MONOD. 1959. The genetic control and cytoplasmic expression of "inducibility" in the synthesis of β-galactosidase by *E. coli*. *J. Mol. Biol. 1:* 165.

ZABIN, I. and A. FOWLER. 1970. β-galactosidase and thiogalactoside transacetylase. *The Lactose Operon*, p. 27. Cold Spring Harbor Laboratory.

Strains

Number	Sex	Genotype	Important Properties
CSH23	F′*lac*$^+$*proA*$^+$,*B*$^+$	Δ(*lacpro*) *supE spc thi*	$i^+z^+y^+$
CSH36	F′*lacI proA*$^+$,*B*$^+$	Δ(*lacpro*) *supE thi*	$i^-z^+y^+$

Method

In order to obtain precision in assaying β-galactosidase, it is necessary to use exponentially growing cells. In the following experiment we will assay two different strains. These have the following genotypes: $i^+z^+y^+$ and $i^-z^+y^+$.

Day 1 Prepare an overnight of each of the two strains in minimal medium 1 × A supplemented with B1, $MgSO_4$, glucose, and IPTG. Also prepare an overnight in the same medium without IPTG for each of the two strains.

Day 2 Subculture each of the four overnights into fresh media of the exact types used in Day 1. Add 4 drops of the overnight to 5 ml fresh media. Continue to aerate at 37° until the cultures are at 2–5 × 10^8 cells/ml (OD_{600} of 0.28–0.70). Cool the cultures to prevent further growth by immersing in an ice bucket containing a mixture of ice and water. After 20 minutes the cultures can be assayed and the bacterial density measured.

Assay: Record the cell density by measuring the absorbance at 600 mμ. Withdraw 1 ml of the culture for this measurement. Immediately add aliquots of the cultures to the assay medium (Z buffer). The final volume will always be 1 ml. If high levels of β-galactosidase are being assayed, add 0.1 ml of culture to 0.9 ml of Z buffer. If low levels are being determined, then add 0.5 ml of culture to 0.5 ml of Z buffer. In this experiment use both of these dilutions for each of the four cultures.

Add one drop of toluene with a pasteur pipette to each tube and immediately vortex for 10 seconds. The toluene partially disrupts the cell membrane, allowing small molecules, such as ONPG, to diffuse into the cell. Allow the toluene to evaporate by placing the tubes on a rotor at 37° for 40 minutes with the top open. (Mild shaking at 37° in a shaker bath is also sufficient.) Place the tubes in a water bath at 28° for 5 minutes. Start the reaction by adding 0.2 ml of ONPG (4 mg/ml) to each tube and shake for a few seconds. Record the time of the reaction with a stop watch. Stop the reaction by adding 0.5 ml of a 1 M Na_2CO_3 solution after sufficient yellow color has developed.

Record the optical density at both 420 mμ and 550 mμ for each tube. Ideally, the 420 mμ reading (yellow color) should be 0.6–0.9. The reading at 420 mμ is actually a combination of absorbance by the *o*-nitrophenol and light scattering by the cell debris. This latter component can be corrected for by obtaining the absorbance at another wave length (550 mμ) at which there is only light scattering (no contribution from *o*-nitrophenol). The light scattering at 420 mμ is proportional to that at 550 mμ. For *E. coli* the OD_{420} (light scattering) = 1.75 × OD_{550}. Using this correction factor we can compensate for the light scattering and compute the true absorbance of the *o*-nitrophenol. For our units we use the following formula:

$$\text{Units} = 1000 \times \frac{OD_{420} - 1.75 \times OD_{550}}{t \times v \times OD_{600}}$$

OD_{420} and OD_{550} are read from the reaction mixture,
OD_{600} reflects the cell density just before assay,
t = time of the reaction in minutes,
v = volume of culture used in the assay, in ml.

These units are proportional to the increase in *o*-nitrophenol per minute per bacterium. They are convenient because a fully induced culture grown on glucose has approximately 1000 units, and an uninduced culture approximately 1 unit. Instead of correcting for the cell debris interference, it is also possible to spin down the debris in a small centrifuge, thus eliminating the 550 mμ reading.

A **sample calculation** follows: Suppose we asay an IPTG-induced culture with an OD_{600} of 0.60. We add 0.1 ml of this culture to 0.9 ml Z buffer + toluene + ONPG, and after 15 minutes stop the reaction. The reading at 420 mμ is 0.900, and at 550 mμ 0.050. Since 1.75 × 0.050 = 0.088 we have:

$$\text{Units} = 1000 \times \frac{(0.900 - 0.088)}{15 \times 0.60 \times 0.1} = \frac{812}{0.9}$$

$$= 902 \text{ units } \beta\text{-galactosidase.}$$

If after several hours there is still no yellow color in the basal level assays, cover them with aluminum foil and allow the reaction to continue overnight. In this case prepare a control with 1 ml Z buffer, but no cells, and incubate along with the basal level cultures, after addition of ONPG. This will serve as a control for the spontaneous splitting of ONPG.

The specific activity of β-galactosidase is defined in terms of units/mg protein. Many investigators define a unit of β-galactosidase as the amount of enzyme which produces 1 mμ-mole *o*-nitrophenol/min at 28°, pH 7.0. Under the above conditions 1 mμ-mole/ml *o*-nitrophenol has an optical density (420 mμ) of 0.0045 using a 10-mm light path. Therefore the above units can easily be converted into these units. Although protein is usually determined directly from extracts or estimated from the dry weight of cells, we can estimate the amount of protein by assuming that 10^9 cells yields approximately 150 μg protein and that an OD_{600} of 1.4 corresponds to 10^9 cells/ml. Using this information, calculate the specific activity of β-galactosidase in terms of mμ-moles *o*-nitrophenol/min/mg protein.

Materials

Day 1

overnight cultures of strains CSH23 and CSH36
10 ml 1 × A medium supplemented with 20 μg/ml B1, 10^{-3} M $MgSO_4$, and 0.4% glucose
10 ml of the same medium + 10^{-3} M IPTG

Day 2

10 ml of each of the media used in Day 1
spectrophotometer
ice bucket
6 ml Z buffer for β-galactosidase assay. Z buffer is per liter:

16.1 g	$Na_2HPO_4 \cdot 7H_2O$	(0.06 M)
5.5 g	$NaH_2PO_4 \cdot H_2O$	(0.04 M)
0.75 g	KCl	(0.01 M)
0.246 g	$MgSO_4 \cdot 7H_2O$	(0.001 M)
2.7 ml	β-mercaptoethanol	(0.05 M)

Do not autoclave! Adjust pH to 7.0.

4 0.1-ml, 5 1-ml and 8 pasteur pipettes
4 test tubes with 0.9 ml Z buffer each
4 test tubes with 0.5 ml Z buffer each
1 ml toluene
2 ml ONPG, 4 mg/ml in 0.1 M phosphate buffer, pH 7.0 (1 × A)
5 ml 1 M Na_2CO_3
vortex
stop watch
waterbath at 28°

Additional Methods

As an alternative to the toluene method, 2 drops of chloroform and 1 drop of 0.1% SDS solution are added to each ml of assay mix. The tubes are then vortexed for 10 seconds. With this procedure more cells are opened and the evaporation step is eliminated.

EXPERIMENT 49

Time Course of β-Galactosidase Induction

What are the steps involved in the synthesis of a protein, and what are the kinetics of this synthesis? This question has been studied in the *lac* operon by numerous investigators. After addition of an inducer (IPTG) to growing *E. coli*, active β-galactosidase molecules are synthesized after a lag of 1.5–2.0 minutes. During this time, the repressor-operator complex dissociates and a specific mRNA is transcribed and translated into β-galactosidase monomers, which are then assembled into active oligomers. In this experiment we follow the β-galactosidase synthesized by a growing culture after the addition of IPTG. A variation of this experiment is shown in Figure 49A. In this experiment aliquots were removed at various times after the addition of IPTG and diluted to prevent further induction. Each diluted sample was then incubated for 20 minutes. In this way, the kinetics of "z messenger" synthesis (capacity to synthesize β-galactosidase) during induction could be followed. It can be seen that the induction of mRNA starts without any measurable lag; or in other words, the maximal rate of transcription is instantaneously achieved (Kepes, 1963).

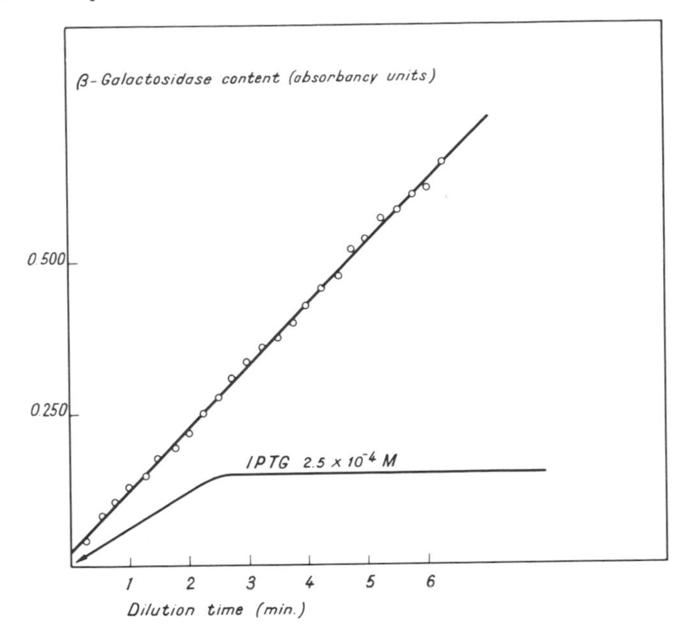

Figure 49A. Kinetics of "z-messenger" synthesis during induction. IPTG (2.5×10^{-5} M) was added to an exponentially growing culture of a *lac*$^+$ *E. coli* strain and 0.1-ml aliquots were removed at various times and diluted in 5 ml of prewarmed culture medium. Each dilution was toluenized 18 minutes after the sampling time and the final amount of β-galactosidase was plotted as a function of the time of dilution. (Reproduced with permission from Kepes, 1963.)

References

CONTESSE, G., M. CRÉPIN and F. GROS. 1970. Transcription of the lactose operon in *E. coli*. *The Lactose Operon*, p. 111. Cold Spring Harbor Laboratory.

KEPES, A. 1963. Kinetics of induced enzyme synthesis: Determination of the mean life of galactosidase-specific messenger RNA. *Biochem. Biophys. Acta 76:* 293.

KEPES, A. 1969. Transcription and translation in the lactose operon of *Escherichia coli* studied by *in vivo* kinetics. *Prog. Biophys. Mol. Biol. 19:* 199.

Strains

Number	Sex	Genotype	Important Properties
CSH62	HfrH	*thi*	$i^+z^+y^+$

Method

Day 1

Prepare an overnight culture of CSH62 in minimal medium ($1 \times$ A) supplemented with B1, $MgSO_4$ (10^{-3} M) and glycerol (0.5%), at 37°. Allow for 1.5 ml of this overnight per student.

Day 2

Growth of cells

Make a 1:20 dilution of the bacteria in fresh minimal medium supplemented as in Day 1, and let the bacteria grow for at least one doubling (1–2 hours). Since each student will receive 30 ml of this suspension, 1.5 ml of the overnight are needed per student. The bacteria should then be put on ice to be used later the same day.

Induction of β-galactosidase

Each group should take 30 ml of an ice-cold suspension of the subcultured overnight into a 125- or 250-ml flask with a bubbler attachment. Air is passed through and the apparatus is put in a 37° waterbath for 30 minutes. (Alternatively, the flask can be set up in a shaking waterbath.) Prepare 12 test tubes with 0.5 ml of Z buffer and 1 drop of toluene. These should be numbered 1–12. Begin the experiment after 30–45 minutes of growth, and label this as zero time. Use the following schedule:

At zero time pipette 0.5 ml of the culture into test tube 1 and vortex.
At 1 minute pipette 0.5 ml of the culture into test tube 2 and vortex.
At 2 minutes add 3 ml 10^{-2} M IPTG to the bacteria (induction).
At 3, 4, 5, 7, 9, 12, 17, 22, 27, and 32 minutes pipette 0.5 ml of the culture into a different test tube and vortex.

Assay for β-galactosidase

Then evaporate the toluene and assay for β-galactosidase exactly as in Experiment 48.

Plot the time after addition of IPTG against the β-galactosidase activity. Compare this graph with those given in the references.

Materials (per group)

Day 1

overnight of CSH62
test tube with 5 ml 1 × A medium supplemented with B1, $MgSO_4$ (10^{-3} M), and glycerol (0.5%)
pasteur pipette

Day 2

1 bubbler apparatus with 125- or 250-ml flask
12 test tubes
15 1-ml and 2 5-ml pipettes
3 ml IPTG 10^{-2} M in H_2O
3 ml ONPG (4 mg/ml)
6 ml 1 M Na_2CO_3
6 ml Z buffer
toluene
vortex
37° waterbath
30 ml 1 × A medium supplemented as in Day 1

Optional

Alternative procedures: Use lactose as an inducer instead of IPTG. With lactose it takes more time until the full rate of β-galactosidase synthesis is attained, since lactose itself is not an inducer. Lactose is recognized by the few β-galactosidase molecules already present in the cell. Not all of it is hydrolyzed. Some lactose molecules are transformed by a galactoside transfer reaction catalyzed by β-galactosidase. The transfer products, 1,6-galactosido-glucose and galactosido-glycerol, are good inducers.

Questions

By extrapolating your curve to the x axis, find how long your cells took to synthesize β-galactosidase after you added the IPTG. The β-galactosidase monomer is approximately 1170 amino acids long (135,000 daltons). If the first active tetramer appears at the time indicated by the x intercept of your graph, what is the rate of chain elongation of this polypeptide, in amino acids per second? Be careful—Does it take four times as long to make the tetramer as the monomer? If transcription and translation are coupled, what is the rate of synthesis of "z message" in bases per second?

EXPERIMENT 50

Assay of *lac* Permease

Lac permease accumulates β-D-galactosides, particularly lactose, in *E. coli*. Some of these, such as IPTG (isopropyl-thio-β-D-galactoside) also enter the cell in the absence of the *lac* transport system, and are then found in the same concentration inside the cell as outside the cell. However, in the presence of a functional *lac* permease these are concentrated more than 50-fold.

One way to study the *lac* transport system is to measure the accumulation of β-D-galactosides against a concentration gradient. Gratuitous substrates which cannot be hydrolyzed by β-galactosidase are widely used for this type of experiment. Here radioactive IPTG is added to a suspension of cells and samples are then applied to a Millipore filter, washed, and then counted in a gas filter or liquid scintillation counter.

An excellent review of the *lac* permease system has recently been published by Kennedy.

References

KENNEDY, E. P. 1970. The lactose permease system of *Escherichia coli*. *The Lactose Operon*, p. 49. Cold Spring Harbor Laboratory.

RICKENBERG, H. V., G. N. COHEN, G. BUTTIN and J. MONOD. 1956. La galactoside-perméase d'*Escherichia coli*. *Ann. Inst. Pasteur 91:* 829.

Strains

Number	Sex	Genotype	Important Properties
CSH23	$F'lac^+ \, proA^+,B^+$	$\Delta(lacpro)$ *supE spc thi*	$i^+z^+y^+$
CSH36	$F' \, lacI \, proA^+,B^+$	$\Delta(lacpro)$ *supE thi*	$i^-z^+y^+$
CSH14	$F' \, lacZ \, proA^+,B^+$	$\Delta(lacpro)$ *supE thi*	$i^-z^-y^+$
CSH26	F^-	*ara* $\Delta(lacpro)$ *thi*	$i^-z^-y^-(\text{Pro}^-)$

Method

Day 1

Use overnight cultures grown in M63 (supplemented with glycerol and B1) of the 4 test strains. Dilute 1:20 into 5 ml fresh medium and continue to grow until an OD_{550} of 1.0 is reached (approximately 4 hours). For strain CSH26 **only**, proline must be present in the medium. Place on ice, since permease is unstable at room temperature. Centrifuge the bacteria at 4°. Resuspend in 5 ml M63 medium containing 150 μg/ml chloramphenicol, to prevent further protein synthesis and induction of permease by IPTG. Samples of 0.9 ml are incubated at 37° with 0.1 ml of ^{14}C-IPTG (10^{-3} M, 500,000 cpm/ml) for 10 minutes in a fresh test tube.

Filter samples of 0.5 ml through membrane filters, using a propipette **(never mouth pipette radioactive solutions).** All "hot" pipettes should be placed in a special container. An additional container is supplied for liquid radioactive waste. Use a fresh pipette for each strain and try to pipette the bacterial suspension on the middle of the filter. Wash with 5 0.2-ml portions of M63 medium containing 150 μg/ml chloramphenicol. The M63 should be at room temperature since washing with ice-cold medium can lead to a loss of accumulated substrates.

Using pincers put the filters in liquid scintillation vials. Add 10 ml Aquasol to each vial. Count in a liquid scintillation counter. Also count the background. For the calculation of permease activity assume that the volume of one *E. coli* cell is 0.5×10^{-12} cm^3 and that an OD_{550} of $1.0 = 5 \times 10^8$ cells/ml. How many counts do you expect in the permease mutant, in which you expect just equilibration of the IPTG? Express the specific activity as % excess ^{14}C-IPTG bound.

Materials (per student)

fresh overnight cultures of strains CSH14, CSH23, CSH26, and CSH36 grown in M63 glycerol medium
4 Millipore filters (0.45 mμ poresize, 2.5 cm diameter)
15 ml M63 medium supplemented with B1 and glycerol (0.5%)
5 ml M63 medium supplemented with proline
20 ml M63 containing chloramphenicol (150 μg/ml)
0.4 ml ^{14}C-IPTG (10^{-3} M, 500,000 cpm/ml) (Calbiochem)
container for radioactive pipettes and liquid waste
4 test tubes
4 centrifuge tubes
4 small test tubes
1 icebucket per several students, crushed ice
50 ml Aquasol or other liquid scintillator
4 0.1-ml, 6 1-ml, and 2 10-ml pipettes and 1 propipette
liquid scintillation counter (if not available, a gas-flow counter is sufficient)
glue (to glue the filters to the planchets)
suction pumps, suction flasks, and filter apparatus

If the ^{14}C-labeled IPTG has a specific activity of 10 mc/mM, dilute it about 100-fold with cold IPTG to have the indicated specific activity. Recall that one curie equals 2×10^{12} disintegrations per minute and there is a counting efficiency of about 90% in the liquid scintillation counter.

EXPERIMENT 51

Assay of *lac* Repressor by Binding to Inducer

Lac repressor can be detected by its binding to a labeled inducer, for example ^{14}C-IPTG. A convenient way to measure the binding is by equilibrium dialysis against inducer. When equilibrium is attained, aliquots are withdrawn from inside and from outside the sac and counted in a liquid scintillation counter.

If the experiment is done with a concentrated extract of wild-type cells, no extra binding is found inside the sac. Wild-type cells produce too few *lac* repressor molecules. It has been estimated that a wild-type cell contains about 10 *lac* repressor molecules, which corresponds to 0.002% of the bacterial cell protein. In order to see an effect in equilibrium dialysis it is necessary to increase the amount of repressor normally present in the cell. This is done by means of two types of genetic manipulations: First, the *i* gene can be mutated to a form that produces more repressor. One mutation of this type, termed i^Q, produces approximately 10 times more repressor than wild type. A further mutation, called i^{SQ} ("Super-Q"), overproduces the repressor by as much as 50-fold. Second, the number of copies of the *i* gene can be greatly increased by incorporating the *lac* region into a prophage chromosome. After induction, the prophage replicates, and a 10- to 20-fold increase in repressor protein is observed. Using the i^{SQ} and a heat-inducible prophage, extracts can be obtained with repressor as 2% of the soluble protein.

In the following experiment cultures of strain CSH46 will be assayed for *lac* repressor both with and without phage induction. CSH46 contains i^{SQ} on a heat-inducible prophage.

Strain

Number	Sex	Genotype	Important Properties
CSH46	F^-	*ara* Δ(*lacpro*) *thi* (λCI857*St*68h80d*lacI*,*Z*)	Pro^- carries $i^{SQ}z^-$ on heat-inducible prophage

Method

Day 1

Heat induction: 10 ml from an overnight culture of strain CSH46 (grown in LB below 34°) is used as inoculum for 100 ml LB in a 500-ml flask. As a control use another 100 ml of LB to grow the same strain without temperature induction. Shake for 1 to $1\frac{1}{2}$ hours at 32°, to an OD_{550} of 3.0–6.0. Heat each flask in a 45° shaking water bath with vigorous aeration for 20 minutes.

If shaking water baths are not available, heat with shaking over an open flame, with a thermometer in the culture. Maintain the temperature between 43° and 45° for 20 minutes. Cool the flasks under cold water and replace on the 32° shaker.

Cool and continue to shake at 32° for $1\frac{1}{2}$ to 2 hours. Then spin down the bacteria in a centrifuge (8000 × *g* for 15 minutes). Discard the supernatant. Weigh the bacteria on a piece of Parafilm and freeze the sample.

Day 2

Breaking the cells: Break the cells by grinding with alumina in a mortar in the cold room for 10–15 minutes at 4°. Use alumina at 2.5 times the weight of the cells. Stop grinding when the material crackles and becomes a shiny paste. Add 2 ml of buffer (5 μg/ml in DNase) per gram of bacteria, and spin down the alumina and cell debris at 10,000 × *g* for 10 minutes. Transfer the supernatant to a new tube with a pasteur pipette.

Equilibrium dialysis: 20 ml TMS buffer which contains ^{14}C-labeled IPTG (10^{-7} M) is used for equilibrium dialysis. The buffer contains about 5000 counts/ml. Tie a knot in one end of a 15-cm strip of #20 dialysis tubing. Put about 200 μl of the extract into the tubing and form as small a sac as possible by knotting the other end. Wear disposable plastic gloves during these operations to avoid contamination of the extract. Put the dialysis sacs into a 125-ml erlenmeyer flask containing the labeled IPTG and TMS buffer. Sacs can be marked by clipping the ends.

After 2 hours or more on a shaker, preferably in the cold, the dialysis sacs are taken out of the flask with a glass rod and the outside of each sac is dried off. Open with scissors, and squeeze the sample into a test tube. Wear gloves for this operation to avoid contamination from radioactivity. Take exactly 100 μl from outside and 100 μl from inside the sac, and pipette them into scintillation vials. Add scintillation fluid and count. Do not forget to measure the background.

Make a protein determination with the rest of the bacterial extract: to 20 μl of extract add 1 ml of Biuret reagent (see Preparation of Materials in Unit VIII). After 30–40 minutes measure the optical density at 546 mμ against 1 ml of a Biuret solution containing 20 μl of the outside buffer.

OD_{546} 0.060 = 10 mg/ml

The specific activity of the repressor is expressed in the following units:

$$\frac{\%\text{ of excess IPTG inside the sac}}{\text{mg protein/ml}}$$

How much more repressor did you find after heat induction?

Buffers

Breaking Buffer (BB): 0.2 M Tris (pH 7.6 at 4°)
0.2 M KCl
0.01 M MgAc
3 × 10^{-4} M DTT
5% glycerol

TMS Buffer: 0.2 M KCl
0.01 M Tris (pH 7.6 at 4°)
0.01 M MgAc
10^{-4} M EDTA
10^{-4} M DTT

Biuret reagent: see Unit VIII, Preparation of Materials

Materials

Day 1

20 ml of an overnight culture of strain CSH46 grown at 30°
2 500-ml erlenmeyer flasks with 100 ml LB each
shaker for the 500-ml flasks in a 32° room
Sorvall or comparable centrifuge
shaking water bath set at 45°

Day 2

6 g levigated alumina (Norton Co., Worcester, Mass.)
5 ml TMS buffer

^{14}C-labeled IPTG, 1×10^{-7} M in 20 ml TMS buffer in a 125-ml erlenmeyer flask
30 cm dialysis tubing, Visking No. 20 (Calbiochem)
2 pairs disposable plastic gloves
small scissors
30 ml Biuret reagent
2 1-ml, 2 10-ml, 2 20-μl, 2 100-μl, and 2 pasteur pipettes
3 small test tubes
2 glass rods
small shaker for the dialysis
liquid scintillation counter
spectrophotometer and cuvettes for measurements at 546 mμ
Aquasol or other scintillation fluid, and 1-dram disposable inserts

Questions

How many (repressor) molecules are there in the uninduced cell if the specific activity of pure repressor is 2000%/mg? How many in the induced cell? Look at the induced cells under a microscope, compared to the control. Can you explain the difference?

EXPERIMENT 52

Assay of the *lac* Repressor by Binding to Operator

There are ten to twenty molecules of repressor normally present in the cell (per gene copy). The binding constant of repressor to the 10 to 20 base pairs that constitute the operator region has been measured *in vitro* to be about 10^{-13} M (Riggs et al.).

We can also make an approximate estimate of the *in vivo* binding constant of repressor for its operator site on the DNA. We shall assume that the relation between repressor and operator (in the absence of inducer, of course) is of the form:

$$K_{eq} = \frac{[O][R]}{[O - R]} \qquad (1)$$

where $[O]$ = concentration of free operator in the cell,
$[R]$ = concentration of free repressor in the cell,
and $[O - R]$ = concentration of operator-repressor complex.

A convenient fact is that one molecule in an *E. coli* cell represents a concentration of about 10^{-9} M. Since there are roughly 10 repressor molecules in a haploid cell (only one of which can be bound to a single operator at a time), we have $[R] = 10^{-8}$ M. The total operator concentration (one per cell) will be 10^{-9} M; this will also be the concentration of free operator in an induced cell. If we assume that the rate of *lac* enzyme synthesis is directly proportional to the concentration of free operator in the cell, the fact that the basal level of β-galactosidase is 1/1000 of the induced level implies that [O] in the repressed state is 1/1000 of the induced free operator concentration, or about 10^{-12} M. Therefore, in the absence of inducer,

$$\frac{\text{free operator}}{\text{total operator}} = \frac{[\mathrm{O}]}{[\mathrm{O-R}] + [\mathrm{O}]} \approx \frac{[\mathrm{O}]}{[\mathrm{O-R}]} = \frac{1}{1000}$$

since $[\mathrm{O-R}] \gg [\mathrm{O}]$.

Substituting in equation (1), we obtain

$$K_{eq} = \frac{[\mathrm{O}]}{[\mathrm{O-R}]}[\mathrm{R}] = \frac{1}{1000}(10^{-8}) = 10^{-11}\ \mathrm{M}. \qquad (2)$$

The binding constant cannot be much weaker than this; the *in vitro* experiments suggest it may be more than an order of magnitude tighter, which is not inconsistent, if the basal level of *lac* enzyme synthesis is due to an escape process rather than simply lack of complete repression at the operator site.

Fortunately, the *lac* repressor also binds very tightly to nitrocellulose filters. At the same time it will also bind either the inducer IPTG (as in the Millipore Filter Assay described in Appendix VI) or any piece of DNA containing *lac* operator. If a repressor-operator complex is bound to the filter, and IPTG is added, this complex dissociates (leaving repressor bound to IPTG on the filter), releasing the operator DNA, which by itself does not bind to the filter. These properties are used in the following assay which demonstrates the specific binding of *lac* repressor to *lac* operator. Using ^{32}P-labeled *lac* DNA it is possible to detect repressor in a crude extract of wild-type (i^+) cells. Using this technique the small fragment of DNA corresponding to the *lac* operator has been isolated (Gilbert).

The experiment requires the preparation of ^{32}P-labeled *lac* DNA, which will be done by the instructor. Each group of students will then measure the amount of this DNA that is bound to nitrocellulose filters by repressor, using extracts from i^+ and i^- cells in the presence and absence of inducer.

References

GILBERT, W. and B. MÜLLER-HILL. 1966. Isolation of the *lac* repressor. *Proc. Nat. Acad. Sci. 56:* 1891.

RIGGS, A. D., S. BOURGEOIS, R. F. NEWBY and M. COHN. 1968. DNA binding of the *lac* repressor. *J. Mol. Biol. 34:* 364.

Strains

Number	Sex	Genotype	Important Properties
CSH46	F^-	*ara* Δ(*lacpro*) *thi* (λ*CI*857*St*68h80d*lacI*,*Z*)	i^+
CSH50	F^-	*ara* Δ(*lacpro*) *strA thi*	i^-
CSH43	F^-	*tonA* Δ(*lac*) *thi* (λh80*CI*857*St*68)	

Method

Preparation of ^{32}P-labeled *lac* Operator DNA (to be done by the advisor a week in advance)

Day 1 Inoculate 10 ml of an overnight culture of strain CSH43 in LB medium into 200 ml of LB medium in a 1-liter flask, and shake at 30°–34° until the OD_{550} reaches 1.0 (about 90 minutes). Add to the culture 10 millicuries of $^{32}PO_4$ (dissolved in a few ml of water), and immediately shift the culture to a shaking water bath at 43° for 20 minutes. Then return the culture to the shaker at 30°–34° for another 2–3 hours. Centrifuge the cells (8000 × *g* for 20 minutes), and very carefully discard the supernatant into a container for liquid radioactive waste. Resuspend the cells in 10 ml of SM medium and proceed exactly as described in Experiment 43 from the DNase treatment on.

Day 2 Purification of phage on CsCl block gradient. Begin equilibrium run.

Day 3 At the conclusion of the equilibrium density gradient in CsCl, the single phage band should be visible. It should contain enough DNA to determine both the OD_{260} (one OD_{260} unit is equivalent to 50 μg/ml) and the radioactivity. There should be more than 25×10^6 cpm, which would be sufficient for 400 binding tests.

Day 4 Extract the DNA from the phage and purify.

Binding of *lac* Operator by *lac* Repressor

Day 1 Harvest 200 ml overnight cultures of strain CSH46 and CSH50, which have been grown in 1-liter flasks at 30–34° overnight, by spinning at 8000 × *g* for 20 minutes. This should yield about 2 grams of cells; resuspend each pellet in about 2 ml of Breaking Buffer. Do not use any DNase! (Why not?) Break the cells by sonication (the muddy suspension will become translucent), spin out the debris at 10,000 × *g* for 10 minutes, and determine the protein concentration by Biuret determination. This will probably be around 30 mg/ml; dilute the sample in Binding Buffer down to about 0.2–0.5 mg/ml. Prepare a series of 9 tubes as outlined below, each containing 1.0 ml Binding Buffer with 5000–10,000 cpm of *lac* ^{32}P-DNA and 20 μg chicken blood DNA (What is this for?). Add the following to each tube:

1. Background control—add nothing
2. 5 μl i^+ extract
3. 10 μl i^+ extract
4. 20 μl i^+ extract
5. 50 μl i^+ extract
6. 50 μl i^+ extract and 10 μl 10^{-1} M IPTG
7. 20 μl i^- extract
8. 50 μl i^- extract
9. 50 μl i^- extract and 10 μl 10^{-1} M IPTG

Use a new disposable pasteur pipette to apply each sample onto its filter. Filter slowly and wash with 1 ml of Binding Buffer. Glue each filter on a planchet and measure its radioactivity. Plot the counts as a function of the amount of extract (μl) added. How large an effect above background can you see?

Buffers

Breaking Buffer: 0.2 M KCl
0.2 M Tris (pH 7.6, 4°)
0.01 M MgAc
3×10^{-4} M DTT
5% glycerol

Binding Buffer: 0.01 M KCl
0.01 M Tris (pH 7.6, 4°)
0.01 M $MgSO_4$
10^{-4} M EDTA
10^{-4} M DTT
5% DMSO (dimethyl sulfoxide)
50 μg/ml BSA (bovine serum albumin)

SM Medium: 6 ml 25% NaCl
0.3 ml 1 M $MgSO_4$
6 ml 1 M Tris (pH 7.5, 4°)
300 ml distilled H_2O
Autoclave; then add 3 ml of a 1% solution of gelatin.

Materials (for binding of *lac* operator)

15 ml BB buffer
0.2 ml 10^{-2} M IPTG
lac o^+ DNA, ^{32}P-labeled 20,000 cpm
100 μg chicken blood DNA (any other DNA which does not contain *lac o* will suffice)
suction pump
filter apparatus (one filtration takes about 4 minutes; 4 filtrations have to be done per student)
4 Millipore filters, pore size 0.45 μ, diameter 2.5 cm
50 ml Aquasol or other liquid scintillator
liquid scintillation or gas flow counter
5 vials for liquid scintillation counting
SM medium
disposable micropipettes, 5-, 10-, 20-, and 50-μl (Yankee, Clay-Adams)

For the preparation of the operator DNA the same equipment is needed as in Experiment 43.

EXPERIMENT 53

Assay of Transacetylase

The transfer of the acetyl group from acetyl coenzyme A to a thiogalactoside acceptor such as IPTG is catalyzed by the enzyme thiogalactoside transacetylase.

$$\text{Acetyl-CoA} + \text{isopropyl-}\beta\text{-D-thiogalactoside} \longrightarrow \text{CoA} + \text{acetylated isopropyl-}\beta\text{-D-thiogalactoside.}$$

This enzyme is coded for by the *a* gene of the *lac* operon. A spectrophotometric assay has been developed which provides a sensitive method for detecting this enzyme in crude extracts. The reagent 5,5′-dithiobis-2-nitrobenzoic acid (DTNB) is used to react with the free CoA liberated in the above reaction. This reaction involves a disulfide interchange producing thionitrobenzoic acid, which is detected by its absorbance at 412 mμ (Figure 53A).

In the following experiment, induced levels of transacetylase will be measured in CSH51 ($i^+z^+y^+a^+$), and also in CSH50, a total *lac* deletion strain. Both strains require proline.

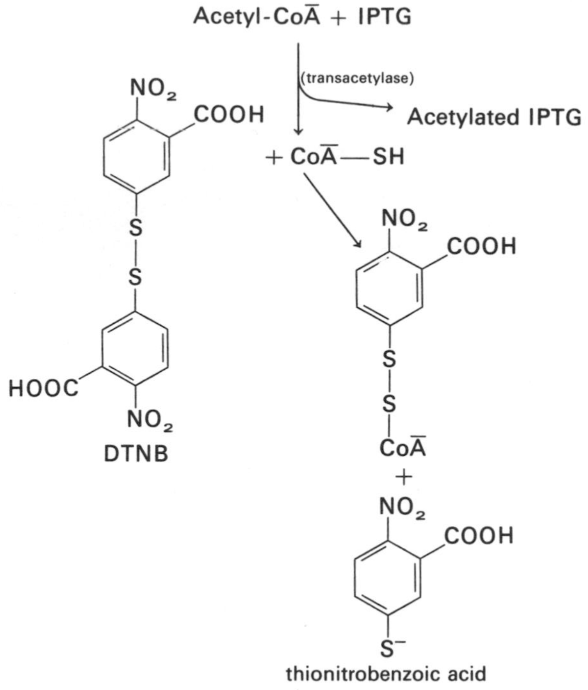

Figure 53A. Assay for transacetylase.

References

Alpers, D. H., S. H. Appel and G. M. Tomkins. 1965. A spectrophotometric assay for thiogalactoside transacetylase. *J. Biol. Chem. 240:* 10.

Michels, C. A. and D. Zipser. 1969. The non-linear relationship between the enzyme activity and structural protein concentration of thiogalactoside transacetylase of *E. coli. Biochem. Biophys. Res. Commun. 34:* 522.

Zabin, I. and A. V. Fowler. 1970. β-galactosidase and thiogalactoside transacetylase. *The Lactose Operon*, p. 27. Cold Spring Harbor Laboratory.

Strains

Number	Sex	Genotype	Important Properties
CSH51	F^-	*ara* Δ(*lacpro*) *strA thi* (ϕ80d*lac*$^+$)	Pro^- Lac^+
CSH50	F^-	*ara* Δ(*lacpro*) *strA thi*	Pro^- Lac^-

Method

Day 1

Preparation of overnights: Subculture an overnight of CSH51 into supplemented 1 × A minimal medium containing 10^{-3} M IPTG and glucose. Use 1 drop of the overnight and 5 ml of medium. Do likewise for CSH50, the control strain for this experiment. Aerate overnight at 37°.

Day 2

Subculture each of the two strains into the same medium, using 3 drops of the glucose-grown overnights and 5 ml fresh medium. Allow to grow until a density of about 4×10^8 cells/ml (OD_{600} = 0.56) is reached. This should take about 4–6 hours. Spin down the cells and resuspend in $\frac{1}{10}$ the original volume of cells in 0.05 M Tris, 0.01 M EDTA at pH 7.9. Sonicate (in a Biosonic Sonicator at 50 watts of power) for 30 seconds, and then heat at 70° for 5 minutes. Spin down the debris and save the supernatant.

Mix 50 μl of the crude extract with 50 μl of the assay medium. Incubate at 25° for 1 hour. Stop the reaction by adding 3 ml of a solution of 25 mg/100 ml DTNB in 0.05 M Tris, pH 7.9. Measure the color at 412 mμ in a spectrophotometer. Also determine the absorbance at 260 mμ. Calculate the transacetylase activity. One unit = OD_{412}/OD_{260}/hour. Do the same for the control strain, which should be a measure of the background.

Repeat this experiment, with and without induction. Compare this to the 1000-fold induction of β-galactosidase. Is there "coordinate induction"?

Materials

Day 1

overnight cultures of CSH51 and CSH50
2 test tubes with 5 ml each of 1 × A medium supplemented with: 50 μg/ml proline, B1, 10^{-3} M $MgSO_4$, 10^{-3} M IPTG, and 0.4% glucose
2 pasteur pipettes

Day 2

2 test tubes with 5 ml each of 1 × A medium supplemented exactly as in Day 1
4 small centrifuge tubes
2 ml 0.05 M Tris, 0.01 M EDTA at pH 7.9
sonicator
waterbath at 70°
3 50-μl, 1 5-ml, and 2 pasteur pipettes
2 small incubation tubes
waterbath at 25°
10 ml 25 mg/100 ml DTNB (5,5′-dithiobis-2-nitrobenzoic acid) in 0.05 M Tris, pH 7.9
spectrophotometer
assay medium: 5 mg/ml acetyl coenzyme A, 250 mg/ml IPTG in 0.05 M Tris, 0.01 M EDTA, pH 7.9

EXPERIMENT 54

Assay for α Complementation

Intracistronic complementation in general has been recently reviewed (Fincham), as has intracistronic complementation in the *z* gene of the *lac* operon (Ullmann and Perrin; see Experiment 20). The experiment described here demonstrates the phenomenon of α complementation *in vitro* (Morrison and Zipser).

If a solution containing β-galactosidase is autoclaved, the enzyme activity is destroyed. However, when the supernatant from this autoclaved solution is incubated with cell extracts of certain z^- deletion mutants, β-galactosidase activity can be recovered. This is approximately 6% of the original activity. During autoclaving, a small polypeptide (MW approximately 7400) is released from the β-galactosidase chain. This peptide is called "auto α." The partially deleted β-galactosidase molecule that complements with auto α is termed the auto α acceptor.

In the following experiment, we will test the extracts of three different nonsense mutants for β-galactosidase activity, and also for auto α activity. The latter test will be done by autoclaving the extracts and then mixing them with extracts of an α acceptor. One of the three nonsense mutations maps within the α region, and should not show α activity.

References

FINCHAM, J. R. S. 1966. *Genetic Complementation*. W. A. Benjamin, New York.
MORRISON, S. L. and D. ZIPSER. 1970. Polypeptide products of nonsense mutations. I. Termination fragments from nonsense mutations in the *z* gene of the *lac* operon of *Escherichia coli*. *J. Mol. Biol. 50:* 359.
ULLMANN, A. and D. PERRIN. 1970. Complementation in β-galactosidase. *The Lactose Operon*, p. 143. Cold Spring Harbor Laboratory.

Strains

Number	Sex	Genotype	Important Properties
CSH23	F′*lac*$^+$ *proA*$^+$,*B*$^+$	Δ(*lacpro*) *supE spc thi*	$i^+z^+y^+$
CSH34	F′*lacZ proA*$^+$,*B*$^+$	Δ(*lacpro*) *supE thi*	$i^+z^-y^+$ (U118, maps within α)
CSH39	F′*lacZ proA*$^+$,*B*$^+$	Δ(*lacpro*) *thi*	$i^+z^-y^+$ (YA536)
CSH21	F′*lacZ proA*$^+$,*B*$^+$	Δ(*lacpro*) *supE nalA thi*	$i^+z^-y^+$ (X90)
CSH12	F$^-$	*lacZ strA thi*	$i^-z^-y^+$ (α acceptor)

Method

Day 1

Prepare broth overnights of the 5 strains to be tested. Inoculate heavily from colonies on a glucose minimal plate without proline, to ensure that the starting colonies all have the episome. It is also possible to use the amber mutants from Experiment 21.

Day 2

Part A

Growth of cells and preparation of extracts: Inoculate 5 ml of each overnight into 100 ml of the growth medium (see Materials) in a 500-ml flask. Aerate at 37° for 4 hours. Spin down and freeze the pellet. Break open the cells by grinding in 1.4 times the wet weight of alumina and extract in 2 times the wet weight of buffer I. Spin down the alumina and save the supernatant, which is the extract.

Part B

Complementation: Autoclave the donor extracts in aluminum foil-covered test tubes at 15–16 lb/in² (122°C) for 20 minutes at slow exhaust. Mix 0.1 ml of each of the 4 α donor extracts with 0.1 ml of the α acceptor extract (CSH12). Use 0.2 ml of each of the extracts alone as a control. Mix well and leave at room temperature for 1 hour. At the end of this time assay for β-galactosidase activity by adding 0.1 ml of these mixed extracts to 0.9 ml buffer (Z buffer), and then proceed exactly as in Experiment 48.

Materials

Day 2

Part A

500 ml growth medium: 1% casamino acids, 10^{-3} M $MgSO_4$, 10^{-3} M IPTG in M9, pH 7.3
5 500-ml erlenmeyer flasks with 100 ml growth medium
5 centrifuge bottles
5 mortars and pestles
alumina
20 ml buffer I, 0.02 M Tris-HCl, 0.1 M 2-mercaptoethanol, 0.01 M EDTA, 0.01 M NaCl, pH 7.2

Part B

5 aluminum foil-covered test tubes
autoclave
10 small test tubes
15 0.1-ml pipettes; 6 1-ml pipettes
3 ml ONPG (4 mg/ml)
10 test tubes with 0.9 ml Z buffer each
5 ml 1 M Na_2CO_3
waterbath at 28°
stopwatch

Questions

Compare the results with those given in Morrison and Zipser. Why do you see less auto α in extracts of late z^- nonsense mutants (X90) than in extracts of wild type? Read Goldschmidt, *Nature*, **228,** 1151, 1970. Does this alter your answer?

UNIT VIII

PROTEIN PURIFICATIONS

INTRODUCTION TO UNIT VIII

Protein Purifications

The purification of a protein is essential for a detailed study of its structure and function. In essence, the particular properties of a protein are utilized to separate it from the numerous other proteins in the cell. This unit offers sample purifications for two proteins of the *lac* operon, β-galactosidase and the *lac* repressor. The purifications differ to some extent from those in the literature because we have tried to include a variety of techniques in adapting them for class exercises.

A purification scheme is usually devised to separate molecular species on the basis of (i) selective solubility, (ii) chromatographic behavior, and (iii) molecular weight. In cases where these are not sufficient, more sophisticated techniques must also be employed. At every stage of a purification it is important to know the **specific activity** of the protein (a measure of its purity) and the **total activity** present in the sample. During the course of a purification the specific activity increases at each step, although the total activity will decrease since some losses are inevitable. Never discard any fractions until you are sure they contain no significant activity. Sample activities for the purification of the *lac* repressor are shown in Table 55A of Experiment 55.

SDS-Polyacrylamide Gels

In recent years, the use of polyacrylamide* gel (or "disc-gel") electrophoresis with sodium dodecyl sulfate (SDS) has proved very important (Shapiro et al.).

* Unpolymerized acrylamide is very toxic and is readily absorbed directly through the skin. Always wear gloves when preparing gels and rinse glassware out thoroughly after use.

(A) The molecular weight of a polypeptide chain in the 10,000–200,000 dalton range can readily be determined (Weber and Osborn). (B) The extent of purity of a protein is qualitatively evident (for contaminants outside of its molecular weight range) (Shapiro et al.). (C) Less than 10 μg of a protein is usually sufficient for these purposes. (D) Gels may also be used on a larger scale to prepare small amounts of pure protein suitable for micro-sequencing and for the production of antibody (Weiner et al.). (E) It is possible to renature proteins eluted from these gels, which is an especially useful technique for the reconstruction of multicomponent systems (Weber and Kuter). (F) The gels are extremely simple to run if you have once gone through the procedure with someone familiar with their use.

We have included here the recipes for pouring and running these gels and the preparation of the reagents. For a full description of how to set up the apparatus and actually run the samples, the instructor and students should refer to the review by Weber (1972).

For those interested in the original sources, a list of references to various purification schemes is given below.

References

lac Repressor:

GILBERT, W. and B. MÜLLER-HILL. 1966. Isolation of the *lac* repressor. *Proc. Nat. Acad. Sci. 56:* 1891.

MÜLLER-HILL, B., K. BEYREUTHER and W. GILBERT. 1971. *Lac* repressor from *Escherichia coli*. *Methods in Enzymology XXI-D:* 483. Academic Press, New York.

RIGGS, A. D. and S. BOURGEOIS. 1968. On the assay, isolation and characterization of the *lac* repressor. *J. Mol. Biol. 34:* 361.

β-Galactosidase:

CRAVEN, G. R., E. STEERS, JR. and C. B. ANFINSEN. 1965. Purification, composition, and molecular weight of the β-galactosidase of *Escherichia coli* K12. *J. Biol. Chem. 240:* 2468.

HU, A. S. L., R. G. WOLFE and F. J. REITHEL. 1959. The preparation and purification of β-galactosidase from *Escherichia coli* ML308. *Arch. Biochem. Biophys. 81:* 500.

KARLSSON, U., S. KOORAJIAN, I. ZABIN, F. S. SJÖSTRAND and A. MILLER. 1964. High resolution electron microscopy on highly purified β-galactosidase from *Escherichia coli*. *J. Ultra. Res. 10:* 457.

Lactose Permease:

FOX, C. F. and E. P. KENNEDY. 1965. Specific labeling and partial purification of the M protein, a component of the β-galactoside transport system of *Escherichia coli*. *Proc. Nat. Acad. Sci. 54:* 891.

JONES, T. H. D. and E. P. KENNEDY. 1969. Characterization of the membrane protein component of the lactose transport system of *Escherichia coli*. *J. Biol. Chem. 244:* 5981.

Thiogalactoside Transacetylase:

ZABIN, I. 1963. Crystalline thiogalactoside transacetylase. *J. Biol. Chem. 238:* 3300.

ZABIN, I., A. KEPES and J. MONOD. 1962. Thiogalactoside transacetylase. *J. Biol. Chem. 237:* 253.

SDS-Polyacrylamide Gels:

SHAPIRO, A. L., E. VINUELA and J. V. MAIZEL, JR. 1967. Molecular weight estimation of polypeptide chains by electrophoresis in SDS-polyacrylamide gels. *Biochem. Biophys. Res. Commun. 28:* 815.

WEBER, K. 1972. Methods in Enzymology.

WEBER, K. and D. J. KUTER. 1971. Reversible denaturation of enzymes by sodium dodecyl sulfate. *J. Biol. Chem. 246:* 4504.

WEBER, K. and M. OSBORN. 1969. The reliability of molecular weight determination by dodecyl sulfate-polyacrylamide gel electrophoresis. *J. Biol. Chem. 244:* 4406.

WEINER, A. M., T. PLATT and K. WEBER. Manuscript in preparation.

See also *The Lactose Operon*, Cold Spring Harbor Laboratory, 1970:

Chapter III—β-Galactosidase and Thiogalactoside Transacetylase, I. Zabin and A. V. Fowler, p. 27.

Chapter IV—The Lactose Permease System of *E. coli*, E. P. Kennedy, p. 49.

Chapter V —The Lactose Repressor, W. Gilbert and B. Müller-Hill, p. 93.

Preparation of Materials

Purifications often involve a considerable amount of physical work, and it is important that all materials for an experiment be ready beforehand. Leaving the preparation of buffers and columns until the last minute can lead to major delays and disastrous results. Make sure you are familiar with the protocol for each stage, with a checklist of all materials. **Careful planning is essential!**

It is convenient to have on hand concentrated stock solutions of commonly used salts. As even reagent grade chemicals usually contain some impurities, each stock solution should be filtered (Millipore DA filters are fine) before it is stored. A partial list is suggested below; this will vary depending on the experiment.

2 M KCl	1 M KH_2PO_4
2 M NaCl	1 M K_2HPO_4
1 M MgAc	0.5 M NaH_2PO_4
1 M Tris HCl pH 7.6 (4°C)	0.5 M Na_2HPO_4
0.2 M EDTA pH 7.0 (20°C)	6 N HCl
	6 N KOH

The preparation of other materials used in protein purifications is described below.

Visking #20 Dialysis Tubing (Union Carbide)

Cut 15 to 20 two-foot lengths of tubing, and boil twice for 15 minutes in 1.5 liter distilled H_2O that is 2×10^{-3} M EDTA (pH 7). Use a 4-liter beaker, and put a 1-liter erlenmeyer flask partly filled with water on top of the tubing to keep it below the surface. Otherwise, the tubing will rise to the surface and dry out as it becomes filled with bubbles during boiling.

Rinse well with plain distilled water between boilings, and always handle with gloves after the first wash. Do not use **any** pointed or sharp objects to stir the tubing: a tiny hole in the membrane will be undetectable until you attempt to dialyze your sample.

Finally, boil once in double-distilled water, 2×10^{-3} M EDTA, rinse, and store the tubing in 2×10^{-4} M EDTA in the cold, in a foil-covered beaker. When tying knots in dialysis sacs, it is helpful to twist the end several times first, and tighten the knot by pulling from the free end toward the center of the sac. Stretching the central portion may enlarge the pores, causing partial loss of the sample. The #20 tubing is the most porous of the Visking series; thus dialysis times can be relatively short. For small samples (as little as 50 μl can be dialyzed in #20 sacs), very rapid equilibrium is attained due to the large surface area available.

Biuret Reagent (Keeps 3 months at room temperature)

Dissolve 1.5 g $CuSO_4 \cdot 5\ H_2O$ + 6.0 g $NaKC_4H_4O_6 \cdot 4\ H_2O$ (sodium potassium tartrate) in 500 ml H_2O. Add with stirring 300 ml of 10% NaOH (CO_2-free). Dilute to 1 liter with H_2O and store in a plastic bottle (not glass). This reagent is used for protein determination, and will detect accurately as little as 0.1 mg of protein. For more sensitive work, the Lowry method (not given here) is recommended.

Method 1: Useful for protein solutions greater than 5 mg/ml, which do not have high concentrations of amino-group cations such as ammonia or Tris (greater than 0.2 M).

Add 1 ml of Biuret reagent to 20 μl of protein solution. As a blank, use 20 μl of whatever buffer the protein is in, plus 1 ml Biuret. Let stand 25–40 minutes at room temperature, and read the optical density at 546 mμ.

$$OD_{546}\ 0.06 = 0.2 \text{ mg protein or } 10 \text{ mg/ml if } 20\ \mu\text{l were used.}$$

Method 2: Useful for dilute protein solutions (0.2–10 mg/ml), or those with a high concentration of amino groups.

Add an aliquot of the sample containing at least 0.1 mg, and preferably 0.2–0.5 mg, to 1 ml of 10% TCA (trichloroacetic acid) in a conical centrifuge tube. Use an equal aliquot of buffer as a control. Centrifuge out the precipitate, discard the supernatant, and resuspend the precipitate in 0.5 ml of water with vortexing.

Add 0.5 ml of Biuret reagent, with more vortexing, then let stand 25–40 minutes at room temperature before reading the optical density at 546 mμ. The correlation between OD_{546} and protein is the same as that in Method 1.

DEAE Cellulose (Whatman DE 52 Preswollen)

Allow about 1 g of resin per gram of cells.

(a) Suspend in water and let settle several times, pouring off the "fines" (very tiny particles) in the supernatant, which only serve to clog up the column. (Usually any resin that does not settle in 20 to 30 minutes is detrimental to the flow rate.)

(b) Wash in 5 volumes 0.5 N HCl for 15–20 minutes.

(c) Neutralize by washing with water on a sintered glass funnel to pH 6.

(d) Wash in 5 volumes 0.5 M NaOH for 15–20 minutes.

(e) Neutralize by washing with water on a sintered glass funnel to pH 8.

(f) Suspend in column buffer and titrate with 6 N HCl to the desired pH. After several hours, check the pH and re-titrate if necessary. Decant the supernatant, and replace with fresh buffer. Repeat this two or three times, until the pH is stable at the desired value.

(g) Pour the column in a slurry with buffer at a total volume of more than three times the volume of the wet resin, and let settle for 5–10 minutes before opening the stopcock. A few column volumes of buffer should be passed through before use. Check the pH and conductivity of the outflowing buffer to verify that the column is equilibrated.

Phosphocellulose (Whatman P-11)

Allow 1 g dry resin (6 ml wet volume) per 5 g of cells. Wash exactly as DEAE cellulose, except with the acid and base washes reversed (base first, then acid), and titrate with 6 N KOH. Equilibration should be started several days before use, as packed resin is about 1 M in phosphate groups. Many generous changes of buffer (especially if it is low in ionic strength) are required, until the pH is stabilized at the desired value.

1 M Tris, pH 7.6 at 4°C. Dissolve 121.1 g (one mole) of Tris Base in about 900 ml of distilled water. Add, with careful stirring, 70 ml concentrated HCl. Chill on ice to a measured 4°–6°C, and titrate with concentrated HCl to pH 7.6. Tris has a high temperature coefficient (0.03 pH units per °C), so use a thermometer. Make up to 1 liter, filter, and store in the cold.

0.2 M EDTA, pH 7.0 at room temperature. Dissolve desired amount (use the sodium salt) in distilled water, and add 50% NaOH, with stirring, to pH 7.0.

You will need about 60–70 ml of 50% NaOH per liter of 0.2 M EDTA. The EDTA will be fairly insoluble until the pH gets above 6. Filter, and store at room temperature.

SDS-Polyacrylamide Gels

Gel buffer: 0.2 M sodium phosphate (pH 7.2) and 0.2% SDS
7.8 g $NaH_2PO_4 \cdot H_2O$ (113 ml of 0.5 M stock)
38.6 g $Na_2HPO_4 \cdot 7H_2O$ (290 ml of 0.5 M stock)
2 g recrystallized SDS (20 ml 10% SDS)
Make up to 1 liter with distilled H_2O.

Acrylamide solution: 22.2 g acrylamide
0.6 g methylenebisacrylamide (MBA)
Make up to 100 ml, filter and store in brown bottle.

Ammonium persulfate (AP): $(NH_4)_2S_2O_8$ at 15 mg/ml made up **freshly**

	Gels	
	5%	10%
acrylamide solution	4.5 ml	9.0 ml
gel buffer	10.0 ml	10.0 ml
AP at 15 mg/ml	1.0 ml	1.0 ml
distilled H_2O	4.5 ml	
TEMED	15 μl	15 μl

De-gas solution at room temperature briefly, then add TEMED, swirl gently and pour gels (1.5–2 ml/tube). Layer a few millimeters of water on top of each gel, to prevent a curved meniscus. This is removed before applying the sample.

Running buffer: Dilute gel buffer 1:1
0.1 M sodium phosphate, pH 7.2
0.1% SDS.

Sample buffer: 0.1 M sodium phosphate, pH 7.2 (Keeps for months at 4°)
0.1% SDS
0.14 M 2-mercaptoethanol
10% glycerol
0.002% bromphenol blue
Add sample to 100 μl of this, heat to 65°C for 10 minutes, and apply to gel.
or Dialyze samples against gel buffer diluted to 20:1, adjusted to 0.2% SDS. Then, per 100–200 μl sample, add:
1 drop glycerol
10 μl 2-mercaptoethanol
5 μl 0.05% bromphenol blue
After applying samples, layer running buffer to the top of the gel tube. Fill the upper and lower reservoirs, and connect to the power supply.

Run: Room temperature at 8 ma/gel
4–5 hours for 5 % gels; 7–8 hours for 10 % gels.

Stain: 0.25 % Coomassie brilliant blue
2.5 g C.B.B.
454 ml methanol
454 ml H_2O
92 ml glacial acetic acid
1000 ml

Filter (on funnel).

Destain: 7.5 % acetic acid
5 % methanol

Storage: 7.5 % acetic acid

All the specialized reagents may be obtained in acceptable form from Bio-Rad:
1. acrylamide
2. N,N′-methylenebisacrylamide (MBA)
3. N,N,N′,N′-tetramethylethylenediamine (TEMED)
4. sodium dodecyl sulfate (SDS)
5. ammonium persulfate (AP)

If any difficulty is encountered with the SDS, it should be recrystallized.

Recrystallization of SDS

1. Add 250 g SDS to 4.0 liters 95 % EtOH at 70°C with stirring (using more SDS will cause great problems later).
2. Filter through Whatman #1 on large preheated (120° oven) Buchner funnel. If it is too cool, it will clog.
3. Allow to cool to room temperature with stirring.
4. Stir overnight in the cold.
5. Put filtrate, a Buchner funnel, and 100 % EtOH into a −20°C freezer for 2 hours.
6. Filter through precooled funnel, rinsing with precooled 100 % EtOH on Whatman #1.
7. Cover crystals with cheesecloth, and dry by sucking through with air.
8. Dry further in a vacuum desiccator over oven-dried (120°) anhydrous $CaCl_2$. The yield is about 70 %.

EXPERIMENT 55

Purification of the *lac* Repressor

The *lac* repressor is normally present at only 10 to 20 molecules per cell, which is about 0.002% of the soluble protein. However, much larger amounts of *lac* repressor can be obtained from strains carrying the *lac* region on a temperature-inducible defective prophage. The strain CSH46, which is used in this experiment, carries such a prophage, with an additional mutation affecting the *i* promoter. This mutation, called i^{SQ}, results in 30–50 times the normal haploid i^{+} level of repressor in the cell. Induction of the prophage gives a further increase of about 20-fold. Thus extracts from this strain contain 1000 times the normal amount of repressor, comprising 1–2% of the soluble protein.

Outline of Purification

A fresh overnight culture of CSH46 is inoculated into LB medium, and grown for 1–2 hours at 34° with good aeration. The *lac* prophage is induced by shifting the temperature of the culture to 43–45°C. (This inactivates its own temperature-sensitive repressor.) Vigorous aeration is required for optimal induction of the phage. After 20 minutes at 45° the culture is cooled to 34°, and growth is continued for 2–3 hours under the maximum possible aeration. The cells are centrifuged and frozen, and may be stored in the freezer for several weeks with no apparent loss of activity.

When the frozen cells are thawed in buffer, they lyse spontaneously and release a large amount of DNA which makes the solution initially very viscous. Treatment with DNase reduces the viscosity and the cell debris may be centrifuged out.

Ammonium sulfate is added to the supernatant to a final concentration of 33% saturation. This step precipitates all the repressor and results in a 5- to 10-fold purification, with the ribosomes and 90% of the other cell proteins remaining in the supernatant. The precipitate is resuspended and dialyzed to remove the ammonium sulfate.

Basic proteins or proteins with strongly basic regions will bind to phosphocellulose. Although the *lac* repressor is an acidic protein it does bind tightly to phosphocellulose. Since 95% of the soluble proteins of *E. coli* do not stick to phosphocellulose, it is a convenient procedure to use in the purification of the repressor.

The sample is applied to a phosphocellulose column equilibrated with a 0.12 M potassium phosphate buffer at pH 7.4; this is followed by a gradient of 0.12 M to 0.24 M potassium phosphate. *Lac* repressor elutes at about 0.17 M potassium phosphate and is greater than 98% pure. The activity peak is pooled and concentrated by precipitation with ammonium sulfate as before.

The precipitate is redissolved in 1 M Tris at pH 7.6 and stored at 4°C. Repressor is very insoluble in low salt, but under these conditions concentrations in excess of 40 mg/ml may be obtained. The yield should be about 70–80%, or roughly 8 mg per 10 grams of cells, with 50% DNA binding activity.* The purity of the repressor can be most easily judged by electrophoresis on SDS-polyacrylamide gels.

Equilibrium Dialysis *vs* Millipore Filter

Experiment 49 and Appendix VI describe two different assays for the *lac* repressor. We employ both during the course of the purification. Equilibrium dialysis can be applied to crude extracts and samples containing high concentrations of other proteins. The Millipore filter assay is faster and more sensitive, and also has a low background relative to the observed effect. As little as 1 μg of repressor can be detected. However, because other proteins also stick to Millipore filters and compete with repressor for sites on the filter, the repressor must be greater than 5% pure. Otherwise the assay is not quantitative.

We therefore use the equilibrium dialysis assay for the early stages of the purification and then switch to the Millipore filter assay at the phosphocellulose column step.

References

GILBERT, W. and B. MÜLLER-HILL. 1966. Isolation of the *lac* repressor. *Proc. Nat. Acad. Sci. 56:* 1891.

MÜLLER-HILL, B., K. BEYREUTHER and W. GILBERT. 1971. *Lac* repressor from *Escherichia coli. Methods in Enzymology XXI-D:* 483. Academic Press, New York.

RIGGS, A. D. and S. BOURGEOIS. 1968. On the assay, isolation, and characterization of the *lac* repressor. *J. Mol. Biol. 34:* 361.

RIGGS, A. D., S. BOURGEOIS, R. F. NEWBY and M. COHN. 1968. DNA binding of the *lac* repressor. *J. Mol. Biol. 34:* 365.

RIGGS, A. D. and S. BOURGEOIS. 1969. On the *lac* repressor-operator interaction and the purification of the *lac* operator. *Biophys. J. 9:* A84.

* Empirically it is found that during the purification some operator binding activity is usually lost.

Strains

Number	Sex	Genotype	Important Properties
CSH46	F^-	*ara* Δ(*lacpro*) *thi* (λCI857*S*t68d*lacI*,*Z*)	Pro^- carries $i^{SQ}z^-$ on heat-inducible prophage z^- is U118, an ochre mutation

Method

Day 1

Growth of bacteria

Inoculate a fresh 100-ml overnight, grown at 30–34°, of CSH46 into 1 liter of LB medium in a 4-liter flask, and shake at 34° for 1–2 hours. By this time the cells will have grown to about $1–3 \times 10^9$/ml (OD_{550} = 3.0–6.0). At this time streak a sample of the culture onto half of a plate. The prophage is now induced by shifting the temperature to 43–45° for 20 minutes. This is most easily done with vigorous shaking of the flask by hand over a bunsen burner with a thermometer in the culture, until the temperature reaches 43°. It should be maintained between 43° and 45° for 20 minutes by occasional warming over the flame. Good shaking is very important for proper induction of the phage!

After 20 minutes, cool the culture under cold water and streak a sample again, on the other half of the plate used above. Replace the flask on the shaker at 34° for $1\frac{1}{2}$–3 hours. The plate is incubated overnight at 30–34°, and will provide a rough measure of the extent of induction (only non-lysogens and uninduced lysogens will survive). Continue to read the OD_{550} (at a 10:1 dilution, since the Zeiss is non-linear above an OD of 1–1.5) every 30–40 minutes. Growth should plateau at about three times the OD_{550} obtained just prior to induction. This phage is lysis defective, which permits protein synthesis to continue for several hours after induction. The cells no longer divide, but instead elongate considerably and look like "snakes" under the microscope. Centrifuge each culture at 8000 × *g* for 20 minutes, and weigh the cells on parafilm and freeze. They may be stored for several weeks if necessary.

The phosphocellulose resin should be washed and de-fined during this time (see Preparation of Materials), equilibrated with the column buffer (0.12 M KP), and stored in the cold. Allow 1 gram (dry weight) of resin per 5 grams of cells. It is important for equilibration to be complete **before** pouring the column (see Day 2). The bed volume of the column should be about equal to the weight of cells used in the prep. This would be about 50 ml for 6 liters of culture, yielding 40–50 g of cells.

Day 2

The purification may begin at any suitable interval after freezing the cells. **All subsequent operations are performed on ice or in a cold room unless otherwise indicated.**

The pH of the phosphocellulose which was prepared on Day 1 should be checked. If it does not agree with the pH of 0.12 M KP (7.4 at 4°), it must be titrated with 1 M HCl or 1 M KOH as appropriate. Let stand for several hours and then recheck the pH. Pour off the top buffer, add fresh 0.12 M KP, and continue to monitor pH. When it is constant for several hours, the column may be safely poured. The phosphocellulose column must be poured and ready by the beginning of Day 3.

Preparation of crude extract

Each group should take their weighed packet of frozen cells (6–8 g) and grind it in a prechilled mortar and pestle with 2.5 times the cell weight of alumina for 15–20 minutes. The sample will be very dry at first, gradually developing a smooth sheen and "crackling" as the cells break.*

Then, add to the paste 5 times the cell weight of breaking buffer (BB) which is 5 μg/ml in crude DNase. Work this in for another 5 minutes, then pour the resulting slurry into a centrifuge tube, and spin at 10,000 $\times$ g for 10 minutes to remove the alumina and cell debris. Decant the clear yellow-brown supernatant into a graduate cylinder, and remove a small aliquot (1.0 ml) for assay.

Ammonium sulfate fractionation

Pour the supernatant into a beaker with a stirring bar (allowing for a further 50% volume increase), and slowly add crystalline ammonium sulfate over a period of 30–45 minutes using 231 mg $(NH_4)_2SO_4$ per ml of starting solution. This will give a final concentration of 33% saturation.† Add 1 ml 1 N NaOH for every 10 grams of ammonium sulfate. Continue stirring for 1–3 hours, and then spin down the precipitate (10 minutes at 10,000 $\times$ g). Decant the supernatant and save. Redissolve the precipitate in 2–4 ml of BB; remove 0.1 ml for assay into about 0.4 ml of TMS buffer.

At this stage, the samples from each group are pooled (the total volume from 50 g of cells should be about 20 ml) and dialyzed against 500 ml distilled water for 30 minutes in #20 dialysis tubing, and then against 500 ml of 0.12 M KP three times (at least 2 hours each). This may be done overnight. Dialysis directly against the phosphate buffer will result in precipitation of an insoluble $Mg(NH_4)PO_4$ complex.

Activity determination of fractions

During this time, assay the three samples obtained so far by equilibrium dialysis against ^{14}C-IPTG (see Experiment 51). These

* Although these cells will lyse when thawed in buffer, it is easier to use alumina when handling small amounts. For batches over 100 g, it is better to allow the cells to break spontaneously during thawing.

† Saturation at 4°C is taken as 70 g per 100 ml of starting solution.

are (i) the crude extract, (ii) the ammonium sulfate supernatant, and (iii) the resuspended ammonium sulfate pellet. Dialyze all three in #20 dialysis tubing in the same flask for at least 3 hours to permit equilibration. Remove the samples and calculate the activities for each.

Assuming that pure repressor has an activity of 2000 %/mg/ml, what fraction of the protein in your crude extract is repressor? What fraction is it in the ammonium sulfate cut? How much activity remains in the supernatant? What factor of purification did you obtain by the ammonium sulfate precipitation? Compare your data to that shown in Table 55A. For a crude extract of 250 ml at 20.0 mg/ml, the total protein is 5000 mg. If the equilibrium dialysis assay yields (per 0.1 ml) 2000 cpm of ^{14}C-IPTG inside the sac and 500 cpm outside, the excess inside is 1500 cpm, or 300 % relative to outside. A protein concentration of 15.0 mg/ml of the dialyzed sample gives a specific activity (S.A.) of 300 %/15.0 mg/ml = 20.0 %/mg/ml. The total activity is obtained from multiplying the specific activity by the total protein:

$$5{,}000 \text{ mg} \times 20.0\,\%/\text{mg/ml} = 100{,}000 \text{ activity units (A.U.} = \text{ml}\,\%).$$

Since pure repressor has an S.A. of 2000 A.U./mg, this sample is 20.0/2000 = 1 % pure, and contains 100,000 A.U./2000 A.U. per mg = 50 mg of repressor.

Day 3

Phosphocellulose column

Remove the dialysis sample (wearing disposable plastic gloves), dry the outside of the sac, cut off the end and empty into a beaker. Squeeze out as much as possible from the sac. Centrifuge the sample to eliminate any remaining insoluble material (10 minutes at 10,000 × *g*). Pour off the supernatant, record its volume, and remove about 0.5 ml for assay. The pH and conductance of the sample must agree with 0.12 M KP and with the effluent from the

Table 55A. Sample Purification of *lac* Repressor from 50 grams of Strain CSH46

	Volume (ml)	Concentration (mg/ml)	Total protein (mg)	Specific activity (%/mg)	Total activity units (A.U.)	Yield	Purity
Crude extract	250	20	5,000	20	100,000	100 %	1 %
Ammonium sulfate precipitate	30	16	480	200	96,000	96 %	10 %
Supernatant	300	15	4,500	1	4,500	<5 %	0.05 %
Phosphocellulose peak	43	1.0	43	2,000	86,000	86 %	>99 %

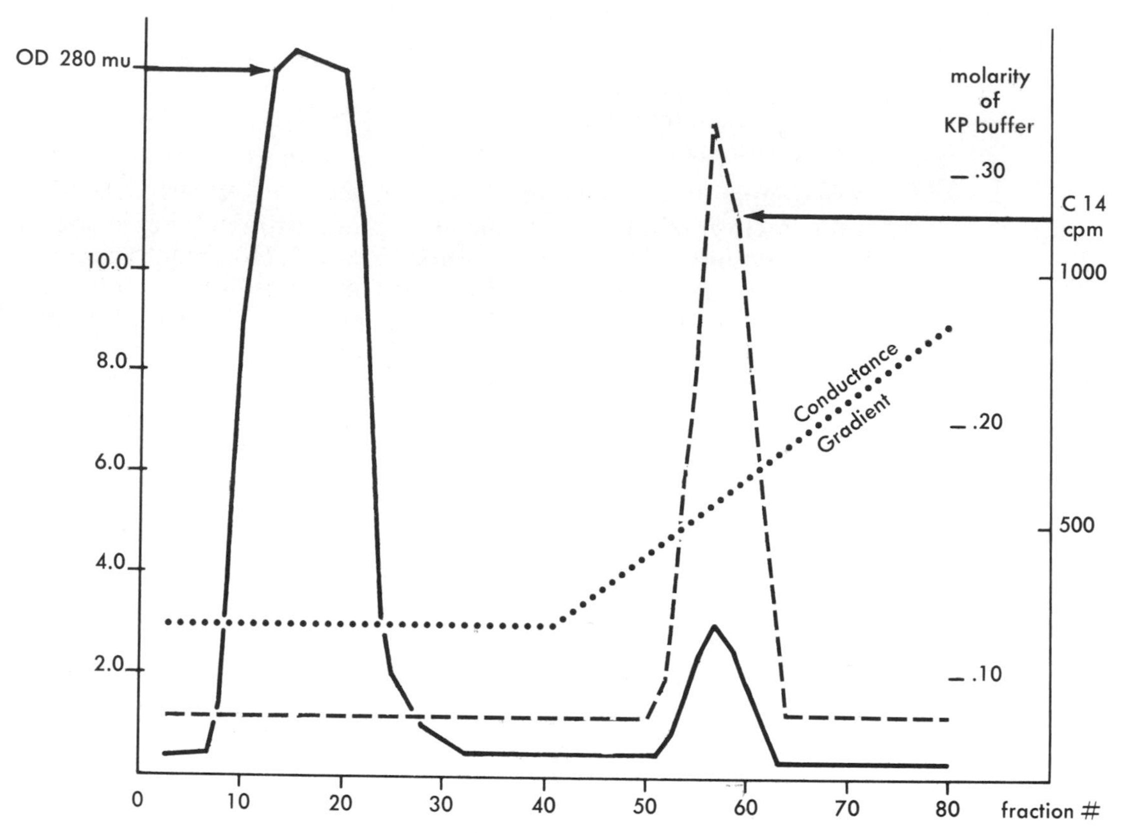

Figure 55A. Phosphocellulose column profile in the purification of *lac* repressor. The dashed line (– – –) indicates *lac* repressor activity as assayed by equilibrium dialysis with ^{14}C-IPTG.

column. Conductivity readings are very temperature dependent, so for reproducibility read all samples on ice. Apply the sample to the column at a flow rate of about one-half bed volume per hour, and wash through with 0.12 M KP until the effluent OD_{280} is below 0.10, relative to 0.12 M KP alone. Collect fractions of 5 ml each.

Begin a gradient of 0.12 M–0.24 M KP, with each buffer 2 times the bed volume of the column, and the same flow rate as for the sample application. At the conclusion of the gradient, wash the column with 2 volumes of 0.24 M KP. Monitor each fraction for OD_{280}, and every second or third fraction for conductance and for repressor activity by the Millipore filter assay. A column profile (see Figure 55A) is obtained which indicates the protein elution, the conductivity gradient, and the position of *lac* repressor, which should be around 0.17 M KP.

Pool the activity peak, remove an aliquot for assay, and concentrate with ammonium sulfate to 33% saturation (231 mg per ml of starting solution). This precipitation may be done overnight.

Day 4 **SDS-polyacrylamide gels**

Spin down the precipitate for 10 minutes at 10,000 × *g*, decant and save the supernatant. Resuspend the precipitate in about 4 ml of 1 M Tris (pH 7.6 at 4°C), add 2 ml of glycerol, and DTT to 10^{-4} M. Each group should take an aliquot to determine the protein concentration and the specific activity by equilibrium dialysis. For pure repressor this would be 2000 to 2500%/mg/ml. Check the supernatant also. If desired an additional aliquot is dialyzed against SDS and buffer in preparation for running on polyacrylamide gels. Prepare six to eight 10% gels, and run a different dilution of the sample on each gel, ranging from 1–100 μg. Run three or four protein standards of known molecular weight on two of the gels, using 5–10 μg of each. Stain for 2–4 hours, then destain. Calculate the mobility and molecular weight of your repressor from the standards. Is it completely pure? This purified repressor may be used in Experiments 56 (Determination of the Binding Constant of the *lac* Repressor for IPTG) and 62 (Repression of β-Galactosidase Synthesis in a Cell-free System).

Buffers

Convenient stock solutions are:
1 M Tris-HCl, pH 7.6 at 4°C
2 M KCl
1 M MgAc (magnesium acetate)
1 M K_2HPO_4
1 M KH_2PO_4
0.1 M DTT (dithiothreitol) (Sigma)
0.2 M EDTA (pH 7.0)

1. Breaking Buffer (BB):
0.2 M Tris (from 1 M stock, pH 7.6 at 4°C)
0.2 M KCl
0.01 M MgAc
3×10^{-4} M DDT
5% glycerol

2. Equilibrium Dialysis Assay Buffer (TMS):
0.2 M KCl
0.01 M Tris (from 1 M stock, pH 7.6 at 4°C)
0.01 M MgAc
10^{-4} M EDTA
10^{-4} M DTT

3. Phosphocellulose Column Buffer (KP):
Correct pH (7.4 at 4°C) is obtained by using $K_2HPO_4:KH_2PO_4 = 5:1$ by volume.
   (a) 0.12 M KP (potassium phosphate)
   10^{-4} M DTT
   5% glycerol

   (b) 0.24 M KP (potassium phosphate)
   10^{-4} M DTT
   5% glycerol

Materials

Day 1

20 g washed phosphocellulose (Whatman P11)
1 liter 0.12 M KP buffer
shaker for 4-liter flasks at 30–34°C
spectrophotometer for reading cell density; cuvettes
centrifuge to accommodate 1 liter per group

Per group:
1 liter sterile LB in a 4-liter flask
1 thermometer
1 bunsen burner

Day 2

50 cm #20 dialysis tubing (Union Carbide)
500-ml flask
2 liters 0.12 M KP
100–200-ml column with stopcock (axial ratio about 10:1)

Per group:
20 g levigated alumina (Norton Co., Worcester, Mass.)
mortar and pestle
50 ml BB
20 g $(NH_4)_2SO_4$, Enzyme Grade (Mann)
50-ml graduate cylinder
100-ml beaker and magnetic stirring bar
magnetic stirring platform
equilibrium dialysis assay materials

Day 3

1 liter 0.12 M KP
1 liter 0.24 M KP
gradient maker (100–200 ml each side)
fraction collector and test tubes
conductance meter
20 g $(NH_4)_2SO_4$, Enzyme Grade
50 Millipore filters (washed)
5 ml ^{14}C-IPTG at 2×10^{-6} M (Calbiochem)
filter apparatus
50 scintillation vials and equivalent scintillation fluid
scintillation counter

Day 4

10 ml 1 M Tris (pH 7.6 at 4°C)
equilibrium dialysis assay materials
SDS-polyacrylamide gel materials
(buffers, gel tubes, gel apparatus, power supply, stain, and destaining apparatus)

Chemicals

^{14}C-IPTG, specific activity 20–30 mc/mM	(Calbiochem)
$(NH_4)_2SO_4$, Enzyme Grade	(Mann)
DNase I (crude), activity 2500 units/mg	(Worthington)
#20 dialysis tubing	(Union Carbide)
Phosphocellulose P-11	(Whatman)
Dithiothreitol (DTT)	(Sigma)
Levigated alumina	(Norton Co., Worcester, Mass.)

EXPERIMENT 56

Determination of the Binding Constant of *lac* Repressor for IPTG

The equilibrium dialysis assay (see Experiment 51) for *lac* repressor depends on the affinity of repressor for the inducer IPTG. This assay can be used to determine the order of magnitude of the binding constant, even with a crude cell extract as a source of repressor. If pure repressor is used, the number of binding sites for IPTG per repressor molecule can also be determined.

If a protein contains n equivalent and independent binding sites for a ligand (L), with an equilibrium binding constant of k_{eq}, we may write

$$\mathrm{P} + \mathrm{L} \rightleftharpoons \mathrm{PL} \quad \text{and} \quad k_{eq} = \frac{[\mathrm{PL}]}{[\mathrm{P}][\mathrm{L}]}$$

where [P] is the concentration of binding sites, and [L] is the concentration of free ligand. Or:

$$k_{eq}[\mathrm{L}] = \frac{[\mathrm{PL}]}{[\mathrm{P}]} = \frac{[\text{occupied binding sites}]}{[\text{free binding sites}]} = \frac{V}{n - V}$$

where V = molecules of ligand bound per protein molecule and n = total number of binding sites per protein molecule. Rearranging this equation:

$$\frac{V}{L} = k_{eq}(n - V)$$

or:

$$\frac{V}{L} = -k_{eq}V + k_{eq}n.$$

A plot of (V/L) against (V) should give a straight line with slope = $-k_{eq}$. If a pure homogeneous protein sample is used, the x intercept will yield the number of binding sites, n. With a sample only partly pure, one cannot obtain n, but the slope will still yield k_{eq}, provided that no other proteins or small molecules in the extract affect the equilibrium binding.

In this experiment, aliquots of a repressor sample are dialyzed overnight (to ensure equilibrium) against ^{14}C-IPTG with different concentrations of unlabeled IPTG. A sample from each dialysis flask is taken before adding the repressor to define the initial IPTG concentration. After equilibrium is reached (IPTG will dialyze through the #20 membrane with a half-time of about 5 minutes), an aliquot of outside buffer defines the final IPTG concentration outside. The excess IPTG concentration inside is calculated from the excess counts inside the sac. This is corrected for the amount of material by dividing it by the protein concentration inside the sac.

From this raw data, a Scatchard plot can be obtained (see Method) as illustrated in Figure 56A. The slope of this line gives the binding constant for IPTG, as described above. The basic assumption here is that the n binding sites per molecule are equivalent and independent. If cooperative interactions are involved, a curve rather than a straight line will be obtained.

If this experiment is coupled with the purification of *lac* repressor, it may be worthwhile to divide the class into smaller groups and have each group determine the binding constant at a different stage of the purification.

Method

Day 1 Prepare 10 125-ml erlenmeyer flasks each with 20 ml TMS assay buffer. To each add 10 μl of ^{14}C-IPTG (2×10^{-4} M) making a final concentration of 10^{-7} M. Then, add aliquots of unlabeled IPTG from a stock at 10^{-3} M, diluting through TMS, to obtain total molar IPTG concentrations in the 10 flasks of:

2×10^{-7}	4×10^{-6}
4×10^{-7}	6×10^{-6}
6×10^{-7}	1×10^{-5}
1×10^{-6}	2×10^{-5}
2×10^{-6}	4×10^{-5}

Remove 100 μl for counting to define the starting concentration. Into each flask put a dialysis sac (#20) containing about 0.2 ml of a crude or partially purified sample of *lac* repressor. Seal each

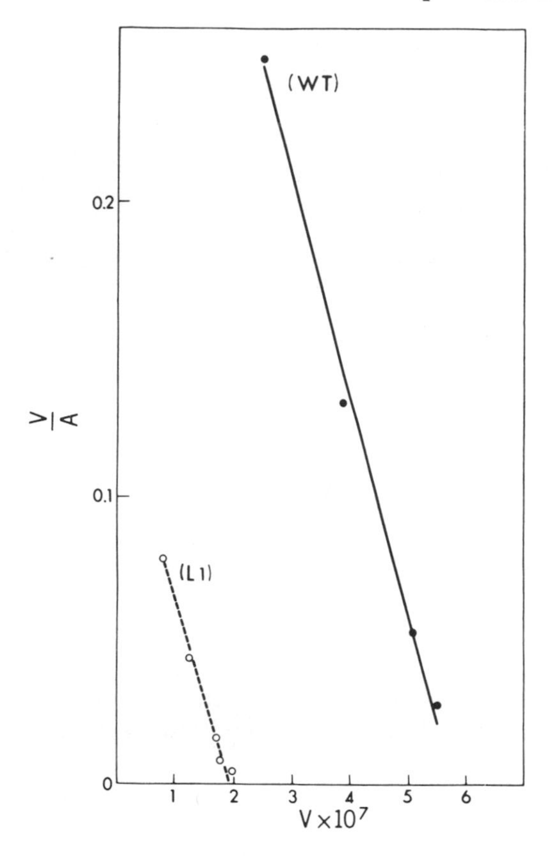

Figure 56A. A sample Scatchard plot to compare the binding constant of *lac* repressor (or IPTG) to that of a mutant repressor which also binds IPTG. This data was obtained by equilibrium dialysis of samples from a crude extract against IPTG at various concentrations. (From Miller, Platt and Weber, *The Lactose Operon*, p. 348.)

flask with Parafilm and tape, and dialyze overnight on a shaker at 4°C.

Day 2

Remove the flasks from the cold, wearing disposable plastic gloves, and remove each sac from its flask with a glass rod. Dry every sac, cut it open, and squeeze the contents into a small (10 × 75 mm) disposable test tube. For each sample determine:

(a) the protein concentration (Biuret determination),
(b) the ^{14}C cpm inside each sac per 100 μl,
(c) the ^{14}C cpm outside each sac per 100 μl.

Using this data, calculate the final IPTG concentrations inside and outside the sac. Subtraction yields the concentration **excess** inside. Divide this by the protein concentration in mg/ml to obtain V, the molar excess of IPTG per mg/ml of protein. The equilibrium concentration of free IPTG is that outside the sac (L).

Plot V/L vs. V. The slope is $-k_{eq}$ (association) and its inverse gives the dissociation constant of IPTG from *lac* repressor, about 1.3×10^{-6} M. Where do small errors in calculation have the largest effect? If you used pure repressor, how many binding sites are there per 150,000 MW species?

Sample Calculation

Starting concentration: 2×10^{-7} M
Starting cpm outside: 1000
Final cpm outside: 800

Final concentration outside (L): $\frac{800}{1000} \times 2 \times 10^{-7}\ \text{M} = 1.6 \times 10^{-7}\ \text{M}$

Final cpm inside: 4000

Final concentration inside: $\frac{4000}{800} \times 1.6 \times 10^{-7}\ \text{M} = 8 \times 10^{-7}\ \text{M}$

Concentration excess: $(8 \times 10^{-7}\ \text{M}) - (1.6 \times 10^{-7}\ \text{M}) = 6.4 \times 10^{-7}\ \text{M}$

Protein concentration: 10 mg/ml

Concentration excess per mg/ml (V): $\frac{6.4 \times 10^{-7}\ \text{M}}{10\ \text{mg/ml}} = 6.4 \times 10^{-8}\ \text{M/mg/ml}$

$$\frac{V}{\text{L}}: \quad \frac{6.4 \times 10^{-8}\ \text{M/mg/ml}}{1.6 \times 10^{-7}\ \text{M}} =$$

$$4 \times 10^{-1}/\text{mg/ml} = 0.4\ \text{ml/mg}$$

Materials

Day 1

10 125-ml erlenmeyer flasks
20 cm #20 dialysis tubing (see Preparation of Materials)
150 μl ^{14}C-IPTG (2×10^{-4} M) (Calbiochem)
1 ml 10^{-3} M IPTG
300 ml TMS assay buffer
disposable plastic gloves
shaker at 4°C

Day 2

disposable plastic gloves
30 scintillation vials and equivalent scintillation fluid
10 small disposable test tubes (10 × 75 mm)
10 100-μl pipettes

Chemicals

^{14}C-IPTG, specific activity 20–30 mc/mM	(Calbiochem)
#20 dialysis tubing	(Union Carbide)

EXPERIMENT 57

Purification of β-Galactosidase

In an uninduced lac^+ *E. coli* cell, *lac* repressor prevents the transcription of *lac* messenger RNA; hence a basal level of only 10 to 20 molecules of β-galactosidase is present. Upon the addition of either lactose or a gratuitous inducer such as IPTG to the growth medium, repression is relieved and the production of β-galactosidase increases 1000-fold, so that the enzyme now constitutes 2% of the protein in the cell. In some strains, β-galactosidase is produced at higher than basal levels even in the absence of inducer. Such strains are called "constitutive," and usually possess mutations, either in the repressor or in the operator, which prevent full repression. Most operator constitutives produce from 1–5% of the full β-galactosidase levels, while an i^- will be 100% constitutive. Intermediate levels may result from a partially functional repressor.

Outline of Purification

A culture of *E. coli* strain CSH36(E7074) which is a full *lac* constitutive, is grown in LB overnight at 37°. (Inoculate from a glucose minimal plate.) The cells are harvested, frozen, and stored in the freezer until ready for usc. If desired, the cells may be grown in glycerol minimal medium to avoid the effects of catabolite repression.

The frozen cells are broken by grinding with alumina and the paste is extracted with buffer. This slurry is centrifuged to remove the alumina and cell debris. Ammonium sulfate is added to the supernatant to a final concentration of 33% saturation, which precipitates most of the β-galactosidase and results in a 5- to 10-fold purification. The ribosomes and 80–90% of the protein remain in the supernatant. The precipitate is redissolved in $\frac{1}{10}$ its initial volume of DEAE column buffer (0.2 M NTM) pH 7.7 and the ammonium sulfate is dialyzed out against the same buffer.

β-Galactosidase elutes from DEAE at a region of the gradient at which few other proteins in the ammonium sulfate fraction elute. Therefore, a DEAE column is particularly suitable for purifying this protein. The sample is applied to a DEAE cellulose column equilibrated with 0.2 M NTM; this is followed by a gradient of 0.20–0.35 M NTM buffer. The β-galactosidase elutes at about 0.25 M NTM and is greater than 95% pure. The activity peak is pooled and concentrated with ammonium sulfate to 33% as before.

The precipitate is redissolved in a small volume of 0.10 M NTM and dialyzed against the same buffer. The dialysate is centrifuged and the supernatant is applied in equal aliquots onto 6 12-ml gradients, which are 5–20% sucrose in 0.10 M NTM buffer. The gradients are run for 10–12 hours at 40,000 rpm, then dripped and assayed. The activity peak is pooled and stored in the freezer.

The enzyme should be more than 99% pure with a recovery of 70–80%, or 30 mg per 10 g of cells. The purity is best judged by electrophoresis on 5% polyacrylamide SDS gels.

References

CRAVEN, G. R., E. STEERS, JR. and C. B. ANFINSEN. 1965. Purification, composition, and molecular weight of the β-galactosidase of *Escherichia coli* K12. *J. Biol. Chem. 240:* 2468.

HU, A. S. L., R. G. WOLFE and F. J. REITHEL. 1959. The preparation and purification of β-galactosidase from *Escherichia coli* ML308. *Arch. Biochem. Biophys. 81:* 500.

KARLSSON, U., S. KOORAJIAN, I. ZABIN, F. S. SJÖSTRAND and A. MILLER. 1964. High resolution electron microscopy on highly purified β-galactosidase from *Escherichia coli*. *J. Ultra. Res. 10:* 457.

Strains

Number	Sex	Genotype	Important Properties
CSH36	F′*lacI proA*$^+$, *B*$^+$	Δ(*lacpro*) *supE thi*	$i^- z^+$

Method

Day 1 **Growth of cells**

Each group inoculates an aliquot from a fresh culture of the *lac* constitutive strain CSH36 into 250 ml of LB in a 1-liter flask. Put on a shaker at 37° overnight.

Day 2

Preparation of crude extract

Centrifuge the cells (20 minutes at 8000 × *g*), weigh and freeze the pellets on parafilm (about 2 grams). **All of the following operations are performed on ice or in a cold room unless otherwise indicated.** After at least an hour at −20°C, put the cells in a prechilled mortar and pestle and grind with 2.5 times the cell weight of alumina for 15–20 minutes. The sample will be very dry at first, gradually developing a smooth sheen and "crackling" as the cells break. Add 5 cell volumes of breaking buffer (BB) and work this in for another 5 minutes. Pour the slurry into a centrifuge tube, and spin out the alumina and cell debris for 10 minutes at 10,000 × *g*. Remove the supernatant, record its volume, and save an aliquot for assay (0.1 ml).

Ammonium sulfate fractionation

At this point, the samples from all the groups are pooled (a total volume of 100 ml for 20 g of cells) and poured into a beaker with a stirring bar, allowing for a 50% volume increase. Crystalline ammonium sulfate is added slowly over a period of 30–45 minutes with gentle stirring using 231 mg $(NH_4)_2SO_4$ per ml of starting solution. This will give a final concentration of 33% saturation.* Add 1 ml 1 M NaOH for every 10 g of ammonium sulfate. Let stirring continue for 1–3 hours, then spin down the precipitate (10 minutes at 10,000 × *g*). Decant and save the supernatant.

Resuspend the pellet in $\frac{1}{10}$ the former volume of **BB**, and remove 0.1 ml for assay. Begin dialyzing the sample against 250 ml of 0.20 M NTM DEAE column buffer (for a 10–20 ml sample) in #20 dialysis tubing. Wear disposable plastic gloves when handling the tubing. The dialysis buffer should be changed 3 times, allowing at least 2 hours for each; one of them may run overnight.

While some groups are doing the ammonium sulfate precipitation, others could be washing and equilibrating the DEAE cellulose with 0.20 M NTM buffer (see Preparation of Materials). Pour the column to be ready at the beginning of Day 3, to obtain a bed volume about equal to the total cell weight.

Activity determination of fractions

Assay the three samples obtained so far in the cell-free β-galactosidase assay. These are (i) the crude extract, (ii) the ammonium sulfate supernatant, and (iii) the redissolved ammonium sulfate pellet. Make a rough trial run first using 10 μl of a 100:1 dilution of each sample in BB. If a strong color develops in less than 5 minutes for any sample, dilute it a further 10:1. Now repeat these assays carefully, using dilutions which yield an OD_{420} of 0.3–0.9 in greater than 10 minutes.

Calculate the specific activity for each sample. If pure β-galactosidase has an activity of 300,000 units/mg, what fraction of the protein in your crude extract is β-galactosidase? How pure is it

* Saturation at 4°C is taken as 70 g per 100 ml of starting solution.

in the ammonium sulfate cut? Does any activity remain in the supernatant? What factor of purification and what yield did you obtain from the ammonium sulfate precipitation?

Day 3

DEAE column

Remove the dialysis sample (wearing disposable plastic gloves), dry the outside of the sac, cut off the end and empty into a beaker. Squeeze out as much as possible from the sac. Centrifuge the dialysate to remove any remaining insoluble material (10 minutes at 10,000 × *g*). Record the volume of the supernatant and remove about 0.2 ml for assay. The pH and conductance of the supernatant must agree with the effluent of the DEAE column and with 0.20 M NTM. Apply the sample to the column with a flow rate of 1–2 column volumes per hour, and wash through with 0.20 M NTM until the effluent OD_{280} is less than 0.10, relative to 0.20 M NTM alone.

Begin a gradient of 0.20 M to 0.35 M NTM, using two column volumes for each half of the gradient and the same flow rate as for the sample application. When the gradient is complete, wash through another 2–3 column volumes of 0.35 M NTM. Monitor the OD_{280} for each fraction sample, and the conductance and β-galactosidase activity for every second or third sample (including those in the flow-through). A column profile is obtained which indicates the protein elution, the conductivity gradient, and the position of β-galactosidase, which should be about 0.25 M NTM. Pool the activity peak, remove an aliquot for assay, and concentrate with ammonium sulfate to 33% saturation as before (231 mg per ml of starting solution). This may be done overnight.

Day 4

Sucrose gradient

Centrifuge the sample at 10,000 × *g* for 10 minutes; decant and save the supernatant (for assay). Resuspend the precipitate in less than 3 ml of 0.10 M NTM and dialyze in #20 dialysis tubing against 500 ml of 0.10 M NTM for 1–2 hours. During this time pour 6 12-ml gradients of 0.10 M NTM in 5% to 20% sucrose. Layer about 0.5–0.8 ml of the sample on each gradient, saving 0.1 ml for assay. The centrifuge run (in a Spinco SW41 rotor) should be about 10–12 hours at 40,000 rpm (200,000 × *g*). When it is convenient, assay: (i) the pooled DEAE peak, (ii) the ammonium sulfate supernatant, and (iii) the sample loaded onto the gradients. What factor of purification has been obtained so far? How good is the recovery of total activity at each step?

Each gradient should be dripped by a different group. Collect about 25 fractions of 0.5 ml each and locate the activity peak by assaying a small aliquot of each fraction. The peak fractions from all 6 gradients are pooled and stored in the cold.

Day 5

SDS-polyacrylamide gels

Each group should assay an aliquot of the pooled sample to determine its specific activity. What is the yield of enzyme, compared to the initial total activity? Run 8 5% SDS-polyacrylamide gels at various dilutions of this sample and the sample before it was applied to the gradient. A good range is to run 5, 20, and 50 μg aliquots for each, and some protein standards on the remaining 2 gels. Stain the gels for 2–4 hours and destain. Is it completely pure? Did the gradient improve the purity?

Buffers

Convenient stock solutions: 1 M Tris-HCl (pH 7.6 at 4°)
1 M NaCl
1 M MgAc

1. BB:

   0.2 M Tris (use 1 M stock, pH 7.6 at 4°)
   0.2 M NaCl
   0.01 M MgAc
   0.01 M 2-mercaptoethanol
   5% glycerol

2. DEAE cellulose column buffer (NTM):
   (a) 0.20 M NaCl
   0.01 M Tris (use 1 M stock, pH 7.6 at 4°)
   0.01 M MgAc
   0.10 M 2-mercaptoethanol
   (b) 0.35 M NaCl instead of 0.20 M NaCl

3. Sucrose gradient buffer (NTM):
   0.10 M NaCl
   0.01 M Tris (use 1 M stock, pH 7.6 at 4°)
   0.01 M MgAc
   0.10 M 2-mercaptoethanol

   Make stocks from this buffer that are 5% and 20% sucrose.

Materials

Day 1

fresh culture of *E. coli* CSH36
250 ml of LB in a 1-liter flask, per group
shaker at 37°

Day 2

200 ml breaking buffer (BB)
30 g $(NH_4)_2SO_4$, Enzyme Grade
β-galactosidase assay solutions
2 liters 0.20 M NTM buffer
20 g DEAE cellulose (DE52)
50–100 ml column with stopcock
Per group:
mortar and pestle
10 g levigated alumina

Day 3 gradient maker (50–100 ml each side)
20 g $(NH_4)_2SO_4$, Enzyme Grade
1 liter 0.20 M NTM and 1 liter 0.35 M NTM buffer

Day 4 10 cm #20 dialysis tubing (Union Carbide)
disposable gloves
1 liter 0.10 M NTM
25 g sucrose
SW41 rotor (or suitable substitute) and 6 centrifuge tubes
gradient maker (6–10 ml each side)

Day 5 β-galactosidase assay solutions
SDS-polyacrylamide gel materials
(buffers, gel tubes, gel apparatus, power supply, stain, and destaining apparatus)

Cell-free β-Galactosidase Assay

Dilute an aliquot of the sample to be assayed at least 100:1 in BB buffer. Add 10–50 μl of this to 1.0 ml of Z Buffer and equilibrate at 28°C. To one tube of Z buffer add an aliquot of BB buffer alone. This is a control for the spontaneous hydrolysis of ONPG. Begin the reaction by adding to each tube 0.2 ml of ONPG at 4 mg/ml, which has also been equilibrated to 28°C. Incubate the samples at 28° (at least 10 minutes for accurate data) until a faint yellow color has developed. Stop the reaction by adding 0.5 ml of 1 M Na_2CO_3, and record the length of time of incubation for each sample.

Read the OD_{420} against the control containing buffer alone. Calculate the total protein in the reaction mixture from the original sample concentration in mg/ml as determined from a Biuret measurement. The specific activity of the β-galactosidase is defined in units/mg, as:

$$\frac{OD_{420} \times 380}{\text{min at } 28^\circ \times \text{mg protein in reaction}}$$

One unit is the amount of enzyme that will hydrolyze 10^{-9} moles/min of ONPG at 28°C. Pure β-galactosidase is usually found to have an activity of about 300,000 units/mg. The number 380 in the above equation is the constant used to convert from the OD_{420} into these units, and depends on the molar extinction coefficient for *o*-nitrophenol, which under these conditions is 4500.

β-galactosidase assay solutions

(a) Z buffer: 0.06 M Na_2HPO_4
0.04 M NaH_2PO_4
0.01 M KCl
0.001 M $MgSO_4$
0.05 M 2-mercaptoethanol
Adjust Z buffer to pH 7.0.

(b) ONPG, 4 mg/ml in distilled H_2O
(c) 1 M Na_2CO_3

Sample calculation: Extract concentration = 10 mg/ml
Assay 10 μl of a 100:1 dilution, thus there is
0.01 ml × 0.1 mg/ml = 0.001 mg protein in reaction
Time = 10.0 minutes
OD_{420} = 0.400

$$\text{Therefore, the S.A.} = \frac{0.400 \times 380}{10.0 \times 0.001 \text{ mg}} = 15{,}200 \text{ units/mg}$$

This sample is 5% β-galactosidase, if pure enzyme is 300,000 units/mg.

UNIT IX

CYCLIC AMP, CATABOLITE REPRESSION, AND CELL-FREE ENZYME SYNTHESIS

INTRODUCTION TO UNIT IX

Cyclic AMP Regulation of Catabolite Operons

Recent studies have shown that a **positive** control, separate from the **negative** repressor-operator control, exists for the *lac* operon in *E. coli* (Zubay et al., 1970; Emmer et al., 1970). The new control system, mediated by cyclic AMP, relates the expression of the *lac* operon to the control of other inducible operons. The *lac* operon, together with other "catabolite-sensitive" operons, requires the presence of cyclic AMP for normal function. Cyclic AMP interacts with a protein factor in the cell called the CAP* factor (termed CRP by some authors) (Zubay et al., 1970; Emmer et al., 1970). This complex activates the promoters of the catabolite operons and stimulates transcription (Perlman et al., 1970; Arditti et al., 1970). Although the exact manner in which this is accomplished is not fully understood, recent evidence suggests that the cyclic AMP-CAP factor complex binds specifically to DNA and perhaps to the catabolite-sensitive promoters themselves (Riggs et al., 1971; Pastan et al., 1971). Experiment 59 describes the isolation of mutants in the cyclic AMP regulation system.

Catabolite Repression

The concentration of cyclic AMP inside the cell is strictly regulated. Cells grown in the presence of certain catabolites, or in the absence of a proper nitrogen source, have lower internal levels of cyclic AMP than normal (Makman and Sutherland, 1965). As a result the derepressed levels of the *lac* enzymes are lower. This phenomenon is called "catabolite repression" (for reviews see Magasanik, 1970; Contesse et al., 1970). The addition of cyclic AMP to the medium can

* CAP is an abbreviation for "catabolite gene activating protein"; CRP for catabolite regulatory protein."

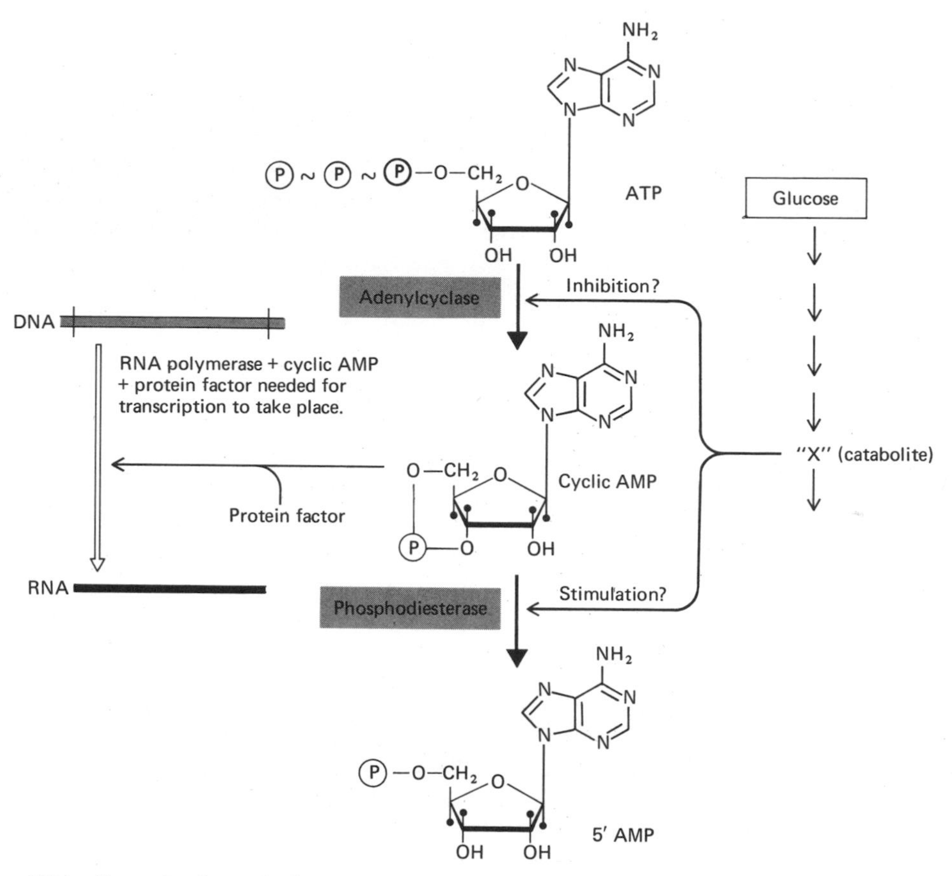

Figure IXA. Control of catabolite-sensitive transcription through cyclic AMP. (Reprinted with permission from J. D. Watson, *Molecular Biology of the Gene*, 2nd Ed., W. A. Benjamin, Inc., New York, 1970.)

relieve catabolite repression (Experiment 58) (Perlman and Pastan, 1968; Ullmann and Monod, 1968). As expected, cells which have lost the ability to make cyclic AMP (due to an alteration in the enzyme adenyl cyclase) cannot synthesize β-galactosidase unless cyclic AMP is supplied externally (Perlman and Pastan, 1969) (Figure IXA).

Cell-free Systems

The development of *in vitro* systems capable of synthesizing specific proteins from an added DNA template allows the direct study of factors involved in the regulation of gene expression, at both the transcription and translation level.

Nirenberg and his collaborators first demonstrated an RNA-dependent cell-free system which synthesized specific polypeptide products (Nirenberg and Matthaei, 1961). RNA extracted from cells (Nirenberg and Matthaei, 1961), synthetic poly ribonucleotides (Matthaei et al., 1962; Lengyel et al., 1962), and viral RNA (Nirenberg and Matthaei, 1961; Ofengand and Haselkorn, 1962) were all successfully used to program these *in vitro* systems.

Cell-free synthesis of a specific protein was first clearly demonstrated by Zinder and his coworkers, who used RNA isolated from the single-stranded RNA phage f2 to direct the incorporation of labeled amino acids into protein. Analysis of tryptic digests of the polypeptide products showed that at least part of the protein synthesized was identical to coat protein (Nathans et al., 1962).

The refinement of these systems was evident several years later when Salser, Gesteland, and Bolle described the synthesis of phage T4 lysozyme with enzymatic activity using a cell-free system programmed with RNA extracted from phage-infected cells (Salser et al., 1967). That this represented *de novo* synthesis was proved with the use of a phage mutant carrying an amber mutation in the structural gene for T4 lysozyme. RNA taken from infected Su^- cells did not stimulate lysozyme synthesis when extracts from Su^- cells were used to support the cell-free system. However, when extracts from an Su^+ strain were employed, lysozyme synthesis was detected (Gesteland et al., 1967). Subsequent to this study the *in vitro* synthesis of alkaline phosphatase monomers directed by *E. coli* mRNA was described (Dohan et al., 1969).

The first verified *in vitro* synthesis of a protein in a coupled system (DNA → RNA; RNA → protein) dependent on added DNA was demonstrated by Zubay and his collaborators (DeVries and Zubay, 1967, 1969; Zubay and Chambers, 1969). They were able to synthesize β-galactosidase in a system completely dependent on added φ80d*lac* DNA. The DNA-dependent synthesis of other enzymes was achieved shortly thereafter by several groups. The cell-free synthesis of T4 α- and β-glucosyl transferase (Schweiger and Gold, 1969a; Gold and Schweiger, 1969; Young and Tissière, 1969) and of T4 lysozyme (Schweiger and Gold, 1969a,b) was reported.

A large number of enzymes have now been synthesized *in vitro*, for example, T7 RNA polymerase (Gelfand et al., 1970; Herrlich and Schweiger, 1971), *E. coli* L-ribulokinase (Zubay et al., 1971; Greenblatt and Schleif, 1971), and *E. coli* galactokinase (Parks et al., 1971; Wetekam et al., 1971).

Uses of Cell-free Systems

In the few years since the first *in vitro* synthesis of an active enzyme in a coupled system, several important uses of cell-free systems have emerged.

The isolation of positive control factors has been achieved both for catabolite-sensitive operons in general (see above discussion of CAP) and for the arabinose enzymes in particular. In the latter case the *araC* gene product, which had previously been shown by genetic studies to mediate positive control of the *ara* operon (Sheppard and Englesberg, 1966), was isolated after an *in vitro* system for the synthesis of one of the arabinose enzymes was developed and shown to be dependent on added *araC*$^+$ extracts (Greenblatt and Schleif, 1971).

Cell-free systems also make possible the isolation of repressors involved in the negative control of gene expression. To facilitate the isolation of the *trp* repressor, a strain was used in which the *lac* operon was fused to the *trp* operon so that β-galactosidase was synthesized under *trp* control. This allowed the use of the more convenient β-galactosidase assay. The *trp* repressor was then detected by its ability to repress the cell-free synthesis of β-galactosidase (Zubay et al., 1972).

In addition to greatly facilitating the isolation of specific factors, cell-free systems allow the direct study of regulation of gene expression. For example, these studies have shown that cyclic AMP is necessary for the expression of the *lac*, *ara*, and *gal* operons (Zubay et al., 1970; Emmer et al., 1970; Parks et al., 1971; Zubay et al., 1971; Greenblatt and Schleif, 1971), enabling the confirmation of *in vivo* results. Similarly, *in vitro* systems have confirmed that induction and repression of the *ara* operon is the result of *araC* protein action (Greenblatt and Schleif, 1971).

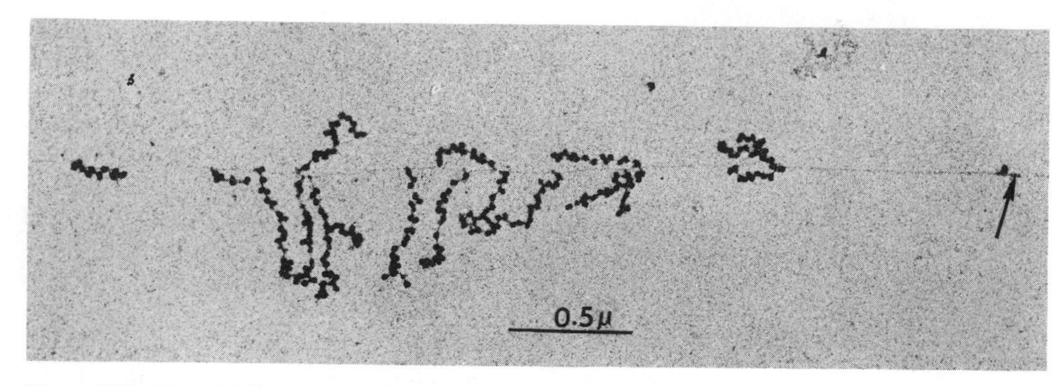

Figure IXB. Genetically active portion of the *E. coli* genome. The arrow indicates a putative RNA polymerase molecule on or very near the initiation site of the active locus. (Reproduced with permission from Miller et al., 1970.)

Because of the increasing use of *in vitro* systems to study protein synthesis and its regulation, we have included a detailed description of such a system (Experiment 60). The effect of cyclic AMP on the *in vitro* synthesis of β-galactosidase is measured in Experiment 61, as is the effect of partially purified *lac* repressor (Experiment 62). The *lac* repressor is obtained from a previous exercise (Experiment 55).

References

ARDITTI, R., L. ERON, G. ZUBAY, G. TOCCHINI-VALENTINI, S. CONNAWAY and J. BECKWITH. 1970. *In vitro* transcription of the *lac* operon genes. *Cold Spring Harbor Symp. Quant. Biol. 35:* 437.

CONTESSE, G., M. CRÉPIN, F. GROS, A. ULLMANN and J. MONOD. 1970. On the mechanism of catabolite repression. *The Lactose Operon*, p. 401. Cold Spring Harbor Laboratory.

DEVRIES, J. K. and G. ZUBAY. 1967. DNA-directed peptide synthesis. II. The synthesis of the α-fragment of the enzyme β-galactosidase. *Proc. Nat. Acad. Sci. 57:* 1010.

DEVRIES, J. K. and G. ZUBAY. 1969. Characterization of a β-galactosidase formed between a complementary protein and a peptide synthesized *de novo*. *J. Bacteriol. 97:* 1419.

DOHAN, F. C., JR., R. H. RUBMAN and A. TORRIANI. 1969. In vitro synthesis of alkaline phosphatase monomers directed by *E. coli* messenger. *Cold Spring Harbor Symp. Quant. Biol. 34:* 768.

EMMER, M., B. DE CROMBRUGGHE, I. PASTAN and R. PERLMAN. 1970. The cyclic AMP receptor protein of *E. coli:* Its role in the synthesis of inducible enzymes. *Proc. Nat. Acad. Sci. 66:* 480.

GELFAND, D. H. and M. HAYASHI. 1970. *In vitro* synthesis of a DNA dependent RNA polymerase coded on coliphage T7 genome. *Nature 228:* 1162.

GESTELAND, R. F., W. SALSER and A. BOLLE. 1967. *In vitro* synthesis of T4 lysozyme by suppression of amber mutations. *Proc. Nat. Acad. Sci. 58:* 2036.

GOLD, L. M. and M. SCHWEIGER. 1969. Synthesis of phage specific α- and β-glucosyl transferases directed by T even DNA *in vitro*. *Proc. Nat. Acad. Sci. 62:* 89.

GREENBLATT, J. and R. SCHLEIF. 1971. Arabinose C protein: Regulation of the arabinose operon *in vitro*. *Nature New Biol. 233:* 166.

HERRLICH, P. and M. SCHWEIGER. 1971. RNA polymerase synthesis *in vitro* directed by T7 Phage DNA. *Mol. Gen. Genet. 110:* 31.

LENGYEL, P., J. F. SPEYER, C. BASILIO and S. OCHOA. 1962. Synthetic polynucleotides and the amino acid code. *Proc. Nat. Acad. Sci. 48:* 282.

MAGASANIK, B. 1970. Glucose effects: inducer exclusion and repression. *The Lactose Operon*, p. 189. Cold Spring Harbor Laboratory.

MAKMAN, R. S. and E. W. SUTHERLAND. 1965. Adenosine 3′,5′-phosphate in *E. coli*. *J. Biol. Chem. 240:* 1309.

MATTHAEI, J. H., O. W. JONES, R. G. MARTIN and M. W. NIRENBERG. 1962. Characteristics and composition of RNA coding units. *Proc. Nat. Acad. Sci. 48:* 666.

MILLER, O. L., JR., B. R. BEATTY, B. HAMKALO and C. A. THOMAS, JR. 1970. Electron microscope visualization of transcription. *Cold Spring Harbor Symp. Quant. Biol. 35:* 505.

NATHANS, D., G. NOTANI, J. H. SCHWARTZ and N. D. ZINDER. 1962. Biosynthesis of the coat protein of coliphage *f*2 by *E. coli* extracts. *Proc. Nat. Acad. Sci. 48:* 1424.

Nirenberg, M. W. and H. Matthaei. 1961. The dependence of cell-free protein synthesis in *E. coli* upon naturally occurring or synthetic polyribonucleotides. *Proc. Nat. Acad. Sci. 47:* 1588.

Ofengand, J. and R. Haselkorn. 1962. Viral RNA-dependent incorporation of amino acids into protein by cell-free extracts of *E. coli. Biochem. Biophys. Res. Commun. 6:* 469.

Parks, J. S., M. Gottesman, R. L. Perlman and I. Pastan. 1971. Regulation of galactokinase synthesis by cyclic adenosine 3′-5′-monophosphate in cell-free extracts of *E. coli. J. Biol. Chem. 246:* 2419.

Pastan, I., B. deCrombrugghe, B. Chen, W. Anderson, J. Parks, P. Nissley, M. Straub, M. Gottesman and R. Perlman. 1971. *Proc. Miami Winter Symp.* (North Holland, Amsterdam).

Perlman, R., B. Chen, B. deCrombrugghe, M. Emmer, M. Gottesman, H. Varmus and I. Pastan. 1970. The regulation of *lac* operon transcription by cyclic adenosine 3′,5′-monophosphate. *Cold Spring Harbor Symp. Quant. Biol. 35:* 419.

Perlman, R. and I. Pastan. 1968. Regulation of β-galactosidase synthesis in *E. coli* by cyclic adenosine 3′,5′-monophosphate. *J. Biol. Chem. 243:* 5420.

Perlman, R. and I. Pastan. 1969. Pleiotropic deficiency of carbohydrate utilization in an adenyl cyclase deficient mutant of *E. coli. Biochem. Biophys. Res. Commun. 37:* 151.

Riggs, A. D., G. Reiness and G. Zubay. 1971. Purification and DNA-binding properties of the catabolite gene activator protein. *Proc. Nat. Acad. Sci. 68:* 1222.

Salser, W., R. F. Gesteland and A. Bolle. 1967. *In vitro* synthesis of bacteriophage lysozyme. *Nature 215:* 588.

Schweiger, M. and L. M. Gold. 1969a. DNA-dependent *in vitro* synthesis of bacteriophage enzymes. *Cold Spring Harbor Symp. Quant. Biol. 34:* 763.

Schweiger, M. and L. M. Gold. 1969b. Bacteriophage T4-DNA-dependent *in vitro* synthesis of lysozyme. *Proc. Nat. Acad. Sci. 63:* 1351.

Sheppard, D. and E. Englesberg. 1966. Positive control in the L-arabinose gene-enzyme complex of *E. coli* B/r as exhibited with stable merodiploids. *Cold Spring Harbor Symp. Quant. Biol. 31:* 345.

Ullmann, A. and J. Monod. 1968. Cyclic AMP as an antagonist of catabolite repression in *E. coli. Fed. Europ. Biol. Soc. Letters 2:* 57.

Young, E. T. and A. Tissière. 1969. *In vitro synthesis* of T4 glucosyl transferase. *Cold Spring Harbor Symp. Quant. Biol. 34:* 766.

Wetekam, W., K. Staack and R. Ehring. 1971. DNA-dependent *in vitro* synthesis of enzymes of the galactose operon of *E. coli. Mol. Gen. Genet. 112:* 14.

Zubay, G. and D. A. Chambers. 1969. A DNA-directed cell-free system for β-galactosidase synthesis. *Cold Spring Harbor Symp. Quant. Biol. 34:* 753.

Zubay, G., D. Schwartz and J. Beckwith. 1970. The mechanism of action of catabolite sensitive genes. *Cold Spring Harbor Symp. Quant. Biol. 35:* 433.

Zubay, G., L. Gielow and E. Englesberg. 1971. Cell-free studies on the regulation of the arabinose operon. *Nature New Biol. 233:* 164.

Zubay, G., D. E. Morse, W. J. Schrenk and J. H. Miller. 1972. Detection and isolation of the repressor protein for the tryptophan operon of *E. coli. Proc. Nat. Acad. Sci.* In press.

EXPERIMENT 58

Cyclic AMP and Catabolite Repression

In the following experiment glucose-6-phosphate is used as the catabolite source. Cells are grown in glycerol and also in glucose-6-phosphate. The cultures are then assayed for β-galactosidase. Those cultures grown up in the presence of glucose-6-phosphate should have lower levels of β-galactosidase due to catabolite repression. Cultures with added cyclic AMP are also assayed, in order to test whether the cyclic AMP can overcome the catabolite repression and restore the lowered β-galactosidase levels to normal (Perlman et al.).

References

PERLMAN, R. L., B. DECROMBRUGGHE and I. PASTAN. 1969. Cyclic AMP regulates catabolite and transient repression in *E. coli*. *Nature 223:* 810.

SILVERSTONE, A. E., R. R. ARDITTI and B. MAGASANIK. 1970. Catabolite-insensitive revertants of *lac* promoter mutants. *Proc. Nat. Acad. Sci. 66:* 773.—This and the following paper provide genetic evidence that the *lac* promoter is the site at which catabolite repression is mediated.

SILVERSTONE, A. E., B. MAGASANIK, W. S. REZNIKOFF, J. H. MILLER and J. R. BECKWITH. 1969. Catabolite-sensitive site of the *lac* operon. *Nature 221:* 1012.

Strains

Number	Sex	Genotype
CSH51	F^-	*ara* Δ(*lacpro*) *strA thi* (φ80d*lac*$^+$)

Method

Day 1

Subculture an overnight of CSH51 into 5 ml of 1 × A medium (supplemented with B1, proline, and $MgSO_4$) + glucose-6-phosphate, and also into 5 ml of 1 × A + glycerol (0.5%). Also subculture 1 drop of CSH51 into these media + 10^{-3} M IPTG.

Day 2

Part A. Catabolite repression

Subculture the cells grown in glycerol + IPTG into fresh medium of the same type (glycerol + IPTG), and do likewise for the culture grown in glucose-6-phosphate + IPTG. Put 3 drops into 5 ml and allow to grow with aeration at 37°. When the cultures are about 2–4 × 10^8 cells/ml (OD_{600} mμ approximately 0.3–0.7) assay for β-galactosidase. Take duplicate samples. Use 0.1 ml cells plus 0.9 ml Z buffer. Compare the induced levels of β-galactosidase in CSH51 cells grown in the two different media.

Part B. Cyclic AMP

Since cyclic AMP can be degraded by the cell, its effect must be measured relatively soon after its addition to the media. Therefore, we will measure the effect of catabolite repression on the differential rate of synthesis of β-galactosidase in growing cultures. IPTG will be added to cells growing in glycerol or glucose-6-phosphate, and the level of β-galactosidase will be measured every 10 minutes for 1 hour in the presence and absence of cyclic AMP. The **increase in β-galactosidase/increase in cell density** will be computed.

For this part of the experiment, use the overnights of CSH51 grown in the absence of IPTG. Subculture each of the two cultures by putting 4 drops into 10 ml of the same medium. Aerate each culture at 37° until a density of about 2 × 10^8 cells/ml (OD_{600} mμ about 0.28) is reached. Add IPTG to a concentration of 10^{-3} M to each tube. Immediately divide each culture into two parts. To one of the glycerol cultures and to one of the glucose-6-phosphate cultures, add cyclic AMP to a final concentration of 3 × 10^{-2} M. Continue to aerate all cultures at 37°. Every 10 minutes take an aliquot from each of the 4 cultures (0.1 ml) and assay for β-galactosidase by adding this to 0.9 ml Z buffer + toluene and immediately disrupting the cells by vortexing. Also measure the turbidity of the culture, either by optical density or klett units. We suggest withdrawing 0.5 ml of the culture and adding it to 0.5 ml of 1 × A medium (stored in an ice bucket) and immediately measuring the optical density.

Take points at 10, 20, 30, 40, 50, and 60 minutes after the addition of IPTG.

Materials

Day 1

overnight culture of CSH51
5 ml of 1 × A supplemented with: B1 (2.5 μg/ml), proline (80 μg/ml), $MgSO_4$ (10^{-3} M), and glycerol (0.5%)
5 ml of same medium + 10^{-3} M IPTG
5 ml of same medium with no IPTG and glucose-6-phosphate (0.2%) instead of glycerol
5 ml of same medium with glucose-6-phosphate and 10^{-3} M IPTG
pasteur pipette

Day 2

Part A

5 ml of each of the 2 types of media used in Day 1 (with IPTG)
2 pasteur pipettes, 2 0.1-ml, and 4 1-ml pipettes
4 test tubes with 0.9 ml Z buffer and 1 drop toluene each
1 ml ONPG (4 mg/ml)
2 ml 1 M Na_2CO_3
waterbath at 28°
spectrophotometer

Z buffer for β-galactosidase assays contains per liter:

$Na_2HPO_4 \cdot 7H_2O$	0.06 M	16.1 g
$NaH_2PO_4 \cdot H_2O$	0.04 M	5.5 g
KCl	0.01 M	0.75 g
$MgSO_4 \cdot 7H_2O$	0.001 M	0.246 g
β-mercaptoethanol	0.05 M	2.7 ml

Do not autoclave! Adjust pH to 7.0.

Part B

10 ml of each of the 2 types of media used in Day 1 (without IPTG)
0.2 ml of 0.1 M IPTG (in H_2O)
cyclic AMP: 1 ml of 0.33 M stock solution in H_2O (0.5 ml of 0.33 M stock solution added to 5 ml of culture gives a final concentration of cyclic AMP of 3×10^{-2} M)
26 0.1-ml, 27 1-ml, and 2 pasteur pipettes
spectrophotometer
ice bucket
waterbath at 28°
24 test tubes with 0.9 ml Z buffer + 1 drop toluene each
24 test tubes with 0.5 ml 1 × A buffer each
5 ml ONPG (4 mg/ml)
12 ml 1 M Na_2CO_3

EXPERIMENT 59

Isolation of Mutants Missing a Factor Necessary for the Expression of Catabolite-sensitive Operons

Cyclic AMP is required in order to obtain maximal expression of the catabolite-sensitive operons. Single-step mutants unable to express all of these operons have been isolated (Perlman and Pastan, 1969; Schwartz and Beckwith, 1970; Perlman et al., 1970). These fall into two categories. One class does not respond to externally added cyclic AMP, whereas cyclic AMP reverses the effects of the mutation in the other class. It has been shown that one group of these mutants lacks the enzyme adenyl cyclase and therefore cannot synthesize cyclic AMP. The inability to grow on lactose, glycerol, maltose, arabinose, and rhamnose is overcome by the addition of 5×10^{-3} M cyclic AMP to the medium.

The second group of mutants lacks a protein factor necessary for the specific stimulation of the catabolite-sensitive operons. This CAP factor (or CRP factor) has been isolated and shown to stimulate the synthesis of β-galactosidase in a cell-free system in the presence of, but not in the absence of, cyclic AMP (Zubay et al., 1970; Emmer et al., 1970). In addition, the transcription of the *lac* operon *in vitro* has been shown to be dependent on CAP and cyclic AMP (de Crombrugghe et al., 1971; Arditti et al., 1970; Perlman et al., 1970). The *cap* (or *crp*) locus maps at approximately 64 min on the *E. coli* chromosome (Perlman et al., 1970; Epstein and Kim, 1971; see Figure 59A).

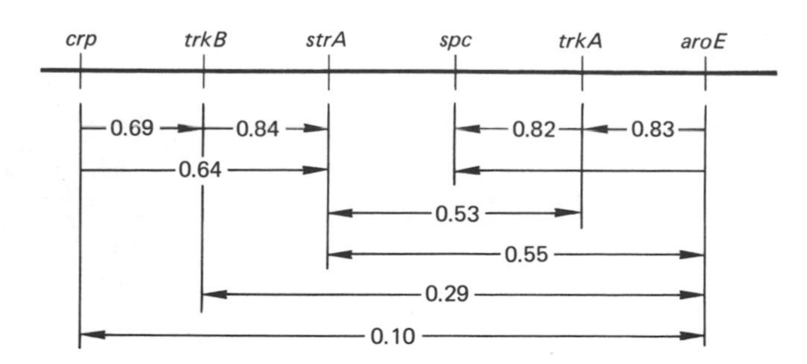

Figure 59A. Detailed map of the *cap* (*crp*) region of *E. coli*. The numbers represent P1 co-transduction frequencies between markers. Arrows point from selected marker to unselected marker; double-headed arrows indicate the average of co-transduction frequencies measured in both directions. Distances indicated at the bottom are arbitrary and not drawn to scale. The *trkA* and *trkB* loci are involved in potassium transport. (Reproduced with permission from Epstein and Kim, 1971.)

In the following experiment we will mutagenize CSH62 ($lac^+ ara^+ mal^+$) and examine colonies on tetrazolium plates containing both maltose and arabinose. Red colonies will be tested for growth on lactose, rhamnose, glycerol, and glucose. Single-step mutants able to grow on glucose minimal medium, but not on maltose, arabinose, rhamnose, glycerol, and lactose, are probably lacking one of the two above-mentioned proteins. We will then test for cyclic AMP restoration of the ability to grow on lactose.

References

ARDITTI, R., L. ERON, G. ZUBAY, G. TOCCHINI-VALENTINI, S. CONNAWAY and J. BECKWITH. 1970. *In vitro* transcription of the *lac* operon genes. *Cold Spring Harbor Symp. Quant. Biol. 35:* 437.

DECROMBRUGGHE, B., B. CHEN, W. ANDERSON, P. NISSLEY, M. GOTTESMAN, I. PASTAN and R. PERLMAN. 1971. Essential elements for controlled *lac* transcription. *Nature New Biol. 231:* 139.

EMMER, M., B. DECROMBRUGGHE, I. PASTAN and R. PERLMAN. 1970. Cyclic AMP receptor protein of *E. coli:* Its role in the synthesis of inducible enzymes. *Proc. Nat. Acad. Sci. 66:* 480.

EPSTEIN, W. and B. S. KIM. 1971. Potassium transport loci in *E. coli* K12. *J. Bacteriol. 108:* 639.

PERLMAN, R. L. and I. PASTAN. 1969. Pleiotropic deficiency of carbohydrate utilization in an adenyl cyclase deficient mutant of *Escherichia coli. Biochem. Biophys. Res. Commun. 37:* 151.

PERLMAN, R. L., B. CHEN, B. DECROMBRUGGHE, M. EMMER, M. GOTTESMAN, H. VARMUS and I. PASTAN. 1970. The regulation of *lac* operon transcription by cyclic adenosine 3′,5′-monophosphate. *Cold Spring Harbor Symp. Quant. Biol. 35:* 419.

SCHWARTZ, D. S. and J. R. BECKWITH. 1970. Mutants missing a factor necessary for the expression of catabolite-sensitive operons in *E. coli. The Lactose Operon*, p. 417. Cold Spring Harbor Laboratory.

ZUBAY, G., D. SCHWARTZ and J. BECKWITH. 1970. Mechanism of activation of catabolite-sensitive genes: A positive control system. *Proc. Nat. Acad. Sci. 66:* 104.

Strains

Number	Sex	Genotype	Important Properties
CSH62	HfrH	*thi*	Lac^+ Str^s

Method

Day 1 Mutagenize CSH62 with nitrosoguanidine, exactly as in Unit III. Grow up overnight at 37° in LB broth.

Day 2 Plate out dilutions of the mutagenized cultures on tetrazolium plates containing both arabinose and maltose. Aim for 500 colonies per plate. Use 10 plates at this dilution. Incubate at 37°.

Day 3 Examine the plates for red colonies. Pick and purify onto the same type of plate.

Day 4 Test the colonies for the ability to grow on minimal plates containing lactose. Also test on minimal plates with the following sugars: arabinose, maltose, glycerol, rhamnose, glucose.

Day 5, 6 Observe the test plates. Strains which are Ara^-, Rha^-, Mal^-, Lac^-, Gly^-, but can grow on glucose minimal plates are candidates for mutants missing a factor necessary for expression of the catabolite-sensitive operons.

We can test for the ability to grow on lactose minimal plates in the presence of cyclic AMP. Spread 0.2 ml of a fresh 0.05 M solution of cyclic AMP in water onto the surface of a lactose minimal plate. Streak the colonies to be tested for growth on this plate. Mutants lacking adenyl cyclase activity will become Lac^+ on lactose minimal plates with cyclic AMP.

Day 6, 7 **Optional Experiments**

Examine the plates and determine which strains respond to cyclic AMP. Those which did not may be lacking the CAP factor. Use any of the following steps to demonstrate that the mutation is a single mutation and also outside the respective operons affected.

1. Cross the *lac* region into an F^- Δ(*lacpro*)*strA* strain (CSH50) by selecting for Pro^+ and examine the recombinants for Lac^+. If a 30-minute interrupted mating is done, neither the gene for CAP nor for adenyl cyclase can be transferred. If the Pro^+ recombinants are Lac^+, this shows that an outside mutation in the CSH62 background caused the Lac^- phenotype.

2. Isolate revertants on one of the sugars and score for the reversion to the ability to grow on the other sugars. If the mutation is a single lesion, than some of these revertants should regain the ability to grow on the other sugars.

3. Spot lysates of a λh80d*lac* phage (Experiment 43) on both the starting strain and the mutants. A CAP^- mutant should not be complemented by the introduction of a second *lac* operon. Spread a few drops of the strain to be tested over the surface of a lactose minimal plate and then apply a drop of a lysate of the phage. Incubate at 30–34°.

Materials

Day 1

exponential culture of CSH62
10 ml 0.1 M citrate buffer, pH 5.5
10 ml 0.1 M phosphate buffer, pH 7.0
nitrosoguanidine, freshly prepared stock solution at 1 mg/ml
5 ml LB broth
2 small centrifuge tubes
desk-top centrifuge
1 1-ml and 5 5-ml or 10-ml pipettes
waterbath at 37°
1 test tube

Day 2

10 tetrazolium plates containing both arabinose and maltose (50 ml of a 20% stock solution of each sugar, added after autoclaving)
3 dilution tubes
5 pasteur pipettes

Day 3

2 tetrazolium plates with arabinose and maltose
12 round wooden toothpicks

Day 4

1 glucose minimal plate
1 lactose minimal plate
1 arabinose minimal plate
1 maltose minimal plate
1 glycerol minimal plate
1 rhamnose minimal plate
12 round wooden toothpicks

Day 5

2 lactose minimal plates
freshly prepared 0.05 M stock solution of cyclic AMP in H_2O (This solution is unstable and should be stored in the freezer. A neutralized solution of cyclic AMP in H_2O or 0.01 M Tris is stable for months if kept frozen.)
1 pasteur pipette

EXPERIMENT 60

Cell-free Synthesis of β-Galactosidase

Recently, an *in vitro* protein synthesizing system capable of synthesizing enzymatically active β-galactosidase has been described (Zubay et al.). Either ϕ80d*lac* DNA or any other transducing phage DNA containing the *lac* operon is used as the template. The composition of the system is shown in Table 60A. It contains various salts, all 20 amino acids, all 4 ribonucleoside triphosphates, an energy generating system, vitamins, transfer RNA, and a crude cell extract containing ribosomes, RNA polymerase, and other enzymes necessary for transcription and translation. The crude extract is prepared from a strain containing a chromosomal *lac* deletion so that no β-galactosidase is present prior to the *in vitro* synthesis. β-Galactosidase is assayed after one hour of protein synthesis at 37°. At least with some *lac* DNA templates, the system responds appropriately to the presence of *lac* repressor, CAP factor, 3′, 5′-cyclic AMP and IPTG. The strain used here for crude extract preparation contains CAP factor, but no *lac* repressor. This cell-free system is very complex and sensitive and careful attention to the described procedures is essential if positive results are to be obtained. It is of general use for many *E. coli* enzymes and has been employed to synthesize enzymes of the *gal* and *ara* operons, among others.

Table 60A. Composition of Cell-free System

Component	Amount per ml incubation mixture
Tris acetate, pH 8.2	44 μmoles
Dithiothreitol	1.4 μmoles
Potassium acetate	55 μmoles
Twenty amino acids	0.22 μmole each
CTP, GTP, UTP	0.55 μmole each
ATP	2.2 μmoles
Trisodium phosphoenol pyruvate	21 μmoles
Ammonium acetate	27 μmoles
3′,5′-cyclic AMP	1.0 μmole
tRNA	100 μg
Pyridoxine HCl	27 μg
Triphosphopyridine nucleotide	27 μg
Flavine adenine dinucleotide	27 μg
Folinic acid	27 μg
p-Aminobenzoic acid	11 μg
Magnesium acetate	14.7 μmoles
Calcium chloride	7.4 μmoles
ϕ80d*lac* DNA	50 μg
S-30	6500 μg protein

Table 60B. Composition of Reaction Mixture

Component	Amount per ml reaction mixture
2 M Tris acetate, pH 8.2	102 μl
5 M Potassium acetate	38 μl
0.1 M Dithiothreitol	68 μl
2 M Ammonium acetate	67 μl
1 M Magnesium acetate	51 μl
Mixture 50 mM in each of 20 amino acids	22 μl
5 mg/ml *E. coli* tRNA (OD_{260} = 100)	100 μl
0.27% Flavine adenine dinucleotide	50 μl
0.27% Triphosphopyridine nucleotide	50 μl
0.27% Pyridoxine-HCl	50 μl
0.11% *p*-Aminobenzoic acid	50 μl
0.27% Folinic acid	50 μl
100 mM UTP, GTP, CTP in 0.01 M Tris-Cl, neutralized with KOH to pH 7.0	27 μl each
100 mM ATP	110 μl
Na_3PEP	34 mg
1 M Calcium chloride	37 μl
Distilled water	95 μl

References

NIRENBERG, M. W. 1963. Cell-free protein synthesis directed by messenger RNA. *Methods in Enzymology*, Vol. VI, p. 17. Academic Press, New York.
ZUBAY, G., D. A. CHAMBERS and L. C. CHEONG. 1970. Cell-free studies on the regulation of the *lac* operon. *The Lactose Operon*, p. 375. Cold Spring Harbor Laboratory.

Strains

Number	Sex	Genotype
CSH44	F^-	*tonA* Δ(*lac*) (λh80*C*I857*St*68) *thi* (λh80*C*I857*St*68d*lac*$^+$)
CSH73	HfrH	Δ*lac* Δ(*ara-leu*) *thi*

Method

Day 1

Growth of cells: For a small scale preparation, cells can be grown and a cell-free extract prepared with minor modifications by the method of Nirenberg. Bacteria for the crude extract are grown in a medium containing, per liter distilled water, 8 g Nutrient Broth and 40 ml of 20% glucose (added after autoclaving). Inoculate each liter of medium with 20 ml of a fresh overnight culture of strain CSH73. Shake the cells at 34°C until they attain an OD_{550} = 1.0 (approximately 5 × 10^8 cells/ml), and then chill by immersing the flask in an ice bath.

Harvest the cells by centrifuging at 6000 rpm for 20 minutes in a Sorvall GSA rotor. Resuspend the pellets from 1 liter of culture in 10 ml of buffer A (10 mM Tris-acetate, pH 7.8, 14 mM magnesium acetate, 60 mM potassium acetate, 0.1 mM dithiothreitol) and centrifuge at 15,000 rpm for 15 minutes using a type 40 rotor in a Spinco Model L ultracentrifuge.

Rewash the pellets with one-half as much buffer A, using the same centrifuging procedure, and then store at −70°C in a deep freeze or use immediately to prepare a crude extract. The yield of cells is about 1 g of wet-packed pellet per liter of growth medium. If a large amount of cell-free extract is to be prepared by one person, it is preferable to use the cell growth method described by Zubay et al. which yields about 5 g of cell pellet per liter of growth medium.

Day 2

Preparation of crude extract: Ordinary glassware may be used for all procedures. Weigh the cell pellet into an unglazed mortar and grind with twice its weight of levigated alumina. When the mixture has become pasty and viscous and a "popping" sound is heard, take up the extract in 1.5 ml of buffer A per gram of cells. Do not add any deoxyribonuclease to the lysate! Centrifuge the extract at 17,000 rpm for 30 minutes in a Spinco type 40 rotor, decant the supernatant and measure its volume. To each ml of supernatant add:

100 μl of 1 M Tris-acetate pH 7.8
20 μl of 0.14 M magnesium acetate
40 μl of 20 mM ATP (neutralized with KOH)
120 μl of 75 mM Na_3PEP (freshly dissolved)
10 μl of 0.1 M dithiothreitol (Cleland's reagent)
20 μl of a solution 500 μM in each of the 20 amino acids
5 μg pyruvate kinase.

Incubate the mixture at 37°C for 80 minutes in a stoppered light-protected vessel, and then dialyze for 8 hours at 4°C against two changes of 50 volumes each of buffer A. Measure the protein concentration of the extract by the biuret procedure, using bovine serum albumin as a standard, and adjust with buffer A to about 22 mg/ml. The extract should be quick-frozen in small portions (about 300 μl each) and stored at −70°C. An appropriate amount of this extract is thawed at 4°C immediately before use. The final yield is 40 mg protein per original liter of culture, sufficient for 6 ml of *in vitro* synthesis reaction. The extracts remain active for months at −70°C.

Day 3

DNA Preparation

Growth of phage: The template λh80d*lac* DNA is obtained from strain CSH44, doubly lysogenic for heat inducible and lysis defective derivatives of the phages λh80 and λh80d*lac*. The cells are grown in medium containing per liter of distilled water: 16 g Difco Bactotryptone, 10 g yeast extract, and 5 g NaCl. A single Lac^+ and temperature-sensitive colony is used to start a 100-ml overnight culture at 33°C. The 100-ml culture is used to inoculate a 5-liter culture for further growth at 33°C. When the OD_{550} reaches 1.0 (about 2 hours), raise the temperature to 43°C for 15 minutes and then continue growth at 35°C for another 4 hours. The most vigorous possible aeration should be maintained throughout and the cells should become greatly elongated.

The culture may be harvested most conveniently with a Sharples or Lourdes continuous flow system or else in several batches using a Sorvall GSA rotor at 6000 rpm for 20 minutes. The wet pellets from a 5-liter culture should weigh about 15 g. The pellets are taken up at 4°C in 70 ml of buffer B (10 mM Tris-chloride, pH 7.5, 5 mM magnesium sulfate, 50 mM sodium chloride) and swirled with 5 ml of chloroform at 37°C. When the solution starts becoming viscous (less than 5 minutes), add deoxyribonuclease to 0.1 μg/ml and continue swirling until the solution liquefies. It is best not to add any more deoxyribonuclease than necessary. Chill the solution to 4°C and remove cell debris by centrifuging in a Sorvall GSA rotor at 10,000 rpm for 20 minutes. Decant and save the supernatant.

Cesium chloride gradients: Distribute the supernatant solution onto 3 preformed cesium chloride block gradients in cellulose nitrate tubes for the Spinco SW25-1 rotor. In order, from the bottom up, each gradient consists of 1.5 ml of density 1.7 cesium chloride (n = 1.392; use refractometer), 2 ml of density 1.5 cesium chloride (n = 1.380), 2 ml of density 1.3 cesium chloride (n = 1.362), and 2 ml of 20% sucrose, all solutions being made up in buffer B. The rotor should be spun at 22,000 rpm for 1.5 hours in the Spinco model L2 ultracentrifuge and allowed to coast to a stop without brake. Remove the phage band near the bottom of each tube by puncturing the tube at the bottom with a 22-gauge hypodermic needle. Leakage around the needle can be prevented by using adhesive tape as a seal and the drip rate can be controlled by using slight negative pressure. The resulting phage solution should be highly opalescent.

Bring the phage solution to a volume of 30 ml with density 1.5 cesium chloride solution in buffer B and adjust to density 1.5. To each of four cellulose nitrate tubes fitted with metal cap assemblies for the Spinco type 40 rotor add 7.5 ml of phage solution and enough paraffin oil to fill the tube. Spin at 25,000 rpm in the Spinco model L ultracentrifuge for 18–24 hours at 4°C and allow to coast to a stop. There should be two visible phage bands. The upper bands can be removed and pooled by puncturing each tube between the bands. The lower bands, which should contain the λh80d*lac* phage, can then be removed and pooled.

Dialyze the phage solutions overnight against 3 changes of 50 volumes each of buffer C (10 mM Tris-chloride, pH 7.5, 50 mM sodium chloride, 1 mM sodium-EDTA) and then adjust with buffer C to an OD at 260 mμ of 10. Add sodium dodecyl sulfate to 0.5% and incubate the solutions 10 minutes at 55°C. Add 4 M potassium chloride to give final concentrations of 0.5 M and put the solutions on ice for 15 minutes. Remove the potassium-SDS precipitates by spinning at 4°C for 5 minutes at top speed in a clinical centrifuge. Carefully decant the supernatant solutions and dialyze against several changes of 10 mM Tris-acetate, pH 8.0, over a period of 24 hours. These final solutions should be stored at 4°C and should contain at least 250 μg/ml DNA (OD_{260} = 5).

Day 4

Synthesis of β-galactosidase: When using *in vitro* protein synthesizing systems of this type, it is necessary to show that synthesis is dependent on added template DNA and is not a result of residual synthesis by unbroken cells. The following experiment should be done to prove that this is indeed the case.

All the components shown in Table 60A, except the crude extract, DNA, and 3′,5′-cyclic AMP, should be mixed according to the protocol shown in Table 60B to form a reaction mixture. Each 100 μl of synthesis mixture should then contain:

- 20 μl reaction mixture
- 30 μl crude extract
- 2 μl 50 mM 3′,5′-cyclic AMP (neutralized with KOH to pH 7.0 in 0.01 M Tris-acetate)
- 28 μl distilled water
- 20 μl λh80d*lac* DNA in 0.01 M Tris-acetate, pH 8.0.

It is best to mix together all the components except the DNA and distribute 80-μl aliquots to small 10 × 75 mm disposable glass culture tubes. Then add to each tube 20 μl of 0.01 M Tris-acetate solution, pH 8.0, containing the appropriate amount of DNA:

1. no DNA
2. 1 μg DNA
3. 2 μg DNA
4. 3 μg DNA
5. 4 μg DNA
6. 5 μg DNA

Seal the tubes with Parafilm and shake gently at 37°C in a water bath. During the incubation a viscous pellet should form in the bottom of the test tube. After 60 minutes at 37°C gently resuspend the pellets in 0.7 ml of β-galactosidase assay mix (0.1 M Na phosphate, pH 7.3, 0.14 M β-mercaptoethanol, 0.35 mg/ml ONPG) and incubate the β-galactosidase assays at 30°C until an appreciable yellow color develops. To each tube add one drop of glacial acetic acid and remove the precipitates by centrifuging at 4°C for 10 minutes at 5000 × *g*. The supernatant of each should be mixed with 0.8 ml 1 M Na_2CO_3 and the OD read at 420 mμ against a distilled water blank.

The synthesis without added λh80d*lac* DNA should give an OD_{420} of about 0.03. The amount of β-galactosidase synthesis observed should be proportional to the DNA concentration. If the system is working well, the amount of β-galactosidase synthesized at 50 μg/ml λh80d*lac* DNA should produce an OD_{420} of at least 1 in 10 hours. If difficulties are encountered, the concentrations of Mg acetate and PEP should be checked since the system is very sensitive to the concentrations of these ions. It is important that the Na_3PEP used be of good quality.

Materials

CTP, GTP, ATP, UTP—of good quality (low in free phosphate)
Na_3PEP—Calbiochem A grade (keep cold and dry)
3′,5′-cyclic AMP—Calbiochem A grade
folinic acid—Calcium leucovorin (Lederle)
levigated alumina—Norton
pyruvate kinase—Calbiochem, rabbit muscle
sodium dodecyl sulfate—Pierce, 99% pure
#20 dialysis tubing prepared as described in Unit VIII, Preparation of Materials

EXPERIMENT 61

Catabolite Repression in a Cell-free System

In vivo experiments have indicated that optimal expression of catabolic operons requires the presence of 3′,5′-cyclic AMP and the protein specified by the *E. coli* gene *cap* (*crp*). The protein specified by the *cap* gene, known as CAP, has now been purified and has been shown to stimulate β-galactosidase synthesis *in vitro* in the presence of 3′,5′-cyclic AMP (Emmer et al., 1970; Zubay et al., 1970).

The crude extract prepared in Experiment 60 was dialyzed, and so contains little, if any, 3′,5′-cyclic AMP. However, the strain CSH73 is cap^+ and the extract does contain CAP factor. This will now be demonstrated by showing that cell-free β-galactosidase synthesis is greatly stimulated by including 3′,5′-cyclic AMP. A series of 100 μl syntheses, identical except for 3′,5′-cyclic AMP concentration, will be run.

References

EMMER, M., B. DECROMBRUGGHE, I. PASTAN and R. PERLMAN. 1970. Cyclic AMP receptor protein of *E. coli:* Its role in the synthesis of inducible enzymes. *Proc. Nat. Acad. Sci. 66:* 480.

ZUBAY, G., D. SCHWARTZ and J. BECKWITH. 1970. Mechanism of activation of catabolite-sensitive genes: A positive control system. *Proc. Nat. Acad. Sci. 66:* 104.

Method

Day 1

Following the procedure in Experiment 60, make up a fresh reaction mixture (these should not be stored). The complete synthesis mixture should contain per 100 μl:

20 μl reaction mixture
30 μl crude extract
20 μl 250 μg/ml ϕ80d*lac* DNA
25 μl distilled water
5 μl 3′,5′-cyclic AMP in the appropriate amount.

Distribute into 75 × 10-mm test tubes 95-μl aliquots of the complete mixture, minus the 3′,5′-cyclic AMP. To each tube add 5 μl of 3′,5′-cyclic AMP solution. A complete range of concentrations, from zero to 5 mM (final concentration) should be tried. Incubate the tubes and assay for β-galactosidase according to the procedures described in Experiment 60.

Questions

The β-galactosidase synthesis should increase with increasing 3′,5′-cyclic AMP concentration until a plateau is reached. Does the concentration of 3′,5′-cyclic AMP giving half-maximal stimulation necessarily indicate the binding K_m of CAP for 3′,5′-cyclic AMP?

EXPERIMENT 62

Repression of β-Galactosidase Synthesis in a Cell-free System

Genetic and biochemical studies have established that β-galactosidase synthesis in *E. coli* is under the negative control of a protein, the *lac* repressor (Jacob and Monod, 1961; Gilbert and Müller-Hill, 1966). The repressor prevents transcription of lactose messenger RNA by binding to a region of the DNA, the *lac* operator (Jacob and Monod, 1961; Gilbert and Müller-Hill, 1967). Galactoside inducers cause induction by binding to the repressor and weakening the repressor-operator interaction. Under ordinary conditions, the ratio of β-galactosidase levels between full induction and full repression is approximately 1000 (Jacob and Monod, 1961).

The isolation of *lac* repressor (see Experiment 55) and the development of a cell-free system capable of synthesizing β-galactosidase using *lac* DNA as template make it possible to verify the action of *lac* repressor *in vitro* (Zubay et al., 1970; deCrombrugghe et al., 1971). To demonstrate that depression of β-galactosidase synthesis is an effect of repressor action and not, for instance, an effect of a nuclease contaminant of the repressor preparation, it is necessary to show that a galactoside inducer completely relieves the observed repression. As an inducer we shall use IPTG.

References

DECROMBRUGGHE, B., B. CHEN, W. ANDERSON, P. NISSLEY, M. GOTTESMAN, I. PASTAN and R. PERLMAN. 1971. Essential elements for controlled *lac* transcription. *Nature New Biol. 231:* 139.

GILBERT, W. and B. MÜLLER-HILL. 1966. Isolation of the *lac* repressor. *Proc. Nat. Acad. Sci. 56:* 1891.

GILBERT, W. and B. MÜLLER-HILL. 1967. The *lac* operator is DNA. *Proc. Nat. Acad. Sci. 58:* 2415.

JACOB, F. and J. MONOD. 1961. Genetic regulatory mechanisms in the synthesis of proteins. *J. Mol. Biol. 3:* 318.

ZUBAY, G., D. A. CHAMBERS and L. C. CHEONG. 1970. Cell-free studies on the regulation of the *lac* operon. *The Lactose Operon*, p. 375. Cold Spring Harbor Laboratory.

Method

Day 1

Make up a fresh reaction mixture according to the method described in Experiment 60. Each 100 μl of synthesis mixture should then include:

20 μl reaction mixture
30 μl crude extract (see Experiment 60)
20 μl φ80d*lac* DNA (250 μg/ml in 0.01 M Tris-acetate, pH 8.0)
2 μl 3′,5′-cyclic AMP (50 mM in 0.01 M Tris, pH 7.0)
18 μl distilled water
5 μl *lac* repressor solution diluted in 0.01 M Tris-acetate, pH 7.4
5 μl distilled water or 0.01 M IPTG.

Mix together all ingredients except repressor and IPTG solutions and distribute 90-μl aliquots to 10 × 75-mm culture tubes. Add repressor and IPTG solutions as follows:

	Repressor (μg/5 μl)	0.01 M IPTG (μl)	Distilled water (μl)
1.	0	0	5
2.	0	5	0
3.	0.05	0	5
4.	0.05	5	0
5.	0.1	0	5
6.	0.1	5	0
7.	0.2	0	5
8.	0.2	5	0
9.	0.5	0	5
10.	0.5	5	0

Incubate all tubes and assay for β-galactosidase as described in Experiment 60.

If repression of β-galactosidase synthesis is not observed in the experiment, the repressor preparation may not be fully active in its ability to complex with the *lac* operator. This possibility can be checked by assaying the operator binding activity (see Experiment 52). If repression is not reversible by IPTG, the repressor preparation should be carefully examined for purity. The repressor should not be diluted through solutions containing glycerol, as glycerol can inhibit *in vitro* protein synthesis in this system.

Materials

Ingredients for cell-free synthesis (see Experiment 60)
IPTG (isopropyl-β-D-thiogalactoside)
lac repressor

Questions

1. Does this experiment prove the hypothesis that correct initiation of lactose messenger RNA synthesis is occurring at the *lac* promoter region? If not, what additional experiments are necessary to prove or disprove this hypothesis?

2. Repression in a cell-free system is much more effective in the presence than in the absence of 3′,5′-cyclic AMP. Is this compatible with the known *in vivo* behavior of the *lac* operon?

APPENDIX I

Formulas and Recipes

1. Minimal Salts

In order to use each of the following (A–C) as growth media, 1 ml of a 1 M solution of $MgSO_4 \cdot 7H_2O$ should be added per liter after autoclaving. In addition, 10 ml of a 20% solution of a carbon source (either a sugar or glycerol) should be added per liter. Vitamins, such as B1, are added to a final concentration of 1 μg/ml and amino acids at 80 μg/ml (in the D, L form; or 40 μg/ml in the L form).

A) M63 Medium

per liter:

KH_2PO_4	13.6 g
$(NH_4)_2SO_4$	2 g
$FeSO_4 \cdot 7H_2O$	0.5 mg

Adjust pH to 7.0 with KOH.

B) M9 Medium

per liter:

Na_2HPO_4	6 g
KH_2PO_4	3 g
NaCl	0.5 g
NH_4Cl	1 g

Add 10 ml of a 0.01 M solution of $CaCl_2$ after autoclaving.

C) Minimal A Medium (1 × A)

per liter:

K_2HPO_4	10.5 g
KH_2PO_4	4.5 g
$(NH_4)_2SO_4$	1 g
Sodium citrate $\cdot 2H_2O$	0.5 g

D) Minimal A Medium (5 × A)

per liter:

K_2HPO_4	52.5 g
KH_2PO_4	22.5 g
$(NH_4)_2SO_4$	5 g
Sodium citrate $\cdot 2H_2O$	2.5 g

2. Minimal Agar Plates

All plates are prepared with the final concentration per liter:

Difco minimal agar	15 g
Salts: (in this case 1 × A)	
K_2HPO_4	10.5 g
KH_2PO_4	4.5 g
$(NH_4)_2SO_4$	1 g
Sodium citrate $\cdot 2H_2O$	0.5 g
$MgSO_4 \cdot 7H_2O$	Add 1 ml from a stock solution of 20 g/100 ml after autoclaving.
B1 (thiamine hydrochloride)	0.5 ml from a 1% stock solution (excess)
Amino acids as required	10 ml from a 4 mg/ml stock solution
Sugar	10 ml from a 20% stock solution
Antibiotics as required	
Streptomycin	2 ml from a stock solution prepared by dissolving 1 g streptomycin sulfate in 17.5 ml sterile H_2O

The salt solution and the agar should be prepared and autoclaved separately at 15 lbs/in^2 for 15 minutes. Therefore the salts are usually prepared in more concentrated form. It is convenient to autoclave 15 g agar in 800 ml H_2O and the salts (at 5 × normal strength) in 200 ml H_2O, or the agar in 500 ml H_2O and the salts (at 2 × normal strength) in 500 ml H_2O. The salts used for the plates required for these experiments is Minimal A Medium, although any of the minimal media listed here will suffice. Mg^{++} and nutrients are prepared and autoclaved separately. Normally, 40 μg/ml of the D,L-amino acid (or 20 μg/ml of the L-amino acid) are sufficient. Vitamins are required in smaller amounts (1 μg/ml). These are then added to the salts after autoclaving.

3. Rich Media

If used in plates, add 15 g agar/liter (except in H medium and R medium).

A) LB Medium

per liter:

Bacto tryptone	10 g
Bacto yeast extract	5 g
NaCl	10 g

B) YT Medium

per liter:

Bacto tryptone	8 g
Bacto yeast extract	5 g
NaCl	5 g

C) 2 × YT Medium (2YT)

per liter:

Bacto tryptone	16 g
Bacto yeast extract	10 g
NaCl	5 g

D) H Medium* for Phage Plate Lysates

per liter:

Bacto tryptone	10 g
NaCl	8 g

Use 12 g agar per liter for plates.

E) R Medium for Phage Lysates

per liter:

Bacto tryptone	10 g
Bacto yeast extract	1 g
NaCl	8 g

After autoclaving add 2 ml 1 M $CaCl_2$ + 5 ml 20% glucose.

Use 12 g agar per liter for plates.

F) B Broth or Tryptone Broth

per liter:

Bacto tryptone	10 g
NaCl	8 g

G) Luria Broth

per liter:

Bacto tryptone	10 g
Bacto yeast extract	5 g
NaCl	0.5 g
1 M NaOH	2 ml

Dissolve; adjust pH to 7.0 with 1 M NaOH. Add 10 ml 20% glucose after autoclaving.

* Not to be confused with the H medium used in Experiment 43.

4. Buffers

A) Phosphate Buffer, pH 7.0

Mix equimolar solutions of:

Na_2HPO_4	61.0 ml
NaH_2PO_4	39.0 ml

B) Citrate Buffer, pH 5.5 (0.1 M)

0.1 M citric acid	4.7 volumes
0.1 M Na_3 citrate	15.4 volumes

C) Saline (for dilutions)

NaCl	8.5 g
H_2O	1 liter

D) Z Buffer for β-galactosidase Assays

per liter:

$Na_2HPO_4 \cdot 7H_2O$	16.1 g
$NaH_2PO_4 \cdot H_2O$	5.5 g
KCl	0.75 g
$MgSO_4 \cdot 7H_2O$	0.246 g
β-mercaptoethanol	2.7 ml

Do not autoclave!
Adjust pH to 7.0.

5. Soft Agar (top agar)

Kept molten at 45°.

A) Minimal "F-top" Agar

per liter:

Difco minimal agar	8 g
NaCl	8 g

B) H-top Agar

per liter:

Bacto tryptone	10 g
Difco minimal agar	8 g
NaCl	8 g

C) R-top Agar

per liter:

Bacto tryptone	10 g
Bacto yeast extract	1 g
Difco minimal agar	8 g
NaCl	8 g

Add after autoclaving 2 ml 1 M $CaCl_2$ + 5 ml 20% glucose.

6. Storage of Strains

A) Stabs. A convenient way to store a large number of strains for several years is to use small airtight stab bottles containing nutrient agar. A colony or a loopful of a fresh culture is stabbed into the medium and the vials are incubated overnight. These are then tightly sealed and stored, preferably in the cold, away from sunlight. Many stab bottles are not airtight. Although molten parafin can be used to seal each vial, it is more convenient to avoid this step. We recommend "1-dram glass vials with rubber-lined plastic screw caps" from Wheaton Glassware, Milville, New Jersey, (Catalog number 24242). These bottles can be autoclaved.

A variety of media are used for stabs. We recommend the following:

per liter:	
Difco nutrient broth (powder)	10 g
NaCl	5 g
Difco agar	6 g

Medium with glucose is often widely used.

per liter:	
Bacto tryptone	10 g
Bacto yeast extract	10 g
Glucose	2 g
Difco agar	6 g

Supplements are generally not added, except for Thy^- strains. In the latter case, thymine at 50 μg/ml is included in the medium. Although some investigators recommend adding 100 μg/ml cysteine to increase the viability, we find that most *E. coli* strains stored in the cold in the above medium remain viable for at least two years.

B) Storage in Glycerol. Cultures can be maintained for several years in liquid form in 40% glycerol at −15°C. A fresh overnight culture, grown in either minimal or rich medium, is used. Add 2 ml sterile 80% glycerol to 2 ml of the culture and put in the freezer at −10° to −20°. This is a particularly useful way of storing unstable strains. Cells are recovered from storage by inoculating fresh medium with a few drops of the glycerol culture.

C) Freeze-drying. Lyophilization or freeze-drying is the surest way to store *E. coli* for a long period of time. An exceptional review by Lapage et al. (1970) thoroughly discusses the methods used in this process.

Reference

Lapage, S. P., J. E. Shelton, T. G. Mitchell and A. R. Mackenzie. 1970. Culture collections and the preservation of bacteria. In *Methods in Microbiology* (J. R. Norris and D. W. Ribbons, ed.) *3A:* 135.

7. Bray's Liquid Scintillator*

Naphthalene	60 g
PPO	4 g
POPOP	0.2 g
Methanol (absolute)	100 ml
Ethylene glycol	20 ml
p-Diozane	to make 1 liter

* (*Anal. Biochem. 1:* 279)

APPENDIX II

Operation of Mating Interrupter

1. The interrupter is designed to shake 13 $\times$ 100 mm culture tubes. The mating cells which are to be separated are pipetted into a tube which contains between 3 and 5 ml of blending liquid. When the cells are going to be plated out after interruption, it is best to use molten agar (0.8 %, kept in tubes at 42–45°C) as the blending liquid. After blending, the cells + agar are poured directly onto a plate to solidify. This reduces the re-formation of mating pairs to a minimum after interruption.

2. Cover the tube with Parafilm M (American Can Co., Neenah, Wisc.). The inner surface of Parafilm (underneath the paper backing) is clean and does not require sterilization.

3. Insert the tube into the Tube Holder.

4. Turn on the shaker for 7–10 seconds.

5. Mating pairs are now interrupted. However, it should be pointed out that after this procedure, cells can form new mating pairs rapidly if left standing in liquid at densities higher than $\sim 10^6$ cells per ml. In order to prevent this, cells should either be plated out directly as described above, or kept at a dilute concentration. It is also helpful to use a male-counterselective agent, such as streptomycin or nalidixic acid, when appropriate strains are being used.

Assembly of Mating Interrupter

1. Drill a hole in the base of the Tube Holder just large enough to fit the drive shaft of a light duty carpenter's sabre saw (e.g. Stanley Model #80456, $\frac{1}{5}$ horsepower).

2. Drill and tap a hole in the base of the Tube Holder for a set screw to fit into the hole in the sabre saw drive shaft (see Figure A1).

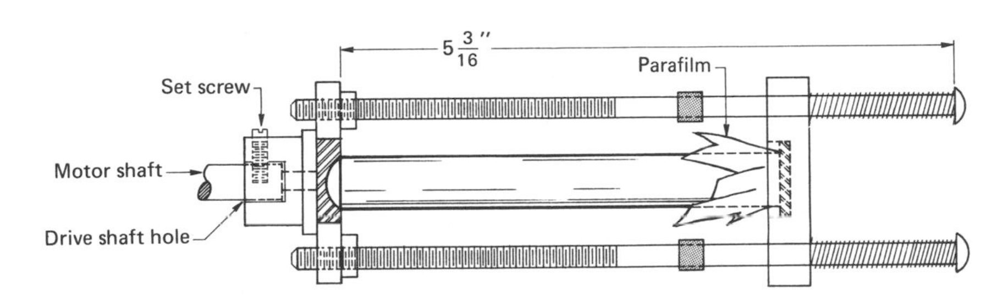

Figure A1.

3. Attach the sabre saw motor, with drive shaft pointing up, to a heavy object, using screws or metal straps. The heavy object may be either:
a) a workbench—this is the simplest method but results in a very noisy operation, or preferably,
b) a heavy lead or iron brick—with this arrangement the motor and brick may be supported on soft rubber padding (and partially enclosed, if desired, as shown in Figure A2) in order to reduce noise.

4. Tighten the nuts on the Tube Holder so that the distance from the nuts to the top of the bolts (under the bolt head) is $5\frac{3}{16}''$ (Figure A1).

5. Attach the Tube Holder to the motor shaft firmly with a set screw.

6. For convenience, plug the motor into an automatic reset timer, such as Time-O-lite Model M-59, Industrial Timer Corp., Parsippany, N.J.

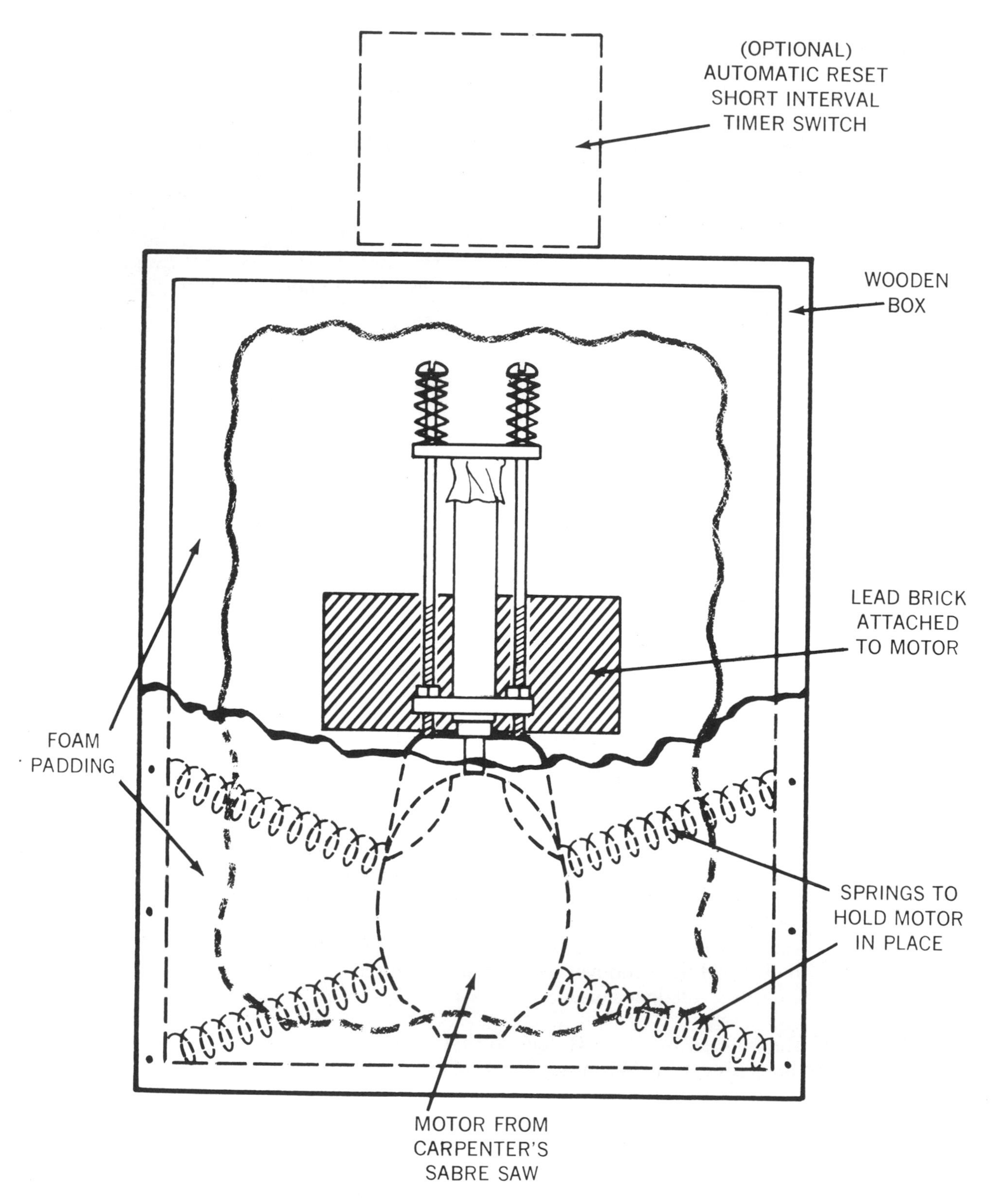

Figure A2. Interrupter mounting.

APPENDIX III

Linkage Map of *E. coli*

(Reprinted with permission from Taylor, 1970, *Bacteriol. Rev. 34:* 155.)

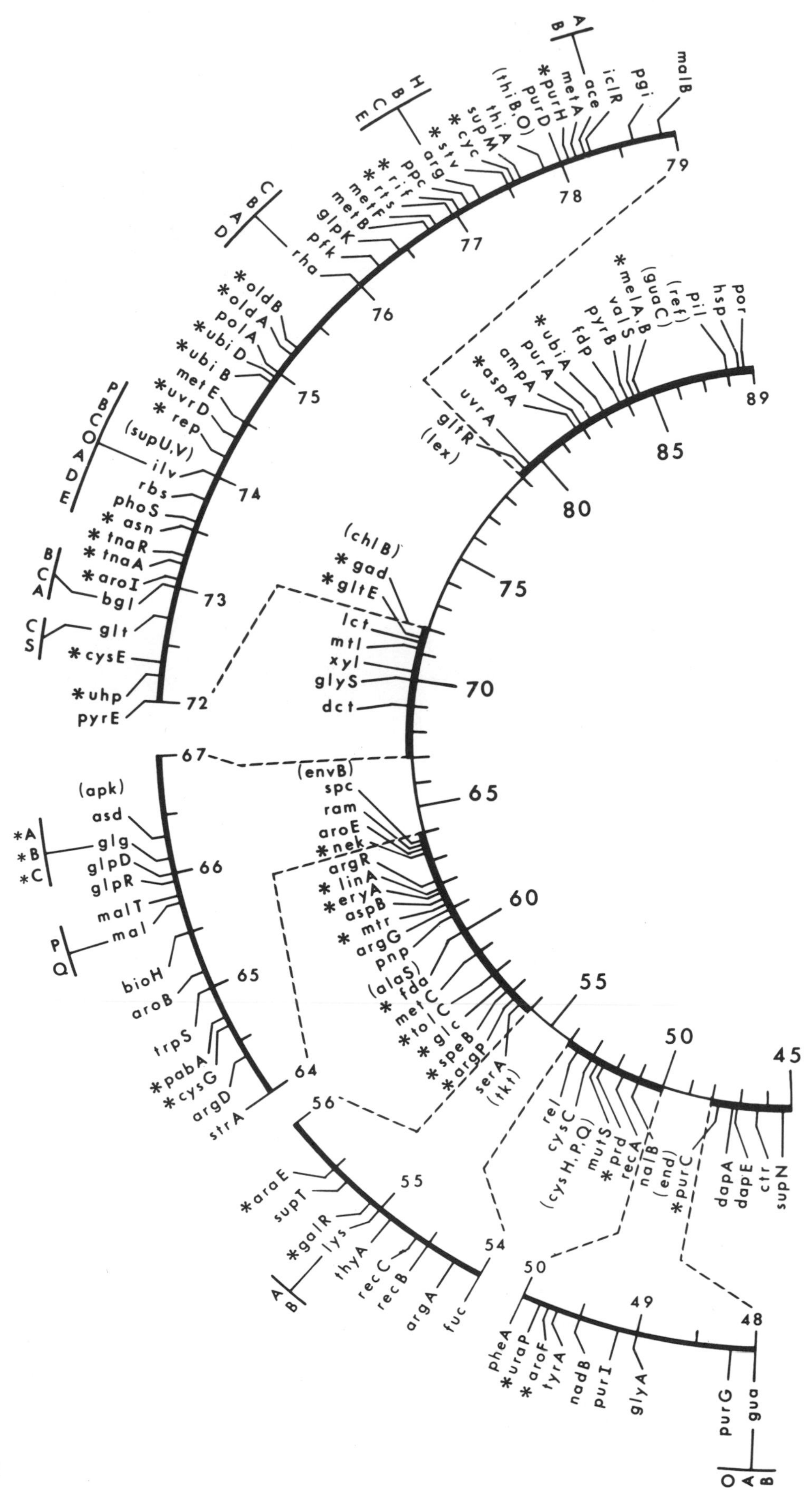

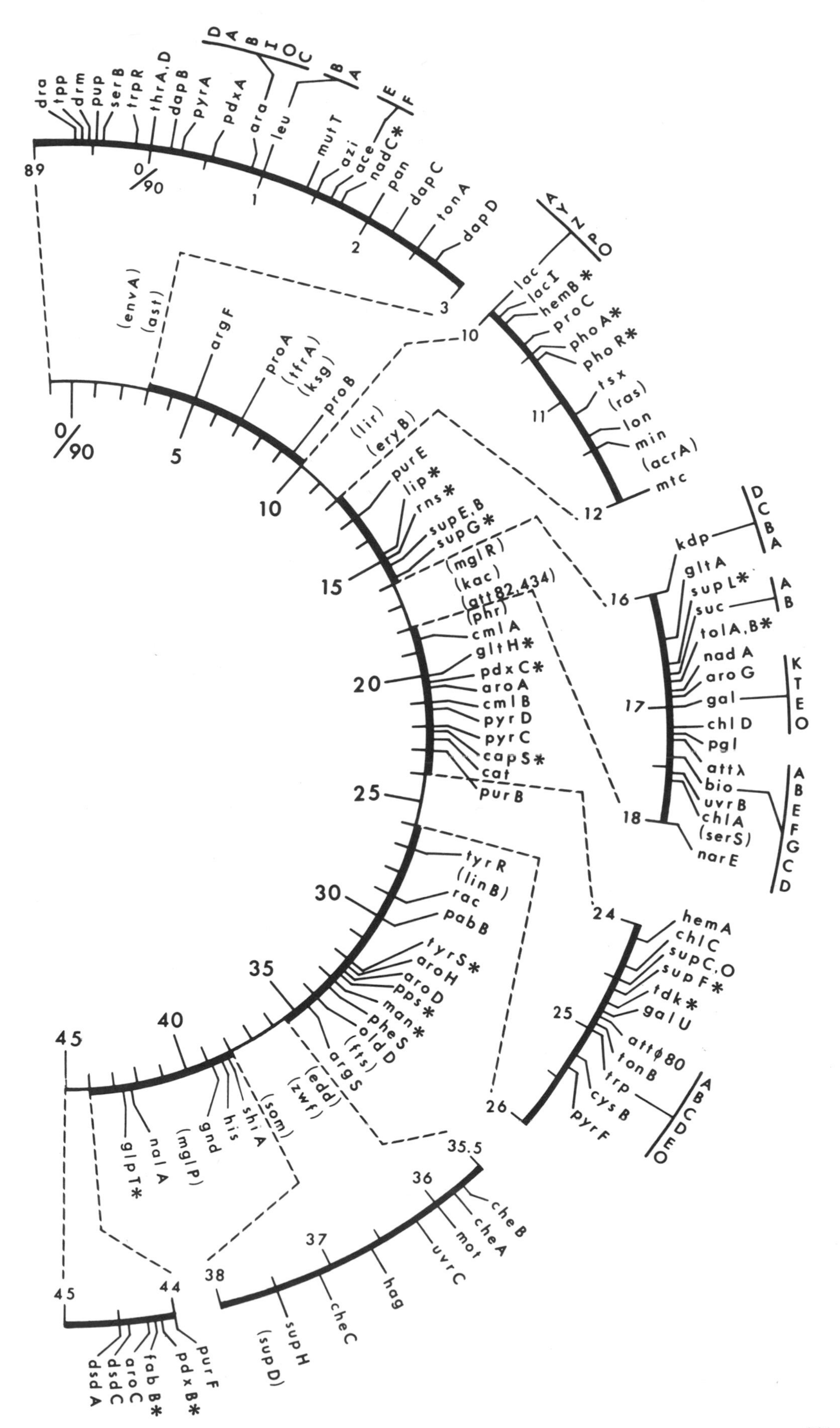

dra
tpp
drm
pup
ser B
trp R
thr A, D
dap B
pyr A
pdx A
ara
D A B I O C
leu
B A
mut T
azi
ace
nad C*
E F
pan
dap C
ton A
dap D
89
0/90
1
2
3
(env A)
(ast)
arg F
pro A
(tfr A)
(ksg)
pro B
(lir)
(ery B)
pur E
lip*
rns*
sup E, B
sup G*
(mgl R)
(kac)
(att 82, 434)
(phr)
cml A
glt H*
pdx C*
aro A
cml B
pyr D
pyr C
cap S*
cat
pur B
0/90
5
10
15
20
25
30
35
40
45
lac
A Y Z P O
lac I
hem B*
pro C
pho A*
pho R*
tsx
(ras)
lon
min
(acr A)
mtc
10
11
12
kdp
D C B A
glt A
sup L*
suc
A B
tol A, B*
nad A
aro G
gal
K T E O
chl D
pgl
att λ
bio
A B E F G C D
uvr B
chl A
(ser S)
nar E
16
17
18
tyr R
(lin B)
rac
pab B
tyr S*
aro H
aro D
pps*
man*
phe S
old D
(fts)
arg S
(edd)
(zwf)
(som)
shi A
his
gnd
(mgl P)
nal A
glp T*
hem A
chl C
sup C, O
sup F*
tdk*
gal U
att φ80
ton B
trp
cys B
pyr F
A B C D E O
24
25
26
35.5
36
37
38
che B
che A
mot
uvr C
hag
che C
sup H
(sup D)
45
44
pur F
pdx B*
fab B*
aro C
dsd C
dsd A

Table A1. List of Genetic Markers of *E. coli*[a]

Gene symbol	Mnemonic	Map position (min)[b]	Alternate gene symbols; phenotypic trait affected	References[c]
aceA	Acetate	78	*icl*; utilization of acetate: isocitrate lyase	26, 223
aceB	Acetate	78	*mas*; utilization of acetate; malate synthetase A	223
aceE	Acetate	2	*aceE1*; acetate requirement; pyruvate dehydrogenase (decarboxylase component)	98, 99
aceF	Acetate	2	*aceE2*; acetate requirement; pyruvate dehydrogenase (lipoic reductase-transacetylase component)	98, 99
acrA	Acridine	(11)	Sensitivity to acriflavine, phenethyl alcohol, Na dodecyl sulfate	149
alaS	Alanine	(60)	*ala-act*; alanyl-tRNA synthetase	150
ampA	Ampicillin	82	Resistance or sensitivity to penicillin	70
apk		(66)	Lysine-sensitive aspartokinase	164
araA	Arabinose	1	L-Arabinose isomerase	129
araB	Arabinose	1	L-Ribulokinase	129
araC	Arabinose	1	Regulatory gene	200, 201
araD	Arabinose	1	L-Ribulose 5-phosphate 4-epimerase	129
araE	Arabinose	56	L-Arabinose permease	67, 156, A
araI	Arabinose	1	Initiator locus	200, 201
araO	Arabinose	1	Operator locus	125
argA	Arginine	54	*argB*, *Arg1*, Arg_2; *N*-acetylglutamate synthetase	90, 111, 215, 226
argB	Arginine	77	*argC*; α-*N*-acetyl-L-glutamate-5-phosphotransferase	87, 88, 133, 226
argC	Arginine	77	*argH*, *arg2*; *N*-acetylglutamic-γ-semialdehyde dehydrogenase	87, 88, 133, 226
argD	Arginine	64	*argG*, Arg_1; acetylornithine-δ-transaminase	112, 226, B
argE	Arginine	77	*argA*, *Arg4*; L-ornithine-*N*-acetylornithine lyase	87, 88, 133, 226
argF	Arginine	5	*argD*, *Arg5*; ornithine transcarbamylase	90, 133, 226
argG	Arginine	61	*argE*, *Arg6*; argininosuccinic acid synthetase	133, 214, 215, 226
argH	Arginine	77	*argF*, *Arg7*; L-argininosuccinate arginine lyase	87, 88, 133, 226
argP	Arginine	57	Arginine permease	132, C
argR	Arginine	62	*Rarg*; regulatory gene	90, 119, 133, 226
argS	Arginine	35	Arginyl-tRNA synthetase	40
aroA	Aromatic	21	3-Enolpyruvylshikimate-5-phosphate synthetase	170, 178, 214
aroB	Aromatic	65	Dehydroquinate synthetase	109, 170, 232
aroC	Aromatic	44	Chorismic acid synthetase	170, 214
aroD	Aromatic	32	Dehydroquinase	170, 214
aroE	Aromatic	64	Dehydroshikimate reductase	170, 215, 232
aroF	Aromatic	50	3-Deoxy-D-arabinoheptulosonic acid-7-phosphate (DHAP) synthetase (tyrosine-repressible isoenzyme)	228

[a] Reprinted with permission from Taylor (1970) *Bacteriol. Rev. 34:* 159–166.
[b] Numbers refer to the time scale on the map. Parentheses indicate approximate map locations.
[c] All references can be found in Taylor, 1970.

Table A1.—*Continued*

Gene symbol	Mnemonic	Map position (min)[b]	Alternate gene symbols; phenotypic trait affected	References[c]
aroG	Aromatic	17	DHAP synthetase (phenylalanine-repressible isoenzyme)	1, 27, 228
aroH	Aromatic	32	DHAP synthetase (tryptophan-repressible isoenzyme)	228
aroI	Aromatic	73	Function unknown	86
asd	—	66	*dap* + *hom*; aspartic semialdehyde dehydrogenase	38, 95, 196
asn	—	73	Asparagine synthetase	34
aspA	—	82	Aspartase	138
aspB	Aspartate	62	*asp*; aspartate requirement	118, 180
ast	Astasia	(4)	Generalized high mutability	244, 245
attλ	Attachment	17	Integration site for prophage λ	118, 184
attϕ*80*	Attachment	25	Integration site for prophage ϕ80	204
att82	Attachment	(17)	Integration site for prophage 82	118, 184
att434	Attachment	(17)	Integration site for prophage 434	118, 184
azi	Azide	2	*pea,fts*; resistance or sensitivity to Na azide or phenethyl alcohol; filament formation at 42C	117, 134, 221, 242
bglA	β-Glucoside	73	β-*glA*; aryl β-glucosidase	192
bglB	β-Glucoside	73	β-*glB*; β-glucoside permease	192
bglC	β-Glucoside	73	β-*glC*; regulatory gene	192
bioA	Biotin	17	Group II; 7-oxo-8-aminopelargonic acid (7KAP) → 7,8-diaminopelargonic acid (DAPA)	1, 51, 65, 182
bioB	Biotin	17	Conversion of dethiobiotin to biotin	1, 51, 65, 182
bioC	Biotin	17	Unknown block prior to 7KAP synthetase	1, 51, 65, 182
bioD	Biotin	17	Dethiobiotin synthetase	1, 51, 64, 65, 182
bioE	Biotin	17	Unknown block prior to 7KAP synthetase	1, 51, 65, 182
bioF,G	Biotin	17	7KAP synthetase	1, 51, 65, 182
bioH	Biotin	66	*bioB*; early block prior to 7KAP synthetase	95, 182, 196
capS	Capsule	22	Regulatory gene for capsular polysaccharide synthesis	139
cat	—	23	*CR*; catabolite repression	219
cheA	Chemotaxis	36	*motA*; chemotactic motility	11, 12
cheB	Chemotaxis	36	*motB*; chemotactic motility	11, 12
cheC	Chemotaxis	37	Chemotactic motility	12
chlA	Chlorate	18	*narA*; pleiotropic mutations affecting nitrate-chlorate reductase and hydrogen lyase activity	1, 174, 175, 224
chlB	Chlorate	(71)	*narB*; pleiotropic mutations affecting nitrate-chlorate reductase and hydrogen lyase activity	174
chlC	Chlorate	25	*narC*; structural gene for nitrate reductase	93, 174, 187
chlD	Chlorate	17	*narD*, *narF*; nitrate-chlorate reductase	1, 224
cmlA	Chloramphenicol	19	Resistance or sensitivity to chloramphenicol	178
cmlB	Chloramphenicol	21	Resistance or sensitivity to chloramphenicol	178

Table A1.—*Continued*

Gene symbol	Mnemonic	Map position (min)[b]	Alternate gene symbols; phenotypic trait affected	References[c]
cmlB	Chloramphenicol	21	Resistance or sensitivity to chloramphenicol	178
ctr	—	46	Mutations affecting the uptake of diverse carbohydrates	231
cyc	Cycloserine	78	Resistance or sensitivity to D-cycloserine	47, D
cysB	Cysteine	25	Pleiotropic mutations affecting cysteine biosynthesis	120, 205, 239
cysC	Cysteine	53	Adenosine 5′-sulfatophosphate kinase	120, 144, 215
cysE	Cysteine	72	Apparently pleiotropic	120
cysG	Cysteine	65	Sulfite reductase	215
cysH	Cysteine	(53)	Adenosine 3′-phosphate 5′-sulfatophosphate reductase	120
cysP	Cysteine	(53)	Sulfate permease and sulfite reductase	120
cysQ	Cysteine	(53)	Sulfite reductase	120
dapA	Diaminopimelate	47	Dihydrodipicolinic acid synthetase	38, E
dapB	Diaminopimelate	0	Dihydrodipicolinic acid reductase	73, E
dapC	Diaminopimelate	2	Tetrahydrodipicolinic acid → N-succinyl-diaminopimelate	E
dapD	Diaminopimelate	3	Tetrahydrodipicolinic acid → N-succinyl diaminopimelate	E
dapE	Diaminopimelate	47	*dapB*; N-succinyl-diaminopimelic acid deacylase	38, E
darA	—	—	*See uvrD*	222
dct	—	69	Uptake of C_4-dicarboxylic acids	124
deo	Deoxythymidine	—	*See dra, drm, pup*, and *tpp*	131
dra	—	89	*deoC, thyR*; deoxyriboaldolase	4, 131, 159
drm	—	89	*deoB, thyR*; deoxyribomutase	4, 131, 159
dsdA	D-Serine	45	D-Serine deaminase	144
dsdC	D-Serine	45	Regulatory gene	144
edd	—	(35)	Entner-Doudoroff dehydrase (gluconate-6-phosphate dehydrase)	167
end	—	(50)	*endoI*; endoduclease I	59
envA	Envelope	(3)	Anomalous cell division involving chain formation	155
envB	Envelope	(65)	Anomalous spheroid cell formation	154
eryA	Erythromycin	62	Resistance or sensitivity to erythromycin	F
eryB	Erythromycin	(11)	High level resistance to erythromycin	9
exr	—	—	*See lex*	
fabB	—	44	Fatty acid biosynthesis	69
fda	—	60	*ald*; fructose-1, 6-diphosphate aldolase	20
fdp	—	84	Fructose diphosphatase	77, 78, 240
ftsA	—	—	*See azi*	221
fts	—	(35)	*fts-9*; filamentous growth and inhibition of nucleic acid synthesis at 42°C	221
fuc	Fucose	54	Utilization of L-fucose	71, 215
gad	—	72	Glutamic acid decarboxylase	135, 137
galE	Galactose	17	*galD*; uridinediphosphogalactose 4-epimerase	3, 29
galK	Galactose	17	*galA*; galactokinase	3, 29
galO	Galactose	17	*galC*; operator locus	29, 30

Table A1.—*Continued*

Gene symbol	Mnemonic	Map position (min)[b]	Alternate gene symbols; phenotypic trait affected	References[c]
galT	Galactose	17	*galB*; galactose 1-phosphate uridyl transferase	3
galR	Galactose	55	*Rgal*; regulatory gene	30, 188
galU	Galactose	25	*UPDG*; uridine diphosphoglucose pyrophosphorylase	198, 93
glc	Glycolate	58	Utilization of glycolate; malate synthetase G	223
glgA	Glycogen	66	Glycogen synthetase	33, 203
glgB	Glycogen	66	α-1,4-Glucan: α-1,4-glucan 6-glucosyltransferase	33, 203
glgC	Glycogen	66	Adenosine diphosphate glucose pyrophosphorylase	33, 203
glpD	Glycerol phosphate	66	*glyD*; L-α-glycerophosphate dehydrogenase	44, 196
glpK	Glycerol phosphate	76	Glycerol kinase	45
glpT	Glycerol phosphate	43	L-α-Glycerophosphate transport system	44
glpR	Glycerol phosphate	66	Regulatory gene	44
gltA	Glutamate	16	*glut*; requirement for glutamate; citrate synthase	13, 68, 100
gltC	Glutamate	73	Operator locus	135, 136
gltE	Glutamate	72	Glutamyl-tRNA synthetase	G
gltH	Glutamate	20	Requirement	135
gltR	Glutamate	79	Regulatory gene for glutamate permease	136
gltS	Glutamate	73	Glutamate permease	136
glyA	Glycine	49	Serine hydroxymethyl transferase[d]	48, 215
glyS	Glycine	70	*gly-act*; glycyl-tRNA synthetase	21
gnd	—	39	Gluconate-6-phosphate dehydrogenase	167
guaA	Guanine	48	gua_b; xanthosine-5′-monophosphate aminase	151, 211, 215
guaB	Guanine	48	gua_a; inosine-5′-monophosphate dehydrogenase	151, 211
guaC	Guanine	(88)	Guanosine-5′-monophosphate reductase	151
guaO	Guanine	48	Operator locus	151, 152
hag	H antigen	37	*H*; flagellar antigens (flagellin)	12
hemA	Hemin	24	Synthesis of δ-aminolevulinic acid	93, 190, 191
hemB	Hemin	10	*ncf*; synthesis of catalase and cytochromes	191
his	Histidine	39	Requirement	214
hsp	Host specificity	89	*hs*, *rm*; host restriction and modification of DNA	23, 39, 128, 235
icl	—	—	*See aceA*	26
iclR	—	78	Regulation of the glyoxylate cycle	26
ilvA	Isoleucine-valine	74	*ile*; threonine deminase	169, 177
ilvB	Isoleucine-valine	74	Condensing enzyme (pyruvate + α-ketobutyrate)	169, 177
ilvC	Isoleucine-valine	74	*ilvA*; α-hydroxy-β-keto acid reductoisomerase	169, 177
ilvD	Isoleucine-valine	74	*ilvB*; dehydrase	169, 177

[d] Enzymatic defect is inferred from studies on the homologous mutant in *S. typhimurium*. Refer to Sanderson (189) for additional references.

Table A1.—*Continued*

Gene symbol	Mnemonic	Map position (min)[b]	Alternate gene symbols; phenotypic trait affected	References[c]
ilvE	Isoleucine-valine	74	*ilvC*; transaminase B	169, 177
ilvO	Isoleucine-valine	74	Operator locus for genes *ilvA*, *D*, *E*	176, 177
ilvP	Isoleucine-valine	74	Operator locus for gene *ilvB*	176, 177
kac	K-accumulation	17	Defect in potassium ion uptake	28
kdpA-D	K-dependent	16	Requirement for a high concentration of potassium	68
ksg	Kasugamycin	(8)	Resistance or sensitivity to kasugamycin (30*S* ribosomal subunit)	210
lacA	Lactose	10	*a*, *lacAc*; thiogalactoside transacetylase	18, 145, 243
lacI	Lactose	10	*i*; regulator gene	49, 145
lacO	Lactose	10	*o*; operator locus	49, 145
lacP	Lactose	10	*p*; promoter locus	49, 116, 145
lacY	Lactose	10	*y*; galactoside permease (M protein)	76, 117, 145
lacZ	Lactose	10	*z*; β-galactosidase	117, 134, 145
lct	Lactate	71	L-Lactate dehydrogenase	163
leuA	Leucine	1	α-Isopropylmalate synthetase	117, 125, 134
leuB	Leucine	1	β-Isopropylmalate dehydrogenase	125
lex	—	(79)	Resistance or sensitivity to X rays and UV light	106
linA	Lincomycin	62	Resistance or sensitivity to lincomycin	F
linB	Lincomycin	(28)	High-level resistance to lincomycin	9
lip	Lipoic acid	15	Requirement	100, 225
lir	—	(11)	Increased sensitivity to lincomycin or erthromycin, or both	9
lon	Long form	11	*capR*, *dir*, *muc*; filamentous growth, radiation sensitivity, and regulation of capsular polysaccharide synthesis	2, 56, 108, 140, 221
lysA	Lysine	55	Diaminopimelic acid decarboxylase	118, 214, E
lysB	Lysine	55	Lysine or pyridoxine requirement	E
malB	Maltose	79	*mal-5*; maltose permease and phage λ receptor site	196, 197
malP	Maltose	66	*malA*; maltodextrin phosphorylase	95, 96, 118, 196
malQ	Maltose	66	*malA*; amylomaltase	95, 96, 118, 196
malT	Maltose	66	*malA*; probably a positive regulatory gene	95, 96, 118, 196
man	Mannose	33	Phosphomannose isomerase	141, 215
melA	Melibiose	84	*mel-7*; α-galactosidase	195
melB	Melibiose	84	*mel-4*; thiomethylgalactoside permease II	173, 195
metA	Methionine	78	met_3; homoserine *O*-transsuccinylase	107, 118, 186, 196
metB	Methionine	77	*met-1*, met_1; cystathionine synthetase	87, 118, 186, 214
metC	Methionine	59	Cystathionase	186, 215
metE	Methionine	75	$met\text{-}B_{12}$; N^5-methyltetrahydropteroyl triglutamate-homocysteine methylase[d]	63, 207, 214
metF	Methionine	77	*met-2*, met_2; N^5, N^{10}-methyltetrahydrofolate reductase[d]	87, 88, 118, 207
mglP	Methyl-galactoside	(40)	*P-MG*; methyl-galactoside permease	82, 185
mglR	Methyl-galactoside	(17)	*R-MG*; regulatory gene	82
min	Minicell	11	Formation of minute cells containing no DNA	36, H
mot	Motility	36	Flagellar paralysis	12

Table A1.—*Continued*

Gene symbol	Mnemonic	Map position (min)[b]	Alternate gene symbols; phenotypic trait affected	References[c]
mtc	Mitomycin C	12	*Mb*, *mbl*; sensitivity to acridines, methylene blue and mitomycin C	110, 161, 213
mtl	Mannitol	71	Utilization of D-mannitol	214
mtr	Methyl tryptophan	61	Resistance to 5-methyltryptophan	103
mutS	Mutator	53	Generalized high mutability	202
mutT	Mutator	1	Generalized high mutability; specifically induces AT → CG transversions	41, 97, 202, 206
nadA	Nicotinamide adenine dinucleotide	17	*nicA*; nicotinic acid requirement	1, 215
nadB	Nicotinamide adenine dinucleotide	49	*nicB*; nicotine acid requirement	118, 218
nadC	Nicotinamide adenine dinucleotide	2	Quinolinate phosphoribosyl transferase	85, I
nalA	Nalidixic acid	42	Resistance or sensitivity to nalidixic acid	94
nalB	Nalidixic acid	51	Resistance or sensitivity to nalidixic acid	94
nar	Nitrate reductase	—	*See chl*	
narE	—	18	Nitrate reductase (*see* also *chl*)	175, 224
nek	—	63	Resistance to neomycin and kanamycin (30*S* ribosomal protein)	10
nic	—	—	*See nad*	
oldA	Oleate degradation	75	*old-30*; thiolase	162
oldB	Oleate degradation	75	*old-64*; hydroxyacyl-coenzyme A dehydrogenase	162
oldD	Oleate degradation	34	*old-88*; acyl-coenzyme A synthetase	162
pabA	*p*-Aminobenzoate	65	Requirement	109, 232
pabB	*p*-Aminobenzoate	30	Requirement	109
pan	Pantothenic acid	2	Requirement	53
pdxA	Pyridoxine	1	Requirement	215
pdxB	Pyridoxine	44	Requirement	54
pdxC	Pyridoxine	20	Requirement	B, D
pfk	—	76	Structural or regulatory gene for fructose 6-phosphate kinase	146
pgi	—	79	Phosphoglucoisomerase	77
pgl	—	17	6-Phosphogluconolactonase	127
pheA	Phenylalanine	50	Prephenic acid dehydratase	170, 214, 215
pheS	Phenylalanine	33	*phe-act*; phenylalanyl-tRNA synthetase	22
phoA	Phosphatase	11	*P*; alkaline phosphatase	60, D
phoR	Phosphatase	11	*R1 pho*, *R1*; regulatory gene	60, D
phoS	Phosphatase	74	*R2 pho*, *R2*; regulatory gene	8, 60
phr	Photoreactivation	(17)	Photoreactivation of UV-damaged DNA (K12-B hybrids)	222
pil	Pili	88	*fim*; presence or absence of pili (fimbriae)	134
pnp	—	61	Polynucleotide phosphorylase	180
polA	Polymerase	75	DNA polymerase	50, 92
por	P1 restriction	89	Restriction of phage P1 DNA	236
ppc	—	77	*glu*, *asp*; succinate, aspartate, or glutamate requirement; phosphoenolpyruvate carboxylase	87, 88, 118
pps	—	33	Utilization of pyruvate or lactate; phosphopyruvate synthetase	25

Table A1.—*Continued*

Gene symbol	Mnemonic	Map position (min)[b]	Alternate gene symbols; phenotypic trait affected	References[c]
prd	Propanediol	53	1,2-Propanediol dehydrogenase	237, J
proA	Proline	7	pro_1; block prior to L-glutamate semialdehyde	46, 214
proB	Proline	9	pro_2; block prior to L-glutamate semialdehyde	46, 214
proC	Proline	10	pro_3; *Pro2*; probably Δ-pyrroline-5-carboxylate reductase	46
pup	—	89	Purine nucleoside phosphorylase	4
purA	Purine	82	ade_k, Ad_4; adenylosuccinic acid synthetase	70, 118
purB	Purine	23	ade_h; adenylosuccinase	205, 211, 214
purC	Purine	48	ade_g; phosphoribosyl-aminoimidazole succinocarboxamide synthetase	151, 211
purD	Purine	78	$adth_a$; phosphoribosylglycineamide-synthetase[d]	211, 214
purE	Purine	13	ade_3; ade_f; Pur_2; phosphoribosyl-aminoimidazole carboxylase	56, 211
purF	Purine	44	*purC*, $ade_{u,b}$; phosphoribosyl-pyrophosphate amidotransferase[d]	211, 214, 215
purG	Purine	48	$adth_b$; phosphoribosylformylglycine-amidine synthetase[d]	211, 218
purH	Purine	78	ade_i; phosphoribosyl-aminoimidazole carboxamide formyltransferase	211
purI	Purine	49	Aminoimidazole ribotide synthetase[d]	217, 218
pyrA	Pyrimidine	0	*cap*, *arg* + *ura*; glutamino-carbamoylphosphate synthetase	17, 214, 215
pyrB	Pyrimidine	84	Aspartate transcarbamylase	17, 214
pyrC	Pyrimidine	22	Dihydroorotase	17, 205
pyrD	Pyrimidine	21	Dihydroorotic acid dehydrogenase	17, 205
pyrE	Pyrimidine	72	Orotidylic acid pyrophosphorylase	192, 214
pyrF	Pyrimidine	25	Orotidylic acid decarboxylase	205
rac	Recombination activation	29	Suppressor of *recB* and *recC* mutant phenotype	K
ram	Ribosomal ambiguity	64	Nonspecific suppression of all nonsense codons	183
ras	Radiation sensitivity	(11)	Sensitivity to UV and X-ray irradiation	227
rbs	Ribose	74	Utilization of D-ribose	215
recA	Recombination	52	Ultraviolet sensitivity and competence for genetic recombination	233
recB	Recombination	55	Ultraviolet sensitivity and competence for genetic recombination	66, 106, 233, 234
recC	Recombination	55	Ultraviolet sensitivity and competence for genetic recombination	66, 234
ref	Refractory	(88)	*refII*; specific tolerance to colicin E2	105
rel	Relaxed	54	*RC*; regulation of RNA synthesis	5, 74
rep	Replication	74	Inhibition of lytic replication of temperate phages	32
rhaA	Rhamnose	76	L-Rhamnose isomerase	88, 172
rhaB	Rhamnose	76	L-Rhamnulokinase	88, 172
rhaC	Rhamnose	76	Regulatory gene	88, 172
rhaD	Rhamnose	76	L-Rhamnose-1-phosphate aldolase	88, 172
rif	Rifampicin	77	DNA-dependent RNA polymerase sensitivity to rifampicin	16, 55, L

Table A1.—*Continued*

Gene symbol	Mnemonic	Map position (min)[b]	Alternate gene symbols; phenotypic trait affected	References[c]
rns	Ribonuclease	15	Ribonuclease I	181
rts	—	77	*ts-9*; altered electrophoretic mobility of 50*S* ribosomal subunit	75
serA	Serine	57	3-Phosphoglyceric acid dehydrogenase	214, 215, 220
serB	Serine	89	Phosphoserine phosphatase	215, 220
serS	Serine	(18)	Seryl-tRNA synthetase	104
shiA	Shikimic acid	38	Shikimate and dehydroshikimate permease	171
som	Somatic	(37)	*O*; somatic (O) antigens	160
spc	Spectinomycin	64	Resistance or sensitivity to spectinomycin	7, 10, 75, 232
speB	Spermidine	57	Putrescine (or spermidine) requirement; agmatine ureohydrolase	M
strA	Streptomycin	64	Resistance, dependence, or sensitivity; "K-character" of the 30*S* ribosomal subunit	75, 118, 183, 196
stv	Streptovaricin	77	DNA-dependent RNA polymerase sensitivity to streptovaricin	241
sucA	Succinate	17	*suc*, *lys* + *met*; succinate requirement; α-ketoglutarate dehydrogenase (decarboxylase component)	100, 101, 215
sucB	Succinate	17	*suc*, *lys* + *met*; succinate requirement; α-ketoglutarate dehydrogenase (dihydrolipoyltranssuccinylase component)	100, 101
supB	Suppressor	16	su_B; suppressor of *ochre* mutation (not identical to *supL*)	24
supC	Suppressor	25	su_C; suppressor of *ochre* mutation (possibly identical to *supO*)	24, 81, 205, 212
supD	Suppressor	(38)	su_I, *Su-1*; suppressor of *amber* mutations	205, 212
supE	Suppressor	16	su_{II}; suppressor of *amber* mutations	68, 205
supF	Suppressor	25	su_{III}, *Su-3*; suppressor of *amber* mutations	83, 212
supG	Suppressor	16	*Su-5*; suppressor of *ochre* mutations	81
supH	Suppressor	38		61, 63
supL	Suppressor	17	Suppressor of *ochre* mutations	62, 63
supM	Suppressor	78	Suppressor of *ochre* mutations	62, 63
supN	Suppressor	45	Suppressor of *ochre* mutations	62, 63, 144
supO	Suppressor	25	Suppressor of *ochre* mutations (possibly identical to *supC*)	62, 63
supT	Suppressor	55		63
supU	Suppressor	74	*su7*; suppressor of *amber* mutations	208
supV	Suppressor	74	*su8*; suppressor of *ochre* mutations	208
tdk	—	25	Deoxythymidine kinase	102
tfrA	T-four	(8)	ϕ^r; resistance or sensitivity to phages T4, T3, T7, and λ	46, D
thiA	Thiamine	78	*thi*; synthesis of thiazole	122, 211
thiB	Thiamine	(78)	Thiamine phosphate pyrophosphorylase	122
thiO	Thiamine	(78)	Probable operator locus for *thiA* and *thiB* genes	123
thrA	Threonine	0	Block between homoserine and threonine	117, 134

Table A1.—*Continued*

Gene symbol	Mnemonic	Map position (min)[b]	Alternate gene symbols; phenotypic trait affected	References[c]
thrD	Threonine	0	*HS*; aspartokinase I-homoserine dehydrogenase I complex	89, 165, 166
thyA	Thymine	55	Thymidylate synthetase	6, 111, 215
tkt	—	(55)	Transketolase	121
tnaA	—	73	*ind*; tryptophanase	84, 168
tnaR	—	73	R_{tna}; regulatory gene	84
tolA	Tolerance	17	*cim*; *tol-2*; tolerance to colicins E2, E3, A, and K	147, 148, 153, 179
tolB	Tolerance	17	*tol-3*; tolerance to colicins E1, E2, E3, A, and K	147, 148, 153, 179
tolC	Tolerance	58	*colE1-i*, *tol-8*, *refI*; specific tolerance to colicin E1	35, 105, 148, N
tonA	T-one	2	*T1*, *T5 rec*; resistance or sensitivity to phages T1 and T5	46, 53, 134, E
tonB	T-one	25	*T1 rec*; resistance to phages T1, ϕ80 and colicins B, I, V; active transport of Fe	91, 204, 230, 239
tpp	—	89	*deoA*, *TP*; thymidine phosphorylase	4, 72, 131
trpA	Tryptophan	25	*tryp-2*; tryptophan synthetase, A protein	113, 238, 239
trpB	Tryptophan	25	*tryp-1*; tryptophan synthetase, B protein	113, 238, 239
trpC	Tryptophan	25	*tryp-3*; indole-3-glycerol phosphate synthetase	113, 238, 239
trpD	Tryptophan	25	*tryE*; phosphoribosyl anthranilate transferase	58, 113, 142, 238
trpE	Tryptophan	25	*tryD*, *anth*, *tryp-4*; anthranilate synthetase	113, 142, 238, 239
trpO	Tryptophan	25	Operator locus	142, 209, 238
trpR	Tryptophan	90	*Rtry*; regulatory gene	37, 114, B
trpS	Tryptophan	65	Tryptophanyl-tRNA synthetase	57, 115
tsx	T-six	11	*T6 rec*; resistance or sensitivity to phage T6 and colicin K	46, 56, 79, 134
tyrA	Tyrosine	50	Prephenic acid dehydrogenase	170, 214, 215
tyrR	Tyrosine	27	Regulatory gene for *tyrA* and *aroF* genes	229
tyrS	Tyrosine	32	Tyrosyl-tRNA synthetase	194
ubiA	Ubiquinone	83	4-Hydroxybenzoate → 3-octaprenyl 4-hydroxybenzoate (OPHB)	42
ubiB	Ubiquinone	75	2-Octaprenylphenol → uniquinone	42, 43
ubiD	Ubiquinone	75	OPHB decarboxylase	43
uhp	—	72	Uptake of hexose phosphates	126
uraP	Uracil	50	Uracil permease	K
uvrA	Ultraviolet	80	*dar-3*; repair of ultraviolet radiation damage to DNA	107, 222
uvrB	Ultraviolet	18	*dar-1,6*; repair of ultraviolet radiation damage to DNA	1, 107, 222
uvrC	Ultraviolet	37	*dar-4,5*; repair of ultraviolet radiation damage to DNA	12, 107, 222
uvrD	Ultraviolet	74	*dar-2*, *rad*; repair of UV radiation damage to DNA	14, 157, 222
valS	Valine	84	*val-act*; valyl-tRNA synthetase	19, 216
xyl	Xylose	70	Utilization of D-xylose	118, 214
zwf	Zwischenferment	(35)	Glucose-6-phosphate dehydrogenase	167

APPENDIX IV

General Mapping Strategy: Mapping Strategy for Two Loci Concerned with Pyridoxine Biosynthesis

Pyridoxine-requiring mutants were originally isolated in a prototrophic Hfr strain, Hfr KL16 (see Figure A3). The two mutants were designated *pdx-13* and *pdx-15*.

Step 1 Mate each mutant with a multiply marked Str^r F$^-$ strain for 90 minutes, without interruption. Select for recombinants listed in Table A2 and examine each class for possible linkage of the donor Pdx$^-$ trait to selected markers. The results shown in Table A2 aren't terribly informative, but they do suggest that both Pdx$^-$ markers are located somewhere in the proximal $\frac{1}{3}$ to $\frac{1}{2}$ of the donor genome. Therefore, recover and purify from each cross a Leu$^+$ recombinant which carries the donor Pdx$^-$ marker and the original His$^-$ and Thy$^-$ markers as well. Use these as recipients in step 2.

Step 2 Choose 2 or 3 prototrophic Hfr strains, such as strains H, OR21, and KL19 illustrated on the diagram, which are known to transfer the *his-thyA* region at reasonable efficiency. Mate each of these

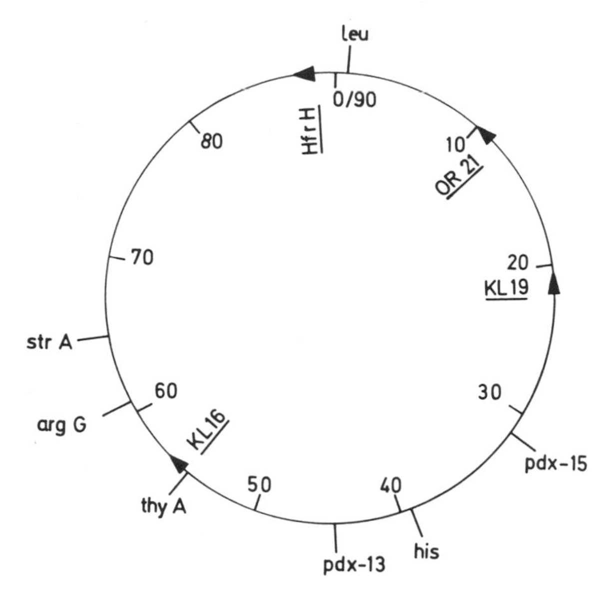

Figure A3.

Table A2.

Hfr (*pdx-13*) and Hfr (*pdx-15*) × F^- (*thyA*$^-$, *his*$^-$, *leu*$^-$, *argG*$^-$, *strA*)

Selected Markers	% of Recombinants that score Pdx^- *pdx-13* Donor	*pdx-15* Donor
Thy^+ Str^r	12	2
His^+ Str^r	26	18
Leu^+ Str^r	2	14
Arg^+ Str^r	5	7

Table A3.

Hfr Donor	Recipient	Recombinants per ml ($\times 10^{-5}$) Pdx^+ Str^r	His^+ Str^r	Thy^+ Str^r
KL19		64.0	81.0	36.0
OR21	*pdx-13* F^-	29.0	47.0	10.0
H		9.8	23.0	2.1
KL19		17.0	12.0	5.8
OR21	*pdx-15* F^-	9.0	5.0	1.4
H		5.9	1.8	0.19

for 60 minutes with the *pdx-13* and *pdx-15* F^- recombinants obtained in step 1. Plate out to determine the number of Pdx^+ Str^r, His^+ Str^r, and Thy^+ Str^r recombinants produced in each cross. See results in Table A3. Since the gradient of marker transmission for any given Hfr is known to be a first-order negative exponential function of distance from Origin, the above data can be plotted on

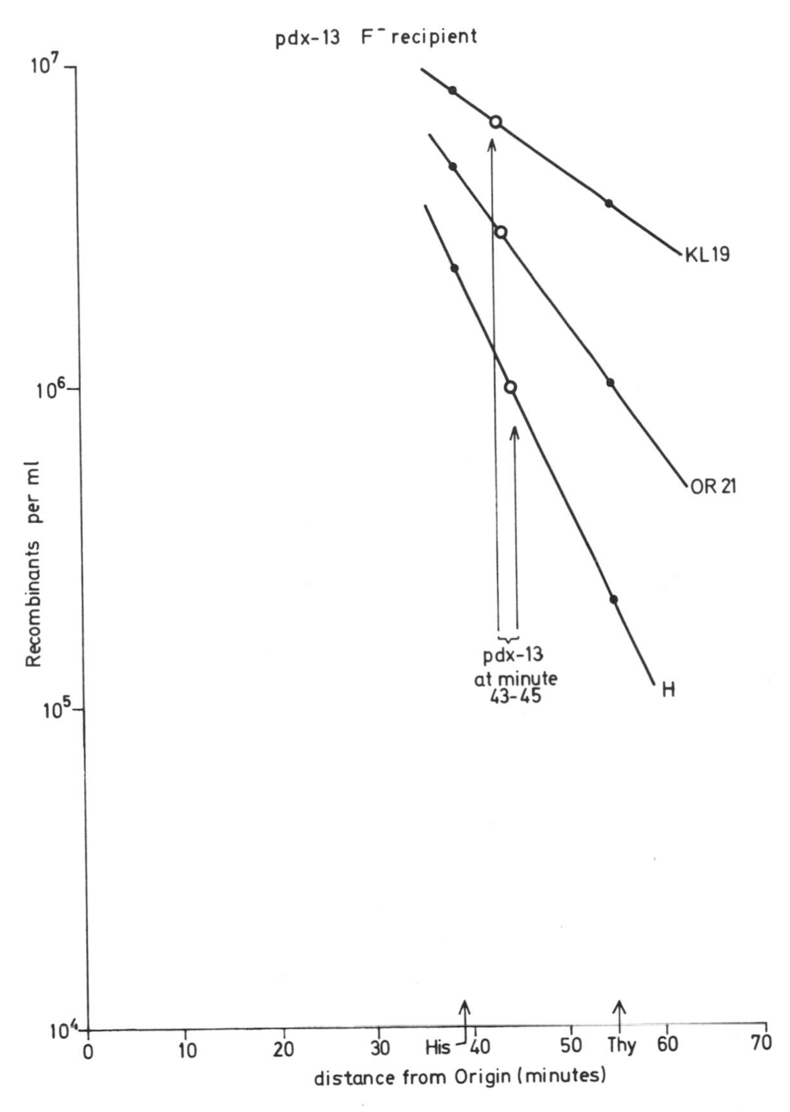

Figure A4.

semilog paper and a straight line can be drawn through the points for His^+ and Thy^+ recombinants. These lines can then be extrapolated to allow plotting the Pdx^+ recombinant yields as illustrated in Figures A4 and A5. The vertical intercept of the latter points with the abcissa provides a good estimate (to within 1–2 minutes) of the locus of the new *pdx* gene.

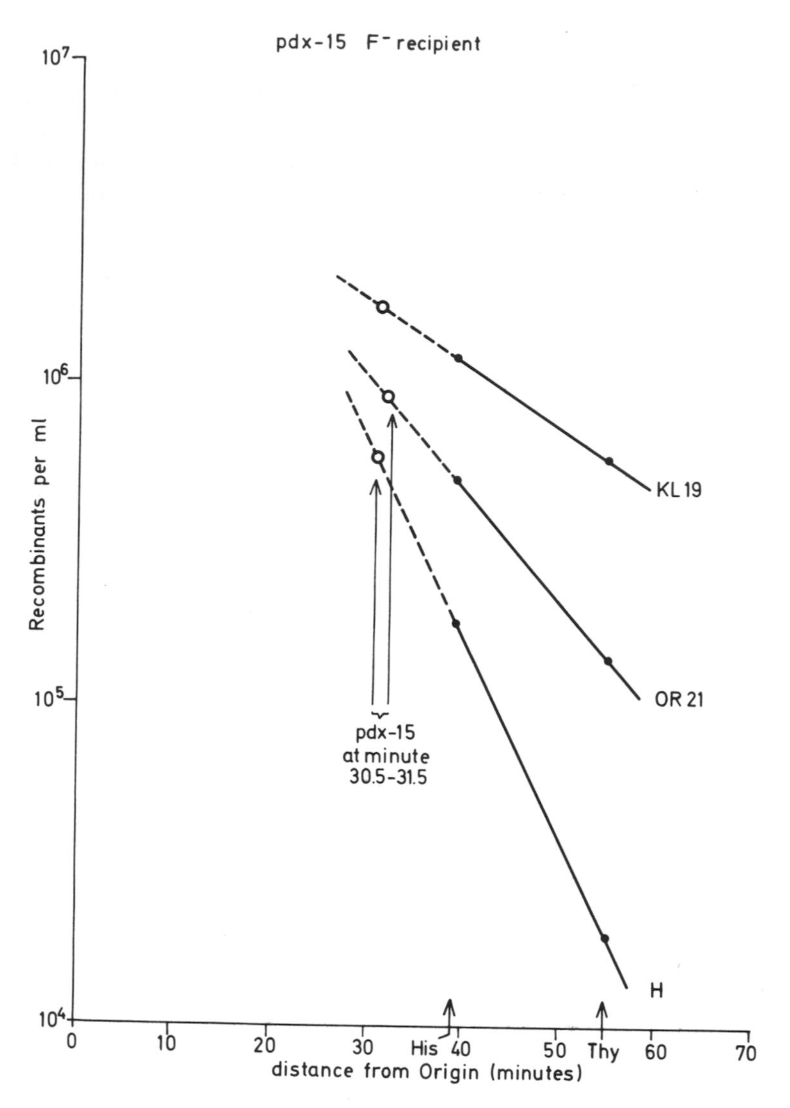

Figure A5.

Step 3 At this stage, precision mapping of the new genes can be done by means of three-factor transduction crosses using markers previously mapped in the 32 min and 44 min regions for *pdx-15* and *pdx-13*, respectively. For some specific examples of gene ordering by transduction, see Tables 2, 3, 4, 6, 7, and 8 in Taylor and Trotter, 1967.

Reference

TAYLOR, A. L. and C. D. TROTTER. 1967. Revised linkage map of *Escherichia coli*. *Bacteriol. Rev. 31:* 332.

APPENDIX V

Lambda Genetic Elements*

Genes**

A, *W*, *B*, *C*, *D*, *E*, *F*† Formation of phage head and probably formation of cohesive DNA ends. Gene *E* determines a major capsid protein. (2, 5, 14)

Z, *U*, *V*, *G*, *T*, *H*, *M*, *L*, *K*, *I*, *J*† Formation of phage tail. Gene *V* determines a major tail protein; *J* determines host range and the antigen with which neutralizing antibody reacts. (2, 14)

int Prophage insertion and excision. (6)

xis Prophage excision. (6)

δ Interference with growth in P2 lysogens (spi^+ phenotype). (7)

exo or *red*α Lambda exonuclease, which promotes general recombination and growth in $recA^-$ and $polA^-$ hosts, but interferes with growth in P1 lysogens. (7, 8)

β or *red*β β protein, which promotes general recombination and resembles λ exonuclease in its effects on growth in various bacterial hosts. (7, 8)

---

* Reprinted with permission from *The Bacteriophage Lambda*, A. Hershey, ed., Cold Spring Harbor Laboratory, 1971.

** Prepared by W. Szybalski and I. Herskowitz. Genes are listed in left-to-right order with respect to the phage map. Numbers in parentheses refer to chapters in *The Bacteriophage Lambda*.

† Essential for plaque formation on standard host.

γ Growth in *recA*$^-$ hosts (*fec*$^+$ phenotype) and interference with growth in P2 lysogens (*spi*$^+$ phenotype). (7)

cIII Establishment of immunity. (12)

N† Positive regulator; required for efficient transcription of genes to the left of *N* and to the right of *cro*. (2, 10, 13)

rex Inhibition of growth of T4*r*II mutants. (11)

cI(*CI*) Lambda repressor. (10, 11, 12, 13)

cro, *tof*, or *fed* Negative control of immunity. Prophage λN^-c$I857cro^-O^-$ re-establishes immunity at 32°C after prolonged growth at 42°C; λN^-c$I857cro^+O^-$ does not. Also responsible for turnoff of transcription of genes *N* through *int*. Lambda cI^+cro^- does not form plaques. (10, 11, 12, 13)

cII Establishment of immunity. (12)

O, *P*† DNA replication. (9)

Q† Positive regulator of late gene transcription. (10, 13)

S, *R*† Cellular lysis. *R* determines the structure of λ endolysin, an endopeptidase formerly thought to be a lysozyme. (2)

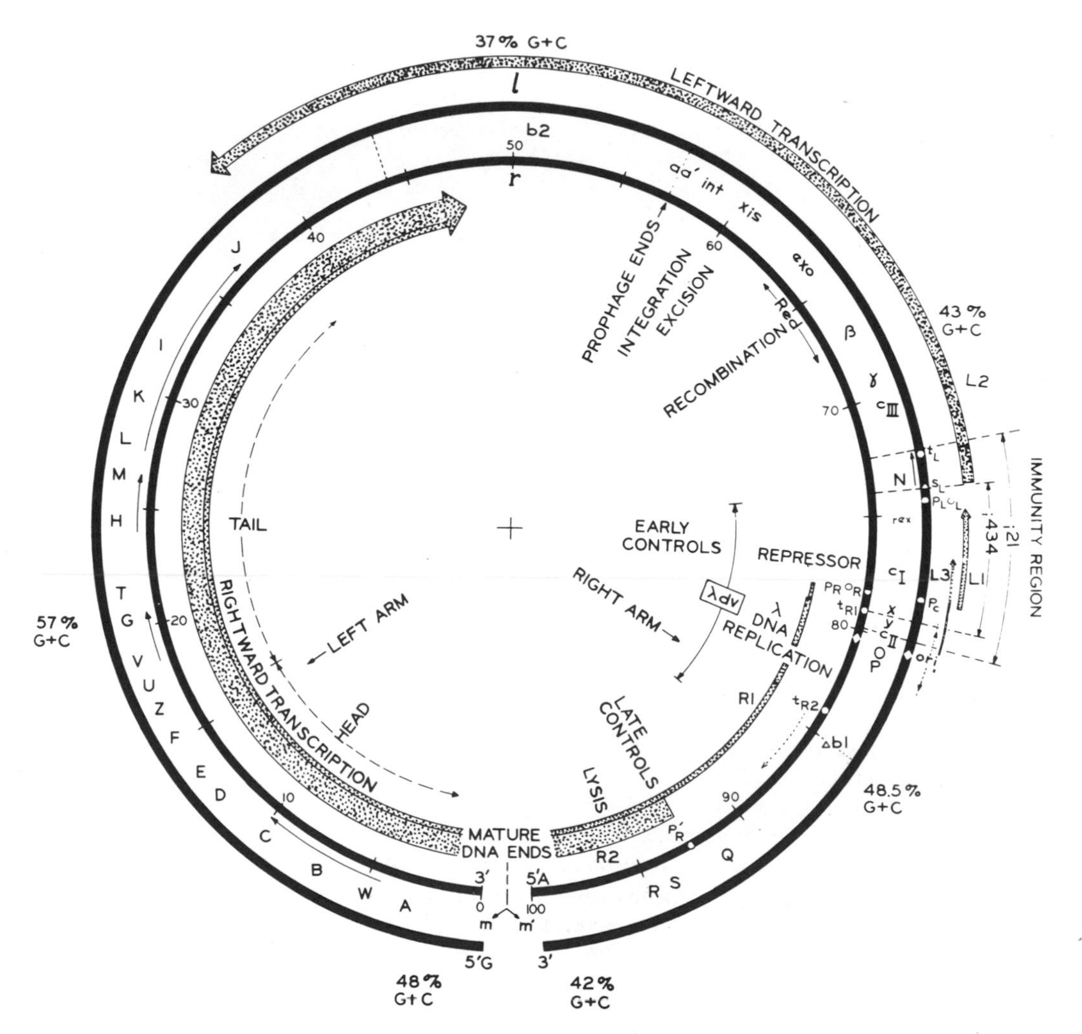

Figure A6. The lambda DNA map. (Reprinted from The *Bacteriophage Lambda*, p. 779, Cold Spring Harbor Laboratory, 1971.)

Recognition Sites‡

m, *m*′† Left and right cohesive ends of the DNA molecule. (3, 5, 8, 9)

att Sites of interaction ("attachment") of phage and bacterial DNAs during lysogenization, called P.P′ (or *a.a*′) in phage DNA and B.B′ (or *b.b*′) in bacterial DNA. (6)

p_L† Early leftward promoter situated within imm^{434} between *rex* and *N*. It is defined by the *sex* and *t*27 mutants, which are deficient in mRNA corresponding to genes *N* through *int*. (10, 11, 12, 13)

p_R† Early rightward promoter situated in the *x* region, defined by polar mutations such as *t*11. Lambda x^- mutants are deficient in mRNA corresponding to genes *cro* through *Q*. (10, 11, 12, 13)

p'_R† Late rightward promoter situated between genes *Q* and *S* required for efficient expression of late genes (*S*, *R*, *A–J*), defined by deletion mapping. Probable site of action of the gene *Q* product. (10, 13)

o_L Operator controlling transcription of genes *N* through *int*; left binding site for the λ repressor; site of the *v*2 (*vir*L) mutation. (10, 11, 12, 13)

o_R Operator controlling transcription of genes *cro* through *Q*; right binding site for the λ repressor; site of the *v*3 and *v*1 (*vir*C and *vir*R) mutations.

ori† Origin of DNA replication, located by several methods in the *c*II–*O* region. Possibly the site of mutations *t*5 and *ti*12. (9)

t Termination signal for transcription. (10, 13, 15)

‡ Sites of *cis*-specific DNA function, identified by mutations or deletions unless otherwise indicated. List prepared by W. Szybalski and I. Herskowitz.

APPENDIX VI

Assay for *lac* Repressor by Millipore Filter

The *lac* repressor will bind to Millipore filters, but IPTG will not. If ^{14}C-IPTG is pre-incubated with an extract containing repressor and then filtered, the only radioactivity associated with the filter is IPTG complexed to repressor. Once this complex is trapped on the filter, the IPTG cannot be washed off. This proves to be a sensitive assay for the *lac* repressor which is more convenient than equilibrium dialysis (see Experiment 51) if the sample is greater than 5% pure. With 1 μg of pure repressor, approximately 100 cpm of ^{14}C-IPTG are trapped on the filter; this is easily visible above the background of 80–120 cpm.

The assay is linear up to about 25 μg (2500 cpm); beyond that the slope decreases as the filter becomes saturated. It cannot be used accurately in a crude extract, since many other proteins also bind to the filter competing out the repressor.

References

RIGGS, A. D. and S. BOURGEOIS. 1968. On the assay, isolation, and characterization of the *lac* repressor. *J. Mol. Biol. 34:* 361.

RIGGS, A. D., S. BOURGEOIS, R. F. NEWBY and M. COHN. 1968. DNA binding of the *lac* repressor. *J. Mol. Biol. 34:* 365.

Method

Boil 25-mm HAWP (0.45 μ) filters twice in distilled H_2O, store in 10^{-4} M IPTG and distilled H_2O at 4°C.

Add 100 μl of the sample (or less, diluted with buffer to 100 μl) to a disposable test tube, then 100 μl of ^{14}C-IPTG at 2×10^{-6} M, making the incubation mixture 1×10^{-6} M (containing about 10,000 cpm). Shake and let sit 5 minutes at room temperature.

a) Put filter in place, with vacuum.

b) Cut vacuum (completely—such that there is *no* suction left), and put entire sample onto filter, using a disposable pasteur pipette.

c) Apply vacuum.

d) Cut vacuum, and rinse with 2 ml of buffer.

e) Apply vacuum until filter is fairly dry (about 30 seconds). Put filter in vial and add scintillation fluid (3 ml of Aquasol in a 1-dram vial; this can be counted as a disposable insert in most normal scintillation vials). Always run 2 duplicate blanks (buffer plus ^{14}C-IPTG) as the first and last samples, and one filter blank (no ^{14}C-IPTG).

The same pasteur pipette may be used for all the samples.

APPENDIX VII

Selected F′ Factors from *E. coli* K12

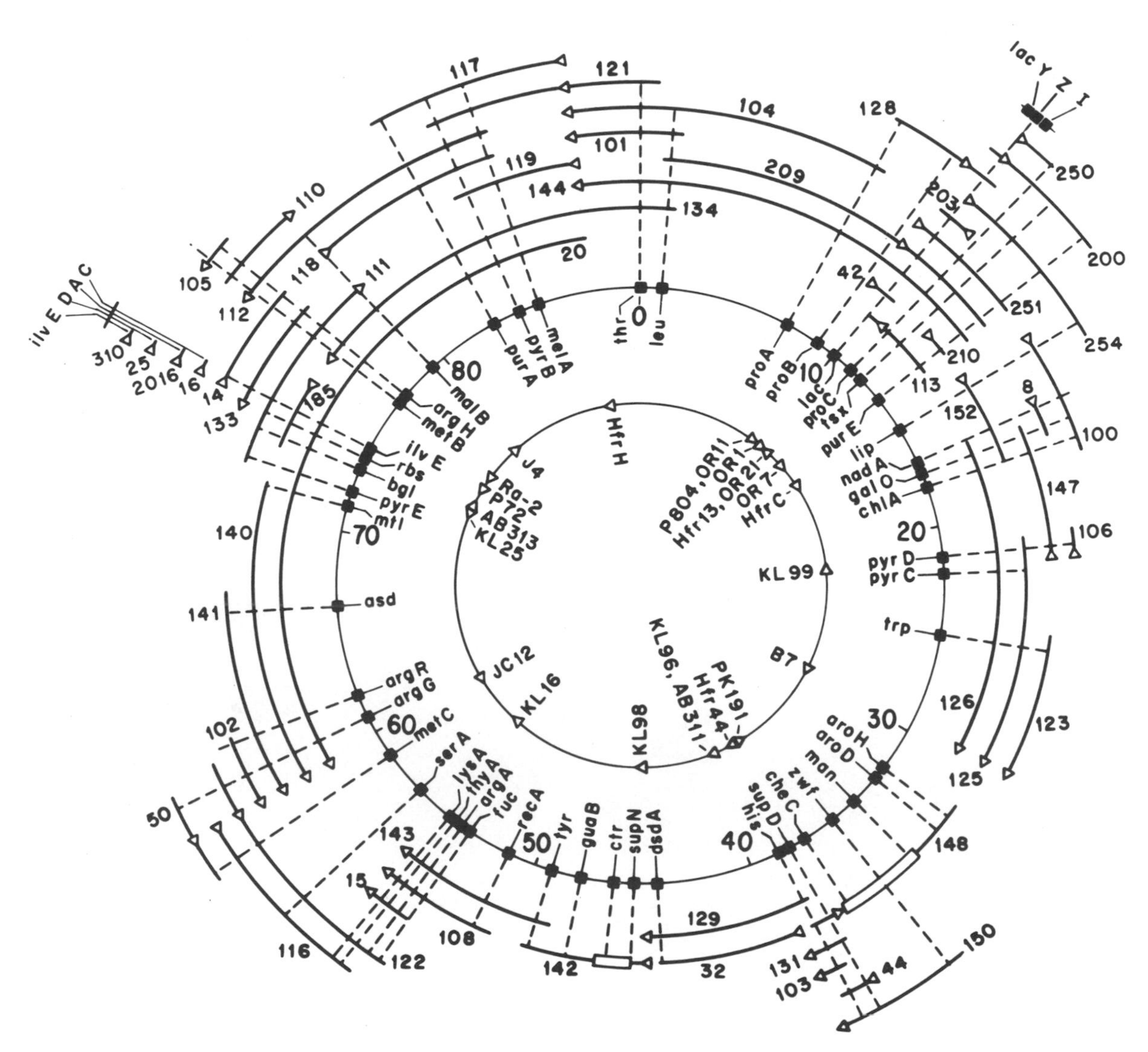

Courtesy of B. Low and J. Falkinham. Reprinted with permission from *Handbook of Microbiology*, (A. I. Laskin and H. Lechevalier, eds.), Chemical Rubber Co., Cleveland, Ohio, 1972.

APPENDIX VIII

Addresses for Laboratory Supplies

1. Chemicals, antibiotics

	J. T. Baker Chemical Co. 122 E. 42nd St. New York, N.Y. 10017
	995 Zephyr Ave. Hayward, Ca. 94544
	Calbiochem P.O. Box 12087 San Diego, Ca. 92112
	P.O. Box 331 5 Remsen Ave. Monsey, N.J. 10952
TONPG, XGal*	Cyclo Chemical P.O. Box 71557 1922 E. 64th St. Los Angeles, Ca. 90001
culture media, dyes	Difco Laboratories Detroit, Mich. 48232
acridine orange, EMS	Eastman Kodak Co. Eastman Organic Chemicals Rochester, N.Y. 14650

* 5-bromo-4-chloro-3-indolyl-β-D-galactoside

hydroxylamine	Fisher Scientific Co. 52 Fadem Rd. Springfield, N.J. 07081 Zeltweg 67 3032 Zurich, Switzerland
folinic acid	Lederle Laboratories Division American Cyanamid Co. Pearl River, N.Y. 10965
phosphonomycin	Merck Chemical Division Merck & Co., Inc. Rahway, N.J. 07065
penicillin (Na), culture media	Nutritional Biochemicals 26201 Miles Rd. Cleveland, Ohio 44128
SDS	Pierce Chemical Co. P.O. Box 117 Rockford, Ill. 61105
polynucleotides, enzymes	Miles Laboratories Research Division P.O. Box 272 Kankakee, Ill. 60901
DNase	Worthington Biochemical Corp. Freehold, N.J. 07728
streptomycin, nalidixic acid, chloramphenicol, rifampicin, Tris	Schwarz/Mann Mountain View Ave. Orangeburg, N.Y. 10962 6308 Woodman Ave. Van Nuys, Ca. 91401
nitrosoguanidine, DTT, 6-azauracil	Aldrich Chemical Co., Inc. 10 Ridgedale Ave. P.O. Box AA Cedar Knolls, N.J. 07927 2098 Pike St. San Leandro, Ca. 94577 Ralph N. Emmanuel Ltd. 264 Water Rd. Alperton, Middlesex HAO 1PY England Aldrich-Europe 2340 Beerse, Belgium

Item	Supplier
IPTG, ONPG, DTT, DNase, RNase, methylglyoxal, 2AP, 6-azauracil	Sigma Chemical Co. P.O. Box 14508 St. Louis, Mo. 63178
ampicillin (Na), kasugamycin	Bristol Laboratories Syracuse, N.Y. 13201
trimethoprim	Hoffman-LaRoche Inc. Nutley, N.J. 07110
D-cycloserine	Eli Lilly & Co. 307 E. McCarty St. Indianapolis, Ind. 46225
chromatography, electrophoresis supplies	Bio-Rad Laboratories 32nd & Griffin Aves. Richmond, Ca. 94804
	220 Maple Ave. Rockville Centre, N.Y. 11570
	Bio-Rad Laboratories GmbH Limestrasse 71 8 Munich 66 West Germany
inorganic & organo-metallic research chemicals	Ventron Corp. Alfa Products P.O. Box 159 Beverly, Mass. 01915
	2098 Pike St. San Leandro, Ca. 94577
	Ventron-Hicol, N. V. Postbus 1151 Rotterdam, Netherlands

2. Glassware

Item	Supplier
	Bellco Glass Inc. 340 Edrudo Rd. Vineland, N.J. 08360
Yankee® pipettes	Clay-Adams, Inc. 141 E. 25th St. New York, N.Y. 10010
side-arm flasks	Metalloglass, Inc. 466 Blue Hill Ave. Boston, Mass. 02121
syringes	American Hospital Supply 2020 Ridge Ave. Evanston, Ill. 60201

cuvettes, disposable micropipettes	Scientific Products 100 Raritan Center Parkway Edison, N.J. 08817
	150 Jefferson Drive Menlo Park, Ca. 94025
screw-cap tubes, disposable pasteur pipettes	Wheaton Glass Co. Laboratory Apparatus Center Millville, N.J. 08332

3. Laboratory equipment

	Ace Scientific Supply Co., Inc. 1420 E. Linden Ave. Linden, N.J. 07036
centrifuges	Beckman Instruments, Inc. Spinco Division 1117 California Ave. Palo Alto, Ca. 94304
	US Highway 22 at Summit Rd. Mountainside, N.J. 07091
	Beckman Instruments RIIC, Ltd. Sunley House, 4, Bedford Park Croydon CR9 3LG Surrey, England
	Ivan Sorvall Inc. Norwalk, Conn. 06856
(International)	Arthur H. Thomas Co. Vine St. at 3rd P.O. Box 779 Philadelphia, Pa. 19105
spectrophotometers	Arthur H. Thomas Co. (see above)
heating blocks	
Sero Block®	Precision Scientific Co. 3737 W. Cortland St. Chicago, Ill. 60647
Multi-Temp-Blok®	Lab-Line Instruments, Inc. Lab-Line Plaza Melrose Park, Ill. 60160
Thermajust®	TechniLab Instruments Pequannock, N.J. 07440

Constantemp®	Roeco Manufacturing Service Box 357 Monterey Park, Ca. 91754
shaking waterbaths	New Brunswick Scientific Co., Inc. 1130 Somerset St. New Brunswick, N.J. 08903
turntables	Fisher Scientific Co. (see No. 1)
	Arthur H. Thomas Co. (see above)
Petroff-Hausser counting chamber	Scientific Products (see No. 2)
UV lamps	Ultra-Violet Products, Inc. San Gabriel, Ca. 91776

4. Miscellaneous supplies

filters	
Whatman	H. Reeve Angel & Co., Inc. 9 Bridewell Place Clifton, N.J. 07014
	H. Reeve Angel & Co., Ltd. 14 New Bridge St. London EC4, England
membrane	Millipore Corp. Bedford, Mass. 01730
	Millipore Ltd. Heron House 109 Wembley Hill Rd. Wembley, Middlesex, England
	Schleicher & Schuell, Inc. Keene, New Hampshire 03431
petri dishes	Falcon Plastics 1950 Williams Drive Oxnard, Ca. 03030
dialysis tubing	Union Carbide Corp. 270 Park Ave. New York, N.Y. 10017
Parafilm®	American Can Co. Neenah, Wisc. 54956

alumina	Norton Co. Worcester, Mass. 01606
polyallomer tubes	Beckman Instruments, Inc. (see No. 3)
Aquasol,® Omnifluor®	New England Nuclear 575 Albany St. Boston, Mass. 02118 206 Professional Bldg. El Cerrito, Ca. 94530
sarcosyl	Geigy Industrial Chemicals Ardsley, N.Y. 10502
disposable gloves	Handguards, Inc. Pittsburg, Ca. 94565